Code Optimization in the Polyhedron Model – Improving the Efficiency of Parallel Loop Nests

Peter Faber

2007-10

Eingereicht an der Fakultät für Informatik und Mathematik der Universität Passau als Dissertation zur Erlangung des Grades eines Doktors der Naturwissenschaften

---

Submitted to the Department of Informatics and Mathematics of the University of Passau in Partial Fulfillment of Obtaining the Degree of a Doctor in the Domain of Science

Betreuer / Advisors:
Priv.-Doz. Dr. Martin Griebl
Universität Passau

Prof. Christian Lengauer, Ph.D.
Universität Passau

Zweitgutachter / External Examiner:
Prof. Albert Cohen
INRIA Saclay Île-de-France
Parc Club Orsay Université

Copyright Notice

ISBN: 978-1-4092-6612-9

Published by lulu.com
Lulu Enterprises, Raleigh, NC, USA

Acknowledgements

This thesis was supported by the DFG through the **LooPo/HPF** project. The text was typeset using the teTEX package and Xemacs; most figures were created using the open source program `xfig`. Most experiments were conducted using the free `HPF` compiler ADAPTOR and the open source GNU-`Fortran` compiler `g77`. The LooPo project makes extensive use of the free software packages `Polylib`, `CLooG`, `PIP`, and `Omega`. Therefore, thanks are due to all authors of the respective packages!

Special thanks are due to all members of the LooPo project – former and present: from the beginning, it has always been a pleasure to be part of this team. I am particularly indebted to Dipl. Inf. Thomas Wondrak. He conducted and analyzed an incredible number of experiments. He also provided valuable assistance during the work on his diploma thesis and later on – not only by doing much of the "dirty work" but also with the valuable discussions we had and with his insightful questions. I am also most grateful to both of my advisors, Christian Lengauer, Ph.D., who gave me the opportunity to work on this thesis, and Priv.-Doz. Dr. Martin Griebl, who went through lengthy discussions with me. He helped me with theoretical and technical aspects of this thesis and always had a word of encouragement for me. Finally, I have to thank my fiancée, Elke Brandstetter, for her continuous support.

Contents

List of Figures

Chapter 1

Introduction

This work was started as part of the DFG project LooPo/HPF. Its goal was to combine parallelization methods based on the polyhedron model with syntax based HPF compilers that carry out low level chores such as generating communication code for distributed memory machines or partitioning of data and computations depending on the number of processors actually available for the task. The main class of programs considered here are numerical computations on large matrices. These computations are usually performed on large parallel computer architectures, often with distributed memory.

HPF The abbreviation HPF stands for High Performance Fortran [Hig97], which is a standardized programming language for parallel computers based on the Fortran standard [Int97]. HPF particularly targets computers with distributed memory. It emerged from different approaches for the automatic generation of parallel code from a sequential program. The main predecessors of HPF are projects like Kali [MR91], the PARADIGM project [SPB93], the Vienna Fortran Compiler [CMZ92], Fortran D [HT92, HKT94], and its successor, the dHPF compiler [AMCS97]. In HPF, computations are distributed across a *processor mesh* by distributing data: computations are only executed at those places that are designated to store the results of these computations. This convention is called the owner computes rule.

When HPF was first introduced, the dominant parallel computer systems where mainly multiprocessors or compute clusters. Both feature distributed memory architectures with high communication costs between different nodes. In the meantime, the situation has gradually changed. Symmetric multiprocessing (SMP) and other shared memory architectures have become increasingly widespread. Moreover, multicore central processing units (CPUs) and graphics processing units (GPUs) featuring advanced programming capabilities that enable parallel processing on the GPU increase the possibilities of high performance computing with relatively low latency communication between processing units (compared to inter-node communication). Yet, clusters are still the dominating parallel architecture, because ever more ambitious applications never cease to demand more processing power.

This hardware development has been accompanied by a development in software engineering. HPF proved to be a very efficient language for expressing regular problems and exploiting synchronous parallelism. But irregular problems cannot be handled well by HPF compilers, since they lead to unstructured communication that can be optimized only insignificantly by the compiler. Thus, inter-node communication is even today largely done with hand-written code that directly calls communication libraries, rather than using compilers. In the late 1990s, other languages have surfaced, such as High Performance Java [ZCF+99], which is based on the popular Java language, but which uses strategies quite similar to HPF for the generation of parallel code. This fact highlights the fundamental merits of the HPF approach. HPF identified basic problems with the generation of efficient communication directives, which are unavoidably inherited by later languages with similar goals.

Another approach is the now widely adopted OpenMP standard [Ope00b, Ope00a]. OpenMP defines language extensions for Fortran and C/C++. The problem of communication generation does not strike OpenMP as much as HPF, since the former is designed only for the now widely available shared memory architectures in which communication can be made much faster and more efficient than inter-node communication, so that programs written in OpenMP may scale well up to some dozen processors. However, the problem does not disappear, so that further scaling is often hindered. Just as with HPF, data locality is a crucial performance factor in programs using OpenMP. Being able to distribute data explicitly is therefore a very helpful feature for software designers and compiler writers alike, since it provides an opportunity to convey semantic information to the compiler and enforce a reasonable data alignment. Unfortunately, these features of HPF where left out in the design of OpenMP – which otherwise resembles HPF in several ways – in favour of a completely hardware managed memory distribution. With its comparatively powerful distribution directives, HPF therefore still represents an invaluable development platform for an automatic parallelizer. In Section 2.1 we will give a quick introduction to the way in which to represent parallel computations in HPF.

Polyhedron Model The polyhedron model was originally developed by Leslie Lamport [Lam74] for automatically parallelizing perfectly nested loop nests. Subsequently, it was taken up by many researchers [Mol82, Qui87, Fea88, Len93] and extended to imperfectly nested loops and non-uniform dependences [Fea92a, Fea92b] and to (possibly unbounded) polyhedra instead of polytopes [GC95, Gri96]. The polyhedron model has been applied in different areas such as generating VLSI algorithms [Mol82], but also in code optimizations such as dead code elimination [Won01], cache optimization [Slo99], and in general a wide range of loop transformations [BCG$^+$03]. The basic idea is to represent different iterations of an n-dimensional loop nest as vectors of $\mathbb{Z}^n$. We will describe the polyhedron model – and some extensions we use in this work – in greater detail in Section 2.3. However, before we review this mathematical model, some basic algebraic definitions are presented in Section 2.2.

Mission Statement The aim of this work is to employ code optimization techniques that can identify different loop iterations to improve the code generation of standard syntax based compilers. In order to model loop nests accurately, these optimization techniques are based on the representation of a program as a set of inter-dependent polyhedra. As a first step, a code placement method based on this fine grained framework is introduced. This transformation is mainly aimed at reducing the number of computations executed and may come at the price of increased memory transfer. In order to reduce this effect, we examine the possibilities of replicated data placement. Both of these transformations may have impact on the final code generation, which we will also inspect. In several places, the code generation itself may influence the ability of later compiler stages to produce efficient code. Since the overall aim of this thesis is to improve automatic code generation, we will also examine further code generation strategies based on our fine grained model that can be employed to improve the resulting target code.

The basic idea behind the code placement method is to optimize the workload in a deep loop nest improve utilization of parallel processors by producing new – transformed – loop nests that contain only parts of the original computation, so that the deeper loop nest is only necessary to combine and scatter the results computed previously. These are important opportunities for optimizations in clusters as well as multicore CPUs [Ram06] that are programmed using parallel computing directives. It may also be used in stream processing compilers [TKA02, GTA06, GR05] and other architectures that feature single instruction multiple data (SIMD) units [Hes06] for improved performance on data-parallel code.

HPF allows for explicit data distributions and directives for marking code segments as independent (and thus executable in parallel) and is supported by stable compilers. Moreover, the optimizations developed in the following chapters are suited quite well for the inclusion within an HPF compiler as a separate phase. Therefore, our foremost interest here is the generation of parallel programs using HPF compilers.

The overall performance of a program often depends not only on the description of the algorithm but also on the compiler and libraries used, as well as on the target machine. We will restrict ourselves to manipulating the first of these points and will therefore sometimes view everything else as one atomic entity. We call this entity (i.e., the processor in the `Fortran` sense) the **system**.

Contents This thesis is organized as follows. The next chapter discusses the data-parallel programming model and compilation techniques – with `HPF` chosen as the example language. It also recalls some basic algebra that will be used throughout the thesis, and describes the framework on which the techniques presented in the subsequent chapters are based. Depending on previous knowledge, the reader might skip sections from this chapter or defer reading them until it appears necessary when definitions and results presented there are used in later chapters. Chapter 3 introduces an optimization strategy based on the polyhedron model that can be used to improve automatically generated loop code – which does not necessarily have to be executed in parallel. Chapter 4 then brings up important points that have to be considered when generating code for parallel programs. In particular, we will shed some light on the placement of calculations on processors. Finally, Chapter 5 will briefly show how all the previously described methods can be combined to produce efficient code, before Chapter 6 concludes with an overview over the lessons learned. Appendix A enumerates the options of the program implementing the optimization strategy presented in chapter 3 and Appendix B contains notational conventions observed throughout this thesis.

Chapter 2

Fundamentals

2.1 Data-Parallel Programmming with Data Distributions

This section gives a brief introduction to the basic features of data-parallel languages. As already pointed out, the generation of data-parallel programs leads to some basic problems and challenges that do not depend on the language used for this task, be it `kali` [MR91], `HPF` [Hig97], `High Performance Java` [ZCF+99] or others. In this work, we choose `HPF` as the language of choice for its easily available and quite mature compilers and the explicit data distribution features and parallel loops constructs. Section 2.1.2 uses `HPF` to sketch some points that should be observed when programming in data-parallel languages. Further introductory material about data-parallel programming (and programming using `MPI`, as well as `Compositional C++`, `Fortran M` and `HPF`) can be found in Foster's *Designing and Building Parallel Programs* [Fos94] and the *High Performance Fortran Handbook* [KLS+94]. Another introduction into parallel programming in general and data-parallel programming in particular, based on `MPI` and `OpenMP` is given in Michael Quinn's *Parallel Programming in C with MPI and OpenMP* [Qui04]. The ultimate source for `HPF` is, of course, the `HPF` standard [Hig97], and an in-depth treatment of how data-parallel compilers work can be found in *High Performance Compilers for Parallel Computing* by Michael Wolfe [Wol95].

2.1.1 Basic HPF Features

In `HPF`, the processors of the parallel machine are supposed to be arranged in meshes, so-called **processor arrays** that can be either user defined or implicitly defined. The actual default arrangement for an *#pdim*-dimensional processor array depends on several factors:

- The compiler – including its run time system – defines a mapping of operations to be executed onto processes.
- A communication library like, e.g., the Message Passing Interface `MPI` [Mes95, Mes97] defines a mapping of these processes to compute nodes.
- The operating system of the parallel computer – if there is one – defines a mapping of processes on a given compute node to the actual CPU that executes the operations.

In general, a compiler will try to use a balanced mapping of these processor arrays to processes. In this work we will mainly use the **ADAPTOR** `HPF` compiler of the Fraunhofer Gesellschaft [BZ94]. The following example illustrates the way in which ADAPTOR distributes physically available processors onto a *#pdim*-dimensional processor mesh.

Example 2.1 *Suppose we have* #cpu *processes given for the execution of our program. Within this program, we may use an #pdim-dimensional processor array. The ADAPTOR compiler then distributes the available processes across this processor array by using the greatest divisor* *of* #cpu *that is smaller than* #cpu *as extent* #cpu[*#pdim*] *of the last dimension of the processor array (i.e.,* #cpu[*#pdim*] *is maximal with* #cpu[*#pdim*]%#cpu $= 0 \wedge$ #cpu[*#pdim*] $<$ max(#cpu, 2)*), while* $\frac{\#\text{cpu}}{\#\text{cpu[}\#pdim\text{]}}$ *processes remain for distribution in* $\#pdim - 1$ *further dimensions according to the same scheme. This means that, if* #cpu *denotes the number of physical processors in the* #pdim*-dimensional processor array, and the shape of the processor array is defined as* (1:ρ[1],...,1:ρ[*#pdim*]), $\rho \in \mathbb{Z}^{\#pdim}$*, this shape is defined by the lexicographic maximum of the vectors whose coordinates multiplied with each other give the number of processors and represent an increasing number sequence*

$$\rho = \max_{\prec}(\left\{(r_1, \ldots, r_{\#pdim})^T \,\middle|\, r_1 \leq \cdots \leq r_{\#pdim} \wedge \#\text{cpu} = \prod_{i=1}^{\#pdim} r_i\right\})$$

For example, for #cpu $= 6$ *and* $\#pdim = 2$*, we have* $\rho = \begin{pmatrix} 2 \\ 3 \end{pmatrix}$.

In this thesis, we will actually only consider placements based on the processor arrangements represented by the HPF processor arrays. Any underlying mapping that is done by the communication library, the operating system, or the parallel machine itself is outside of our area of influence; the underlying infrastructure may or may not choose to distribute the resulting processes in a way that actually corresponds to the processor array.

The basic way of creating a parallel program in HPF is to define an alignment of variable arrays to processor arrays. Different iterations of a loop that is marked with the INDEPENDENT directive are then executed according to the owner computes rule: a processor executes only those iterations of the loop that write to local array elements, i.e., array elements that are declared to be owned by that particular processor. However, even if all write accesses in a loop nest may be local, there may still be read (and write!) accesses necessary to perform the given computation that refer to non-local data. In addition, even a single statement may define several different write accesses and therefore, both local *and* non-local write accesses. Thus, usually *one* write access is picked as the one that is localized. On distributed memory machines, auxiliary variables have to be introduced that serve as buffers for (possibly) non-local write or read accesses.

HPF supports several regular distribution schemes; accesses to regularly distributed arrays can be implemented in an efficient manner, if the reference to corresponding subscript function is itself regular (with compilers differing in the notion of a regular subscript function; we will go into further detail in Section 2.1.2 and Chapter 4.1).

Distribution of an array dimension onto a dimension of a processor array is defined via the DISTRIBUTE directive:

```
!HPF$ DISTRIBUTE <distributee> ( <dist-format-list> ) [ ONTO <processors-name> ]
```

The elements of <dist-format-list> may be one of:

```
BLOCK  [ ( <int-expr> ) ]
CYCLIC [ ( <int-expr> ) ]
*
```

where <int-expr> is an integer expression. This directive specifies that the dimension of <distributee> that corresponds to a <dist-format> in <dist-format-list> is distributed across different processors, if the <dist-format> is not an asterisk (*). The i-th element of <dist-format-list> that *does* specify a distribution (via BLOCK or CYCLIC) selects a distribution format across the i-th dimension of the processor array specified by <processors-name>. If no processor array is given, a default processor array is used whose dimension matches the number of non-asterisk-elements of <dist-format-list>.

An asterisk may appear as an element of <dist-format-list>. In this case, the <distributee> is not distributed along that dimension: array elements that only differ in that dimension are stored on the same processor.

Example 2.2 *Consider the following program fragment:*

```
!HPF$ PROCESSORS P(2,2)
!HPF$ DISTRIBUTE A(BLOCK,CYCLIC) ONTO P
```

In this example, the first dimension of A is distributed across a processor array P according to a `BLOCK` *distribution (contiguous elements are stored on the same processor) in the first dimension, and according to a* `CYCLIC` *distribution (round-robin fashion) in the second dimension. Figure 2.1 illustrates this situation. The dotted lines separate different array elements, and the different shadings represent the processors that store the respective elements.*

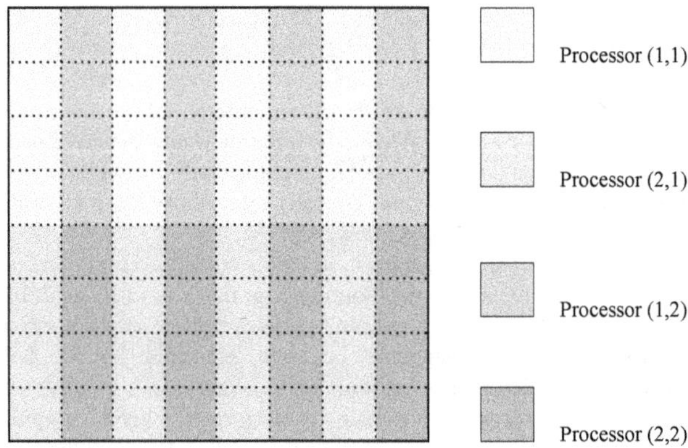

Figure 2.1: Distribution of A onto P as defined in Example 2.2.

The mapping of array elements to a $\#pdim$-dimensional processor array can be modeled by $\#pdim$ functions that map array elements to processors.
The most general regular distribution scheme that can be obtained using `DISTRIBUTE` directives is the `BLOCK-CYCLIC` distribution that is selected via a `CYCLIC(`b`)` directive. The `BLOCK-CYCLIC` distribution (also called `CYCLIC(`b`)`) corresponds to a placement function

$$\Phi_{\texttt{BLOCK-CYCLIC}} : i \mapsto \lfloor \tfrac{i}{b} \rfloor \% \#\text{cpu}\,\texttt{[}j\texttt{]}$$

where $\#\text{cpu}\,\texttt{[}j\texttt{]}$ is the number of available processors in that dimension (j). The `CYCLIC` distribution can be viewed as a special case of a `BLOCK-CYCLIC` distribution, in which the block size b is 1.

Simpler distributions are `BLOCK(`b`)` and `BLOCK` distributions. In the first case, the corresponding placement function is

$$\Phi_{\texttt{BLOCK}} : i \mapsto \lfloor \tfrac{i}{b} \rfloor$$

and the user is responsible for asserting that this distribution scheme does not need more processors than available. In the second case, b depends on the shape of the **`<distributee>`** and the number of processors (the elements of the **`<distributee>`** will be partitioned into equally sized blocks with a single block assigned to one processor).

In Example 2.2, **`<distributee>`** is the array A. In `HPF`, not only arrays can be distributed, but also so-called **templates**. Templates can be viewed as arrays that do not occupy any memory. A template can be used as a helper object in order to obtain a certain placement for an actual array. The general idea is to distribute a template onto a processor array and align an actual array to that template via a linear integer function (cf. Figure 2.3 on page 9). Of course, more complicated combinations – such as an array A aligned to some template T, which is aligned to an array B that is finally distributed to some user-defined processor array C are allowed in the `HPF` standard.

However, we will only consider the case that an array is simply aligned to a template which is then distributed onto some processor array. But let us first look at the way an alignment is defined in HPF.

An **alignee** (an array or a template) is aligned to an **align target** by an ALIGN directive:

```
!HPF$ ALIGN <alignee> ( <align-dummy-list> )          &
      WITH <align-target> ( <align-subscript-list> )
```

Example 2.3 *The* ALIGN *directive*

```
!HPF$ ALIGN B(i,j) WITH A(2*i,j+1)
```

specifies that array B has to be distributed across the processor array so that element $B(i,j)$ is stored on the same processor as $A(2*i,j+1)$. The resulting placement of B is shown in Figure 2.2. Arrows from array B to A indicate alignments – in order to keep the representation clear, alignments are only shown for the first column and the first row of B.

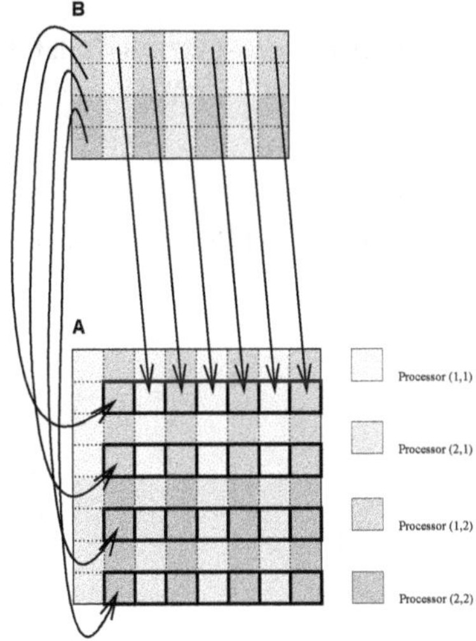

Figure 2.2: Alignment of B with A as defined in Examples 2.2 and 2.3.

The only expressions allowed as a subscript in <align-dummy-list> are **align dummies** (or a single asterisk). An align dummy is an identifier that is only used to identify a certain dimension of an integer vector space $\mathbb{Z}^{n_{tgt}}$ that may or may not correspond to a dimension of the alignee. It is allowed at most once in <align-dummy-list>, so that there is a injective relation between dimensions of the alignee and the align dummies in <align-subscript-list>. Note that <align-dummy-list> may contain align dummies that do *not* occur in <align-subscript-list>. Such an align dummy marks a dimension of <alignee> that does not have any relation to any dimension of <align-target>. In other words: this dimension is collapsed – all array elements that only differ in their coordinates in that dimension are stored on the same processor (just as with asterisks in <dist-format-list> in the DISTRIBUTE directive), and, again, such an align dummy can be replaced by an asterisk.

An element of <align-subscript-list> is a linear function in at most one align dummy and constant expressions. The HPF standard even allows rather complicated functions that are still viewed as constants (for example the application of intrinsic integer functions on constants).

However, existing HPF compilers are usually more restrictive in the expressions they support.[1] An align dummy may appear at most once in `<align-subscript-list>`. This guarantees a simple representation of the corresponding distribution as a linear function: for each dimension of the align target (possibly an array or processor array, usually a template) that is subscripted with an expression $a \cdot i + k$ containing the align dummy i from the alignee's subscript list, an element i in that dimension of the alignee is colocated with the align target element given by

$$\Phi_{\texttt{ALIGN}} : i \mapsto a \cdot i + k$$

i.e., if the align target is a processor array and the alignee is an array, the placement function for that array in the respective processor dimension is given directly by $\Phi_{\texttt{ALIGN}}$.

Again, if an align dummy is used only in an element of `<align-subscript-list>`, but not in `<align-dummy-list>`, there is no relation between the corresponding dimension of `<align-target>` and any dimension of `<alignee>`. This means, the dimension is replicated: all elements of `<align-target>` – possibly a processor array – that differ only in their coordinates in that dimension hold the same element of `<alignee>`. In other words: all the processors that differ only in that dimension store copies of the same elements of `<alignee>`. And – just as with `<align-dummy-list>` – an asterisk is equivalent to the use of an align dummy that does not occur anywhere else in the ALIGN directive.

The align target may be a distributee of some DISTRIBUTE directive. In this case, the placement function is given by the composition of the two placement functions

$$\Phi_{\texttt{ALIGNDIST}} : i \mapsto \left\lfloor \frac{a \cdot i + k}{b} \right\rfloor$$

with b as above.

The following examples give an overview of the possibilities of data placement in HPF.

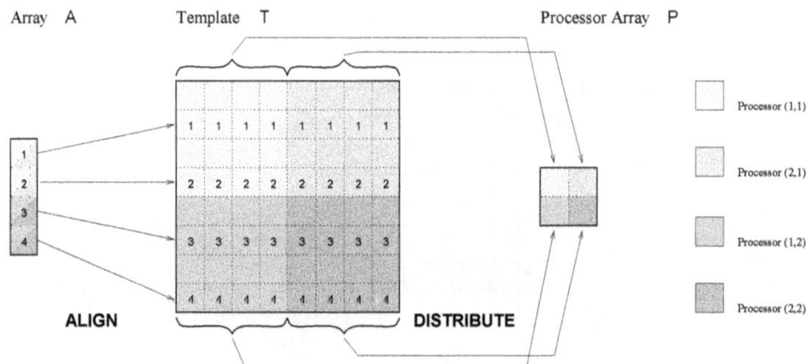

Figure 2.3: Alignment of array A with template T and distribution onto processor array P as in Example 2.4.

Example 2.4 *Figure 2.3 depicts the alignment of an array A to a template T that is* BLOCK-BLOCK *distributed onto a processor array P according to the following declarations:*

```
DOUBLE PRECISION A(1:4)
!HPF$ PROCESSORS P(2,2)
!HPF$ TEMPLATE T(8,8)
!HPF$ DISTRIBUTE T(BLOCK,BLOCK) ONTO P
!HPF$ ALIGN A(i1) WITH T(2*i1,i2)
```

[1]E.g., ADAPTOR does not allow further variables in an output dimension of an align subscript function, if it contains an align dummy.

The elements of A are distributed along the first dimension of P, but replicated along the second dimension of P. Which processor gets to own a copy of the respective array element, is indicated by the shading: each array element is stored on two processors.

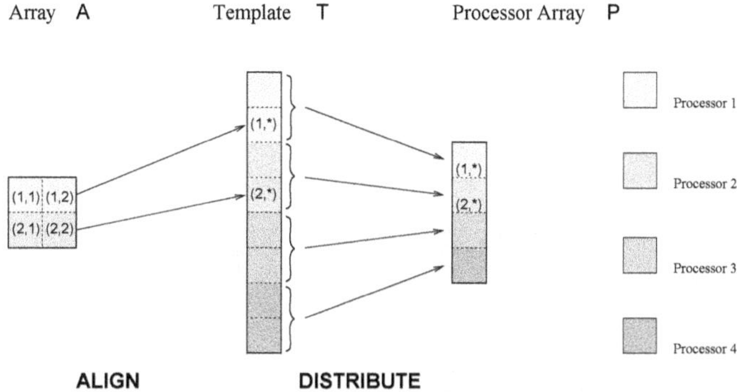

Figure 2.4: Alignment of array A with template T and distribution onto processor array P as in Example 2.5.

Example 2.5 *On the other hand, an alignment of a two-dimensional array to a one-dimensional processor array may look as follows:*

```
DOUBLE PRECISION A(2,2)
!HPF$ PROCESSORS P(4)
!HPF$ TEMPLATE T(8)
!HPF$ DISTRIBUTE T(BLOCK) ONTO P
!HPF$ ALIGN A(i1,i2) WITH T(2*i1)
```

In this case, we have to decide, whether the first or second dimension of A should be aligned (and distributed); the other one is collapsed. In this example, the first dimension is distributed; of course, i_2 could also be replaced by an asterisk in the `ALIGN` *clause. Figure 2.4 visualizes this alignment.*

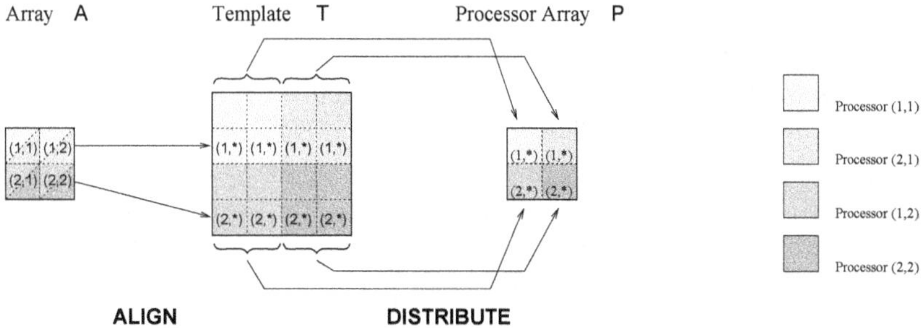

Figure 2.5: Alignment of array A with template T and distribution onto processor array P as in Example 2.6.

Example 2.6 *It is also possible to combine the collapsing of an array dimension with replicated storage. Figure 2.5 shows the alignment corresponding to the following directives:*

```
DOUBLE PRECISION A(2,2)
!HPF$ PROCESSORS P(2,2)
!HPF$ TEMPLATE T(4,4)
!HPF$ DISTRIBUTE T(BLOCK,BLOCK) ONTO P
!HPF$ ALIGN A(i1,*) WITH T(2*i1,*)
```

As mentioned earlier, the restriction for an align dummy to be used at most once in any of `<align-dummy-list>` or `<align-subscript-list>`, respectively, guarantees a simple placement relation. However, it also inhibits some more complicated placement relations, as the following example shows.

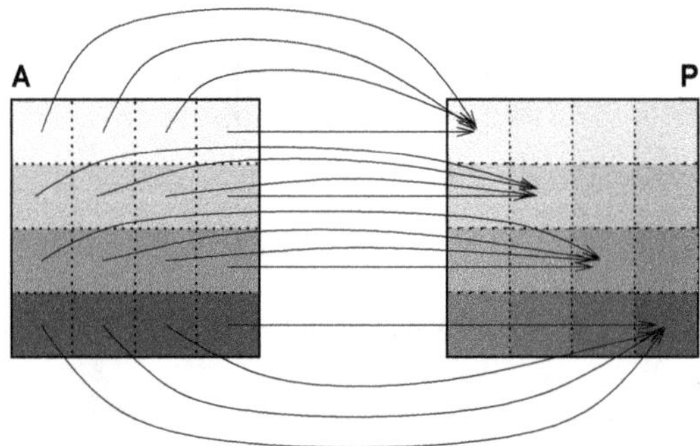

Figure 2.6: Illegal HPF alignment of Example 2.7: only diagonal elements of the processor array P store any part of array A.

Example 2.7 *Figure 2.6 represents an alignment of a two-dimensional array A onto a two-dimensional processor array P according to a placement function*

$$\Phi : \mathbb{Z}^2 \rightarrow \mathbb{Z}^2 : (i_1, i_2) \mapsto (i_1, i_1)$$

The arrows and the shadings indicate the processor to which an array element is mapped. This alignment cannot be described in HPF, *because the same align dummy would have to be used several times. The placement can only be approximated by replicating the second dimension of A on all processors of the same row or column of the processor array.*

As noted earlier, the usual case is that an actual array is aligned to a template, which is distributed onto some processor array of the same dimensionality. In Section 4.1, we will not assume any specific distribution format, but will base our observations on alignments to BLOCK distributed templates, since the regular nature of these distributions lends itself to a more predictable behaviour than a CYCLIC based distribution.

As we have seen, the lowest level of abstraction from the actual CPUs that is at the user's disposal in HPF is the processor array. However, the actual mapping of some user defined processor array to the processes that execute the given code on some CPU core depends not only on the compiler but also on many other factors such as the operating system. Usually, an HPF compiler will at least expect to have as many processes available at run time as there are elements in the processor array. However, we will mostly deal with loops whose bounds depend on values that are only known at run time. And since the compiler cannot influence the number of processes at run time (which could be used to adapt the number of processor arrays for which some given processor is responsible), it is impractical to consider single elements of these HPF processor arrays. Instead, the lowest level of abstraction we will use here is the level of templates: templates may be of arbitrary size, since they can be distributed across a processor array such that the run time system will guarantee that the template elements are evenly distributed across the actually available processes. Therefore, we can use templates as an HPF implementation of a **processor mesh** – a rectangular mesh of points in $\mathbb{Z}^{\#pdim}$ with each point representing a processing unit (may that processing unit be a process, an actual CPU core, or something completely different). Since these processing units – the points in $\mathbb{Z}^{\#pdim}$ – build the lowest level of abstraction from actual CPU cores, we call these points the *physical* processors. The implementation of a physical processor in HPF is an element of an HPF template.

Note that, for a BLOCK distribution of a template, different template extents will enforce different denominators in a placement function that defines a mapping to processes as $\Phi_{\texttt{BLOCK}}$ above, since the block size b in $\Phi_{\texttt{BLOCK}}$ is calculated from the size of the template in the distributed dimension. Therefore, we model the alignment to templates T_1, T_2 with different extents as relations between the respective array elements and different processor meshes of physical processors. This liberates us from having to represent placements using non-linear expressions: the level on which we compare the place of execution of certain calculations is the points of the processor mesh – implemented by template elements: two calculations are executed in the same place, iff they are placed on the same physical processor (the same template element).

In order to be able to talk about placements that only differ in the processor mesh to which array elements (and executions of operations) are mapped, we have to be able to consider different processor meshes. Therefore, we never consider a single processor mesh alone, but the space of *all* processor meshes available. From now on, we will not only consider a *#pdim*-dimensional space representing the processor mesh for a single template, but the *#pdims*-dimensional space of *all* templates of a program fragment put together (with *#pdims* being the sum of all distributed dimensions of templates included).

Since the advent of Fortran 90, Fortran supplies the programmer with a convenient way of addressing regular array sections, which are sometimes used by the run time system of an HPF compiler in order to produce efficient communication primitives. Array sections are commonly represented in the **triplet notation** (the subscripts in this case are called **subscript triplets**):

```
A(lb:ub:st)
```

represents the array section from *lb* to *ub*, where only every *st*-th element is used. The notation has its direct correspondence in a loop

```
DO i=lb, ub, st
  ... A(i) ...
END DO
```

The reverse, however, is not necessarily true: any loop nest in which the bounds of an inner loop depend on the current value of an outer loop counter, *cannot* be represented as triplet, since there is no way to represent this correlation. Consequently, an SPMD program cannot use a subscript triplet for an array section that is to be stored locally in Example 2.7: the local section of the first array dimension can be calculated as a subscript triplet, however, the second dimension is then either the complete dimension or nothing at all, depending on the coordinate of the local processor in the same dimension as was already used for the first array dimension. The corresponding loop nest would introduce an `if`-statement or a loop that depends on the current value of an outer loop. The reason is that $\mathrm{im}(\Phi)$, as a k-dimensional subspace of $\mathbb{Z}^{\#pdim}$ (here, $k = 1$, $\#pdim = 2$), cannot be generated by k unit vectors of $\mathbb{Z}^{\#pdim}$.

Of course, it would be possible to generate loop nests corresponding to more complex local sections, e.g. for communication generation. However, many compilers restrict themselves to these simple cases, and the HPF standard only requires support for these simpler alignment strategies. For example, in Fortran – and in most other languages –, it is not possible to (statically) reserve storage for any array section that cannot be described in triplet notation. Therefore, it is easier to support only alignment directives that lead to these simpler variable declarations.

2.1.2 Guidelines for Advanced Programming in Data-Parallel Languages

The guidelines presented in this section are not intended as an introduction to HPF but rather as a collection of good coding practices for writing data-parallel code that can be compiled into efficient target programs, e.g., by an HPF compiler. We will adopt these practices in our code generation strategies, which can then be implemented in a fully automatic code generation module.

The main task of an HPF compiler is to transform a program written for a single memory image into another program that uses local addressing where data is locally available and explicit communications where it is not. The techniques for this transformation may differ from compiler to compiler, depending on the degree in which a semantic analysis is performed. Here, we review some basic heuristics a programmer should follow in order to enable a completely syntax oriented compiler to generate efficient code for computations on array elements in a loop nest.

Note that the compiler techniques are usually only bound to the use of arrays in loops, and therefore not actually dependent on the use of HPF, so that most of these heuristics will not only apply to the compiler we used predominantly, ADAPTOR, or even HPF, but also to other compilers and languages – such as HP Java [ZCF+99] and other approaches – that use similar techniques in order to transform shared memory programs to distributed memory programs.

The basic way in which an data-parallel compiler reduces the workload in a parallel loop is by reducing the range of values to which the loop variable is bound [CK89]. The values that a loop

```
DO i1=l1,u1,s1
  ⋮
  S
  ⋮
END DO
```

enumerates can be changed by changing the values of l_1, u_1, or s_1:

```
DO i1=l1',u1',s1'
  ⋮
  S
  ⋮
END DO
```

Workload is reduced, if

$$\begin{aligned} l'_1 &> l_1, \text{ or} \\ u'_1 &< u_1, \text{ or} \\ s'_1 &> s_1 \end{aligned}$$

In addition, $s_1 \% s'_1 = 0$ and $(l'_1 - l_1)\% s_1 = 0$ have to hold in order not to enumerate values in the second loop that are not enumerated in the original one. If we restrict ourselves to such changes, the resulting set of i_1-values for which a statement S in the loop body is executed is always restricted to a lattice in $\mathbb{Z}$, i.e., a finite set of equidistant integers. Of course, one could introduce more sophisticated control structures in order to be able to enumerate more complicated subsets of the integer numbers. However, the plain modification of loop bounds and strides is the main technique for syntax-directed restructuring compilers to reduce the workload, so that we can safely assume that the quality of the resulting code will be directly connected to the possibility of creating loops of that form with modified values for l'_1, u'_1, and s'_1.

In order to avoid communication in each iteration of a parallel loop nest, which would probably become quite inefficient, HPF disallows any dependence between different iterations of INDEPENDENT loops.[2] This restriction enables the compiler to execute any part of the workload that was determined as above at any time – possibly in parallel. Additionally, it enables the compiler to hoist

[2] Actually, parallel loops can also be expressed as FORALL loops in HPF. These loops define possible dependences

all communication between different iterations of these loops out of the outermost INDEPENDENT loop of the loop nest.

Since the compiler must be allowed to hoist all communication outside all INDEPENDENT loops, dependencies between code fragments enclosed in INDEPENDENT loops have to be expressed in one of the following two ways:

1. Place one INDEPENDENT loop nest containing such a program fragment below the other in the program text.
2. Place both INDEPENDENT loop nests inside a sequential loop that is itself *not* enclosed by an INDEPENDENT loop.

In other words, only synchronous programs [Len93] can be written in this way.

Pitfall: Different Homes within a Single Loop Nest

As discussed in Section 2.1.1, the compiler is guided in its decision as to which values to choose for the new bounds and strides of a loop by the distribution and alignment of data. The result of a computation has to be stored in some array element – of course, different iterations of a parallel loop have to store the result in different places (since the loop nests have to be executed in a synchronous fashion). However, the exact loop range depends on the distribution of the array whose **home** (i.e. placement) is used as placement of the computation.

Example 2.8 *Consider the loop initializing an array B with the value $A - 2 \cdot n$.*

```
!HPF$ TEMPLATE T(0:m)
!HPF$ DISTRIBUTE T(BLOCK)
!HPF$ ALIGN B(i) WITH T(i)
!HPF$ INDEPENDENT
      DO i1=0,m
        B(i1)=A-2*n
      END DO
```

Since B is distributed across template T, which is BLOCK *distributed, the loop bounds can be reduced so that only those i_1-values are enumerated that initiate a write to a local portion of B on any given processor. This may result in code of the form:*

```
      BLOCKSIZE=((m+1-1)/SQRT_NCPU)+1
      DO i1=0+BLOCKSIZE*MY_ID, MIN(m,BLOCKSIZE*(MY_ID+1)-1)
        B(i1)=A-2*n
      END DO
```

However, if B is not distributed across the elements of T but, for example, stored on a single processor:

```
!HPF$ TEMPLATE T(0:m)
!HPF$ DISTRIBUTE T(BLOCK)
!HPF$ ALIGN B(i) WITH T(0)
!HPF$ INDEPENDENT
      DO i1=0,m
        B(i1)=A-2*n
      END DO
```

between iterations of the loop, if the target statement syntactically succeeds the source statement in the loop body. However, these dependences are *not* necessarily equivalent to the sequential execution. We do not consider these loops here, since they do not provide any additional feature for our purposes.

the situation is different – said processor has to do the complete initialization:

```
IF (MY_ID==0) THEN
  DO i1=0,m
    B(i1)=A-2*n
  END DO
END IF
```

Of course, the same holds if B is stored in a replicated fashion. The only difference is that, in this case, all *processors have to do the initialization.*

Example 2.9 reveals the problems occurring when several differently distributed write accesses appear in the same loop body: since most compilers use only one home, i.e., only one placement function, for a complete nest of parallel loops – i.e., loops that are marked as `INDEPENDENT`–, the compiler may have to generate quite complicated communication code, for it is not necessarily clear which write access should determine the placement.

Example 2.9 *If we mix assignments to arrays that are distributed according to different placements, e.g., one distributed array, the other one replicated:*

```
!HPF$ TEMPLATE T(0:m)
!HPF$ DISTRIBUTE T(BLOCK)
!HPF$ ALIGN B1(i) WITH T(*)
!HPF$ ALIGN B2(i) WITH T(i)
!HPF$ INDEPENDENT
      DO i1=0,m
!       Statement S1:
        B1(i1)=A-2*n
!       Statement S2:
        B2(i1)=A-2*n
      END DO
```

the generated loop might look as follows:

```
      DO i1=0,m
!       Statement S1:
        B1(i)=A-2*n
        IF(MY_ID==i/m) THEN
!         Statement S2:
          B2(i)=A-2*n
        END IF
      END DO
```

I.e., the compiler chose to distribute the loop body in the same way as the first write access (the one to B_1), which happens to be replicated; therefore, the complete loop body is executed for every i_1-value. However, if the compiler had chosen to align loop executions according to Statement S_2, things would have looked different: the calculation would have been aligned to B_2, and a broadcast of B_1 would have had to be created after calculation.

```
BLOCKSIZE=((m+1-1)/SQRT_NCPU)+1
DO i1=0+BLOCKSIZE*MY_ID, MIN(m,BLOCKSIZE*(MY_ID+1)-1)
  B2(i1)=A-2*n
  TMP(i1)=A-2*n
END DO
CALL BROADCAST_SENDRECV(B1,TMP,0,MY_ID)
```

More complicated communication structures than in Example 2.9 may arise, if completely differently distributed arrays are assigned in the same loop.

Example 2.10 *Consider this example:*

```
      INTEGER p1,p2
      REAL A(100)
      REAL B(100,100)
!HPF$ DISTRIBUTE A(BLOCK)
!HPF$ DISTRIBUTE B(BLOCK,BLOCK)
!HPF$ INDEPENDENT, NEW(p2)
      DO p1=1,100
        A(p1)=p1
!HPF$ INDEPENDENT
        DO p2=1,100
          B(p2,p1)=p1+p2
        END DO
      END DO
```

In this example, the home for the p_1 loop cannot be determined. Let us suppose, we have four processors; then the data layout may look as follows:

Processor	*1*	*2*	*3*	*4*
Part of A	A(1:25)	A(26:50)	A(51:75)	A(76:100)
Part of B	B(1:50,1:50)	B(51:100,1:50)	B(1:50,51:100)	B(51:100,51:100)

The loop counter values we have to enumerate for A and B, respectively, do not necessarily have anything in common with each other. Of course, the same holds for communication generated if array A or B is read in the loop nest.

However, if we fission the loops into

```
!HPF$ INDEPENDENT
      DO p1=1,100
        A(p1)=p1
      END DO
!HPF$ INDEPENDENT, NEW(p2)
      DO p1=1,100
!HPF$ INDEPENDENT
        DO p2=1,100
          B(p2,p1)=p1+p2
        END DO
      END DO
```

the compiler is perfectly able to generate very efficient code – in this case even without any communication.

Note that the HPF standard actually obliges the user to specify the NEW directive for p_2 in the above example. The NEW directive tells the compiler that the given variable is written anew in each iteration of the loop body before it is used in that loop, and that its final value is *not* used anywhere outside the loop nest. In other words, it asserts that the compiler may hold private copies of the variable on different processors, which corresponds to OpenMP's PRIVATE directive. Otherwise, we could not use that variable as loop counter of the nested loop (without asserting that the final value is propagated correctly etc.). However, most compilers actually do not heed this restriction, but use private copies as loop counters anyway.

```
!HPF$ INDEPENDENT
      DO i1=1,n
!HPF$ INDEPENDENT
        DO i2=m,m
          A(i2,i1)=B(i1)
        END DO
      END DO
```

```
!HPF$ INDEPENDENT
      DO i1=1,n
        DO i2=m,m
          A(i2,i1)=B(i1)
        END DO
      END DO
```

```
!HPF$ INDEPENDENT
      DO i1=1,n
        i2=m
        A(i2,i1)=B(i1)
      END DO
```

Figure 2.7: Possibilities of implementing an assignment to an array with a constant value in a subscript dimension: parallel loop, sequential loop, and assignment

As a rule of thumb, a compiler will probably be able to generate the most efficient code from an `INDEPENDENT` loop nest, if the write accesses in its body all refer to the same template coordinates in the same iteration.

Pitfall: Enumeration of Pathological Value Ranges

An `INDEPENDENT` loop whose counter does not appear in the subscript function of the array access that defines where the compiler places the computation of the body of the loop nest is likely to irritate the compiler. Although the `HPF` standard allows such `INDEPENDENT` loops that assign several times to the same memory cell, the result of the calculation is not well defined in this case. In order to yield the same results as a sequential compiler, a compiler may chose to execute the corresponding loop sequentially, which will at the very best not lead to any speedup.

Now, let us suppose the opposite: an access to an array that has a completely loop invariant component. This, seemingly, is quite a simple case:

```
!HPF$ DISTRIBUTE A(BLOCK,BLOCK)
!HPF$ INDEPENDENT
      DO i1=1,n
        A(m,i1)=i1
      END DO
```

In this case, it may be quite important how the constant value (m) is expressed in the subscript $A(m, i_1)$. This subscript defines which processor(s) execute(s) the loop body (note that, although A is distributed in two dimensions, only one of those dimensions is actually enumerated by an `INDEPENDENT` loop). If the compiler cannot deduce which processors to activate, it will have to enumerate the loop nest sequentially. Of course, the straight-forward way to express such an assignment is the one above – to use the value directly in the array subscript. However, in automatic code generation we may have to generate code for some assignment

```
      A(i2,i1)=i1
```

where i_2 just happens to be constant for the copy of the assignment statement at hand. It can be tedious to decide that, for this copy, we should replace the value of i_2 by the constant entry, while, for another one, we should replace i_1. Therefore, it may be easier to use an assignment to i_2. Figure 2.7 shows different possibilities of how to implement this assignment.

Both the sequential loop (Figure 2.7, middle) and the assignment (Figure 2.7, right) have the drawback that the compiler might not be able to determine whether such an assignment to i_2 changes the processor set on which to execute the loop body. Therefore, the complete loop nest may get enumerated sequentially. However, the `INDEPENDENT` loop can be translated in just the same way as any other `INDEPENDENT` loop – if the loop counter is used in the array subscript.

The safest way to implement such an assignment is to extract it from the parallel loop nest, if possible. Otherwise, the compiler may not be able to determine that the assignment to i_1, and with it the array access `B(i1)`, is loop independent. If we help the compiler with this analysis by extracting this assignment from the parallel loop, we assert that both communication generation and the determination of active processors stay unaffected by that assignment.

Simple Array Subscripts and Loop Bounds

As we have just seen, array subscript functions should be as *simple* as possible – for both write and read accesses. Of course, it is highly compiler dependent, what function can be considered *simple*.

One point that should be easy for *any* compiler to handle is the case in which the subscript consists of a single constant – like the constant m above. This is because the value of this expression cannot change during execution of the parallel loop nest. However, the use of variables that change – even potentially – during execution may inhibit from the generation of efficient communication code, since constant values are always easy to incorporate into communication statements, while changing values are not. *All* loop counters enumerated in loops that are marked `INDEPENDENT` or are themselves placed in the body of a parallel loop nest will be seen by the compiler as potentially changing their values, i.e., the compiler assumes that several iterations of the loop will be executed. Therefore, if it is a priori clear that such a loop only executes at most once, the use of its loop counter should be avoided. In this case, one should try to use as many constants (integer constants and variables that are known to stay constant) as possible.

As we have already seen, only *synchronous* programs (i.e., programs with sequential loops at the outer loop level) can be expressed well in `HPF` (and other data-parallel languages). This is because program execution is usually sequential up to the begin and after the end of a parallel loop nest. Therefore, a strategy to produce expressions that can be handled well by data-parallel compilers is to use as many constants and loop counters of loops that enclose as large a portion of the loop nest as possible. Moreover, we can build a coarse hierarchy of subscript functions:

1. Function calls or variables that can change during the execution of the independent loop nest lead to a subscript function that cannot be analyzed at all at compile time. A compiler may implement an inspector-executor scheme [MP87] or some other kind of run time resolution [CK89].

2. Depending on the compiler, even affine functions in loop counters may be too complicated to analyze – for example, ADAPTOR will not generate efficient communication for subscript functions that contain the same loop counter in different subscript dimensions (e.g., the subscript function $\mathbb{Z}^2 \rightarrow \mathbb{Z}^2 : (i_1, i_2) \mapsto (i_1, i_1)$).

3. Subscript functions that generate communication patterns that can be expressed by a sequence of subscript triplets – like a shift (e.g., $\mathbb{Z} \rightarrow \mathbb{Z} : i_1 \mapsto i_1 - n$) or a permutation – do lead to communication but, usually, this communication can be done very efficiently.

4. If the array elements accessed in a parallel loop are already stored locally, even copies between memory elements may be avoided, and the overhead for being able to execute the loop body in parallel instead of sequentially is minimized.

We will take a closer look at these levels of simplicity in Section 4.1. It should be noted, however, that not only subscript functions may influence the efficiency of the generated communication code, but also the loop bounds and strides.

As stated earlier, the basic idea behind communication optimization for this kind of compilers is to hoist communication statements out of all parallel loops – since there is no dependence between different iterations of parallel (`INDEPENDENT`) loops, this is always legal. However, since the sending side of the communication has to know which array elements to send (and to which receiver), communication can only be hoisted outside a loop if the areas of arrays that a processor should send can be deduced.

2.2 Some Algebra

In this section, some basic algebraic concepts are recalled that sometimes differ in their exact definition in the literature. The gentle reader may at first want to skip this section and come back if some definitions or properties used later on in this thesis seem unclear. For further information we refer the reader to basic literature such as Usmani [Usm87] and the handbook by Bronstein and Semendjajev [BS91].

In the remainder, we will work mostly with rational numbers, and sometimes with integer numbers, although the following definitions and theorems could as well be defined on *any* field or commutative ring, respectively. However, only the field $\mathbb{Q}$ and the commutative ring $\mathbb{Z}$ are of importance in this work.

Matrix Forms For integer numbers r and s, any r-dimensional linear subspace $\mathfrak{U}$ of $\mathbb{Q}^s$ can be defined as the solution of a linear equation system:

$$\mathfrak{U} = \{\mu \in \mathbb{Q}^s \mid M_{\mathfrak{U}} \cdot \mu = 0\}$$

for a suitable matrix $M_{\mathfrak{U}} \in \mathbb{Q}^{(s-r)\times s}$. An approved method to find such a matrix is to write down the linear constraints in a matrix and transform it into *echelon form*, which reduces the number of non-zero rows in the matrix to r. A special case of the echelon form is the *reduced echelon form.*

Definition 2.11 (Reduced Echelon Form) *Let $M = (m_{i,j})_{i\in\{1,\ldots,r\},j\in\{1,\ldots,s\}} \in \mathbb{Q}^{r\times s}$ be a $r\times s$ matrix. M is said to be in* **reduced echelon form**, *if it is in echelon form, i.e.:*

$$(\forall i : i \in \{1,\ldots,r\} : (\exists k_i : k_i \in \{k_{i-1}+1,\ldots,s\} : (\forall j' : j' \in \{1,\ldots,k_i-1\} : m_{i,j'} = 0))) \quad (2.1)$$

(with $k_0 = 0$), and the first non-zero entry of each row is 1, and is the only non-zero-entry in its column:

$$\left(\forall i : i \in \{1,\ldots,r\} : \left(\forall i' : i' \{1,\ldots,r\} : m_{i',k_i} = \begin{cases} 1 & \text{if } i = i' \\ 0 & \text{if } i \neq i' \end{cases}\right)\right) \quad (2.2)$$

(with the k_i from Formula (2.1)).

Note that, for every $M \in \mathbb{Q}^{r\times s}$, there is a matrix $U \in \mathbb{Q}^{r\times r}$ so that $U \cdot M$ is in reduced echelon form. Moreover, the reduced echelon form of a matrix is unique, i.e., for each matrix $M \in \mathbb{Q}^{r\times s}$, there is exactly one matrix $M' \in \mathbb{Q}^{r\times s}$ in reduced echelon form so that $(\forall x : x \in \mathbb{Q}^s : M \cdot x = 0 \Leftrightarrow M' \cdot x = 0)$.

Note further that we can also define a reduced echelon form of a matrix over the commutative ring $\mathbb{Z}$ instead of the field $\mathbb{Q}$. However, in that case, we cannot guarantee the first non-zero element of a row to be 1. Instead, the corresponding condition is that the greatest common divisor (gcd) of all entries in a row be 1.

Generalized Inverses Singular matrices are not invertible, i.e., there may be several pre-image points for each image point of the corresponding linear mapping. However, one can always choose *some* pre-image point as a representative for a solution, *if* the given point for which we want a pre-image is actually part of the image of the linear mapping. For example, the point $(1,2)$ is clearly *not* an image point of the function $\Psi : \mathbb{Q}^2 \to \mathbb{Q}^2 : (i_1, i_2) \mapsto (i_1, i_1)$. However, in the case that $\nu \in \text{im}(\Psi)$, i.e., $\nu[1] = \nu[2]$, for example $\nu = (1,1)$, *any* point $\nu' \in \mathbb{Z}^2$ that has the same first coordinate, $\nu'[1] = \nu[1]$, is a valid pre-image point – in our example any $(1, i_2)$ for $i_2 \in \mathbb{Q}$.

A *generalized inverse* can be used to pick some possible pre-image point for a given image point. Formally, a **generalized inverse** is defined as follows.

Definition 2.12 (Generalized Inverse) *Let $A \in \mathbb{Q}^{r\times s}$. A matrix $X \in \mathbb{Q}^{s\times r}$ is said to be a* **generalized inverse** *of the matrix A if and only if X satisfies the property:*

$$A \cdot X \cdot A = A \quad (2.3)$$

This definition means that the associated function $\Psi_X : \mu \mapsto X \cdot \mu$ of X maps all image points μ of the function Ψ_A associated with A to some point ν in the pre-image of Ψ_A so that $\Psi_A(\nu) = \mu$. This is the definition we will use, although for different purposes, sometimes other properties defining generalized inverses are adopted in the literature [Usm87, p. 84]. Note that Definition 2.12 does not necessarily define a unique matrix X for some arbitrary matrix A. If it is irrelevant which exact matrix to use, we denote a generalized inverse of A with A^g. The following algorithm can be used to obtain a particular unique generalized inverse.

Algorithm 2.2.1 [*GeneralizedInverseMin*]:
Input:
A : matrix $A \in \mathbb{Q}^{r \times s}$.
Output:
X : generalized inverse $X \in \mathbb{Q}^{s \times r}$ of A.
C : matrix $C \in \mathbb{Q}^{(r-\mathrm{rk}(A)) \times r}$ representing conditions for a vector μ to be in the image of A: $(\exists \nu : \nu \in \mathbb{Q}^s : A \cdot \nu = \mu) \Leftrightarrow C \cdot \mu = 0$.

Procedure:

Let $A' = L \cdot A$ be the echelon form of A;
/* Solve the inhomogeneous equation system for each column of L */
/* as solution vector */
$X := 0$;
for $j = 1$ **to** r
  /* For each row of A', solve for a different column entry */
  for $i = \mathrm{rk}(A)$ **to** 1, **step** -1
    $k := \min(\{k' \mid A'[i, k'] \neq 0\})$;
    $X[k, j] := \frac{L[i,j] - A'[i,\cdot] \cdot X[\cdot,j]}{A'[i,k]}$;
  endfor
endfor

$$C := \begin{pmatrix} L[\mathrm{rk}(A)+1, \cdot] \\ \vdots \\ L[r, \cdot] \end{pmatrix};$$

return (X, C);

Theorem 2.13 *Algorithm 2.2.1 calculates a generalized inverse X of the given matrix $A \in \mathbb{Q}^{r \times s}$. For each given vector μ, $X \cdot \mu$ is the vector whose componentwise absolute value in reverse order is lexicographically minimal in the set of vectors $\{\nu \mid A \cdot \nu = \mu\}$.*
Matrix C represents a condition for a vector μ to be in the image of A (i.e., only for vectors $\mu \in \mathbb{Q}^r$ that satisfy $C \cdot \mu = 0$, there is at least one vector $\nu \in \mathbb{Q}^s$ so that $A \cdot \nu = \mu$): $(\exists \nu : \nu \in \mathbb{Q}^s : A \cdot \nu = \mu) \Leftrightarrow C \cdot \mu = 0$.

Proof:

The proof proceeds in two steps. First, we verify that Algorithm 2.2.1 computes a generalized inverse of A. Then we examine the additional properties of the resulting matrices.

Generalized Inverse X Algorithm 2.2.1 actually performs a simple backward substitution. For each step j, it calculates a solution χ_j for the equation system

$$A' \cdot \chi_j = L[\cdot, j]$$

which is solvable iff the equations $A'[\cdot,i]\chi_j = L[i,j]$ are solvable for all $i \in \{1,\ldots,r\}$. The unique solution for the i-th component of χ_j is then

$$\frac{L[i,j] - A'[i,\cdot] \cdot \chi_j}{A'[i,k]}$$

since $(\forall l : l \in \{1,\ldots,k-1\} : \chi_j[l] = 0)$, and thus the values of components $1,\ldots,k-1$ of χ_j are irrelevant, while the values of components $k,\ldots,s$ are already chosen so that the corresponding linear equations are satisfied. However, this does not hold for rows consisting entirely of zeros in A' whose corresponding row in L does not consist entirely of zeros. These represent equations that define a condition for a vector μ to be part of the image of A. This means there may be no solution for a χ_j with $A' \cdot \chi_j = L[\cdot,j]$, which is equivalent to $A \cdot \chi_j = \iota_j$; however, there is always a solution for $A \cdot \chi_j = \nu$ with $\nu \in \text{im}(A)$. Put it differently, these rows represent conditions for $L[\cdot,j]$ to be in the image of A. So, the calculated solution does not hold for any $\nu \in \mathbb{Q}^s$, but only for some $\nu \in \text{im}(A)$. Since the columns of A represent a base of $\text{im}(A)$, we can only solve for an X with $A' \cdot X \cdot A = L \cdot A$.

Therefore, each column j of X, separately, represents a solution to obtain column j of $L = L \cdot I_{r,r} \cdot A$ and, since in matrix multiplication, the result matrix can be built by consecutively calculating its columns from the corresponding columns of the right hand side matrix, X is a solution for $A' \cdot X \cdot A = L \cdot I_{r,r} \cdot A$, and thus $A \cdot X \cdot A = A$.

By setting X to 0 at the beginning, we implicitly choose the value 0 for an entry of χ_j – and thus of X – that is not explicitly defined by a constraint. In the case that $\text{rk}(A) < r$, there is at least one constraint that contains several variables that are not defined in earlier steps of the algorithm. In that case, only the lowest numbered component of χ_j that occurs in that constraint is defined – the other components are left at their default value 0. Thus, if several components of a vector ν with $A \cdot \nu = \mu$ can be freely chosen, Algorithm 2.2.1 chooses all but the least numbered component to be zero – the minimal absolute value possible. Therefore the absolute value of the result vector is always minimal with respect to a lexicographic order in which the precedence of the components of vectors is reversed.

Conditions represented by C The role of C becomes clear when we look at the condition

$$\begin{array}{rcl} A \cdot \nu & = & \mu \\ & \Leftrightarrow & \\ \underbrace{L \cdot A}_{=A'} \cdot \nu & = & L \cdot \mu \quad \text{(with } L \text{ as in Algorithm 2.2.1)} \end{array}$$

Read as a system of equalities, the first $\text{rk}(A)$ equalities

$$\begin{pmatrix} A'[1,\cdot] \\ \vdots \\ A'[\text{rk}(A),\cdot] \end{pmatrix} \cdot \nu = \begin{pmatrix} L[1,\cdot] \\ \vdots \\ L[\text{rk}(A),\cdot] \end{pmatrix} \cdot \mu$$

always have a solution for ν, given any μ (and vice versa). Yet, the rest of the equalities

$$\underbrace{\begin{pmatrix} A'[\text{rk}(A)+1,\cdot] \\ \vdots \\ A'[r,\cdot] \end{pmatrix}}_{=0} \cdot \nu = \begin{pmatrix} L[1,\cdot] \\ \vdots \\ L[\text{rk}(A),\cdot] \end{pmatrix} \cdot \mu$$

also has to be met. Since this part of A' is a zero matrix, this is possible if and only if

$$\underbrace{\begin{pmatrix} L[1,\cdot] \\ \vdots \\ L[\operatorname{rk}(A),\cdot] \end{pmatrix}}_{=C} \cdot \mu \;=\; 0$$

which is exactly the definition of C. ✓

By iterating through the j-loop, Algorithm 2.2.1 calculates a generalized inverse of a matrix $A \in \mathbb{Q}^{r\times s}$ by solving the inhomogenous equation system

$$A \cdot X \;=\; I_{r,r}$$

for each column of $I_{q,q}$. I.e., for each j, we solve the equation system

$$A \cdot \chi_j \;=\; \iota_r$$

and thus obtain χ_j as column j of the solution X (since each column of X only depends on its corresponding column of $I_{r,r}$ and on matrix A, we can compute each column independently of the others) [BS91]. Since we always choose 0 for a coordinate that remains undetermined in the equation system, we obtain a unique generalized inverse that maps a vector $\mu \in \operatorname{im}(A)$ to a vector ν such that ν has as many zero coordinates as possible (in as highly numbered dimensions as possible) [Fab97]. We denote this generalized inverse by A^{go}. Note, however, that this specific property that makes A^{go} unique is not actually needed in this work. We will only use this as our implementation of a generalized inverse.

Example 2.14 *The matrix M_Ψ defining the mapping $\Psi : \mathbb{Q}^2 \to \mathbb{Q}^2 : \begin{pmatrix} i_1 & i_2 \end{pmatrix}^T \mapsto \begin{pmatrix} i_1 & i_1 \end{pmatrix}^T$ above is:*

$$M_\Psi = \begin{pmatrix} 1 & 0 \\ 1 & 0 \end{pmatrix}$$

The corresponding unit matrix is $I_{2,2}$, and $L = \begin{pmatrix} 1 & 0 \\ -1 & 1 \end{pmatrix}$, i.e., we solve

$$\underbrace{\begin{pmatrix} 1 & 0 \\ 1 & 0 \end{pmatrix}}_{=A} \cdot X \;=\; \begin{pmatrix} 1 & 0 \\ 0 & 1 \end{pmatrix}$$

$$\Leftrightarrow$$

$$\underbrace{\begin{pmatrix} 1 & 0 \\ 0 & 0 \end{pmatrix}}_{=A'} \cdot X \;=\; \underbrace{\begin{pmatrix} 1 & 0 \\ -1 & 1 \end{pmatrix}}_{=L\cdot I_{2,2}=L}$$

Algorithm 2.2.1 assigns the echelon form of M_Ψ to A'. The rest of the algorithm is devoted to calculating X column by column – iterating through column j.
For this task, we only have to solve the equation system

$$A' \cdot X[\cdot,j] = L[\cdot,j]$$

by Gaussian elimination. This is performed for each row i of $X[\cdot,j]$, stepping from the last row to the first one; for each row, we perform the following steps:

1. *Assign the number of the first column of $A'[i,\cdot]$ to k – in our example, this yields $k = \bot$ (undefined) for $i = 2$, and $k = 1$ for $i = 1$, since the only non-zero entry of A' is $A'[1,1]$.*

With the assignment to k, we choose the row of $X[\cdot, j]$ that we want to define in this $\begin{pmatrix} i \\ j \end{pmatrix}$-iteration. If – contrary to this example – there were several *non-zero entries in the i-th row of A' (with only zero-entries below those entries), we could actually choose k from a set of column numbers. In the algorithm, we fill all other rows of $X[\cdot, j]$ that we could have chosen for k with a default value of 0 (by assigning $X = 0$ at the beginning). Always choosing the same value for k in different j-iterations asserts that it is always a complete row of X that is filled with such a default value. And choosing 0 as this default value ensures that this default value does not influence the value of $X[\cdot, j]$ in other rows; a different default value* might *lead to other values – but does not necessarily have to.*

2. *If k is well defined – which is the case in our example if $i = 2$, independent of the j-iteration, we leave X as it is and proceed to the next iteration. In particular, this means that $X[\cdot, j]$ stays 0 where we originally defined it to be 0. Note that k is undefined iff row $A'[i, \cdot] = 0$. Note further that, in this case, the corresponding row $L[i, \cdot]$ does not necessarily have to be 0 – in our example, we even have $L[2, \cdot] = \begin{pmatrix} -1 & 1 \end{pmatrix}$. This means that we have an additional condition to meet, namely that, given a vector μ, this vector has to satisfy*

   $$-1 \cdot \mu[1] + 1 \cdot \mu[2] = 0$$

   in order for our resulting generalized inverse X to "work", i.e., in order for $X \cdot \mu$ to return a vector such that $A \cdot X \cdot \mu = \mu$. Here, this condition is that the first component of μ equals the second component of μ, which is not astonishing, since the mapping Ψ, from which we obtained A', is defined as $\begin{pmatrix} i_1 & i_2 \end{pmatrix}^T \mapsto \begin{pmatrix} i_1 & i_1 \end{pmatrix}^T$, i.e., both output coordinates are equal.

3. *If k is well defined – which, in our example, is the case if $i = 1$, independent of the j-iteration, the equation that has to hold to satisfy the equation system reads*

   $$X[k, j] \cdot A'[i, k] + (\textstyle\sum l : l \in \{k+1, \ldots, s\} : \ A'[i, l] \cdot X[l, j]) = L[k, j]$$

   Since X is filled with 0 for all entries not defined yet, this can be satisfied with the assignment

   $$X[k, j] = \frac{L[i, j] - A'[i, \cdot] \cdot X[\cdot, j]}{A'[i, k]}$$

   Since k is only well defined for $i = 1$ in this example, this equation defines the first row of X. The values for the different j-iterations are:

j	$X[1, j]$
1	$\frac{1-0\cdot 0}{1} = 1$
2	$\frac{0-0\cdot 0}{1} = 0$

So, in our example, the algorithm first sets X to 0. Then it leaves the second row of X untouched, but changes its first row to $\begin{pmatrix} 1 & 0 \end{pmatrix}$. The result is a matrix

$$X = \begin{pmatrix} 1 & 0 \\ 0 & 0 \end{pmatrix}$$

We see that

$$M_\Psi \cdot X \cdot M_\Psi = \begin{pmatrix} 1 & 0 \\ 1 & 0 \end{pmatrix} \cdot \begin{pmatrix} 1 & 0 \\ 0 & 0 \end{pmatrix} \cdot \begin{pmatrix} 1 & 0 \\ 1 & 0 \end{pmatrix} = \begin{pmatrix} 1 & 0 \\ 1 & 0 \end{pmatrix}$$

which fits exactly with the definition of a generalized inverse.

Linear Relations Viewed as sets, functions are a special case of relations. However, their representation as matrices may differ. We always represent a linear function $\Psi : \mathbb{Q}^r \to \mathbb{Q}^s : \nu \mapsto$

$M_\Psi \cdot \nu$ by its defining matrix M_Ψ. However, sometimes we may have to represent more general relations that still are solutions to linear equation systems:

Definition 2.15 (Linear Relation) *Let $\mathfrak{R}$ be a ring. A* **linear relation** $\Lambda \subseteq \mathfrak{R}^r \times \mathfrak{R}^s$ *is a relation such that*

$$\Lambda = \left\{ (\upsilon, \nu) \in \mathfrak{R}^r \times \mathfrak{R}^s \,\middle|\, M_\Lambda \cdot \begin{pmatrix} \upsilon \\ \nu \end{pmatrix} = 0 \right\}$$

for a suitable matrix $M_\Lambda \in \mathfrak{R}^{q \times (r+s)}$ with

$$(\forall i : i \in \{1, \ldots, q\} : (\exists j, k : j \in \{1, \ldots, r\}, k \in \{r+1, \ldots, r+s\} : M_\Lambda[i,j] \neq 0 \wedge M_\Lambda[i,k] \neq 0)) \tag{2.4}$$

Equation 2.4 may come as a bit of a surprise: it actually only states that each row in M_Λ really has to define a relation between some element of the domain $\mathfrak{R}^r$ and the image $\mathfrak{R}^s$ of the relation with $\mathfrak{R}^r$ and $\mathfrak{R}^s$ being unrestricted $\mathfrak{R}$-moduli. This does not mean that we disallow degenerate relations. It may be a helpful representation in an implementation of the algorithms presented in this thesis but does not really restrict the relations we work with in any way. We will only consider $\mathfrak{R} = \mathbb{Q}$ or $\mathfrak{R} = \mathbb{Z}$. Note that in the definition of a (non-scoped) linear relation, it is actually not necessary to use a matrix of rational numbers even for $\mathfrak{R} = \mathbb{Q}$, since we can always multiply the condition $M_\Lambda \cdot \begin{pmatrix} \upsilon \\ \nu \end{pmatrix} = 0$ with the least common multiple (lcm) of the denominators:

$$\begin{array}{rcl}
\left(\sum i : i \in \{1, \ldots, r\} : \frac{m_i}{d_i} \cdot \upsilon[i]\right) + \left(\sum i : i \in \{1, \ldots, s\} : \frac{m_{r+i}}{d_{r+i}} \cdot \nu[i]\right) & = & 0 \\
& \Leftrightarrow & \\
\mathrm{lcm}(d_1, \ldots, d_s, d_{s+1}, \ldots, d_{r+s}) \cdot & & \\
(\left(\sum i : i \in \{1, \ldots, r\} : \frac{m_i}{d_i} \cdot \upsilon[i]\right) + \left(\sum i : i \in \{1, \ldots, s\} : \frac{m_{r+i}}{d_{r+i}} \cdot \nu[i]\right)) & = & 0
\end{array}$$

For the same reason, a linear relation based on rationals also represents a linear relation based on integers by virtue of convention: we may just restrict ourselves to the integer solutions to the above equations. However, the inclusion of scoping information – i.e., the information which dimension is restricted by which bounds – into a linear relation raises the need for distinguishing rational from integer numbers.

We distinguish between the representation of a linear function as a function and its representation as a linear relation above, although, of course, any linear function also has a representation as linear relation.

Example 2.16 *Any homomorphism*

$$\Psi : \mathbb{Q}^r \rightarrow \mathbb{Q}^s : \upsilon \mapsto M_\Psi \cdot \upsilon$$

can be represented as a linear relation $\Lambda \subseteq \mathbb{Q}^r \times \mathbb{Q}^s$, where $M_\Lambda \in \mathbb{Q}^{r \times (r+s)}$ has the form

$$\left(\begin{array}{c|ccc} & -1 & & \\ M_\Psi & & \ddots & \\ & & & -1 \end{array} \right)$$

In Section 2.4 we will discuss the issue of different representations in greater detail.

2.3 The Polyhedron Model

In contrast to traditional code analysis and transformations in which loops are viewed as unpredictable control structures, the polyhedron model considers different values of loop counters (or

indices) in different loop iterations as a subset of an integer vector space with the loop bounds as restrictions in the extents of the corresponding dimensions of the vector space [Fea92a, Fea92b, Len93, Ban93, Ban94]. This so-called **index space** of a loop nest – and thus the index space IS(S) of a statement S – can be expressed as a polyhedron in $\mathbb{Z}^{n_{src}+n_{blob}}$ where n_{src} is the number of loops and n_{blob} the number of loop independent (**structural**) **parameters** (also called **blobs**) of the source program fragment considered. In contrast to the usual conventions in the polyhedron model, we view IS(S) as a subspace of the index space of a complete program fragment, i.e., as a subset of a vector space whose dimensions are spanned by *all* indices in the program fragment. This enables a simple mapping between index space dimensions and corresponding program variables in an implementation. Another possible representation would be to use only loop indices of embracing loops, reducing the complexity of matrix computations. As we will later see, one may easily convert from the first representation to the second one if necessary. The elements of the index space are called **index vectors**. Both indices and parameters are scalar variables in the source program. The difference between indices and parameters is that indices may change their respective value during program execution (albeit in a well defined, regular manner), while parameters do *not* change their values. In addition to the symbolic constants defined in the program fragment considered, we always consider two special parameters:

- m_c is known to be 1; this enables us to use so-called homogeneous coordinates in order to represent affine expressions as *linear expressions* in a vector space that has an additional dimension for this parameter.
- m_∞ is known to be infinity – we will discuss this parameter later in detail.

In general, we call an expression **linear** if it is a linear expression in the indices and parameters (including the special parameters above) of the program fragment under consideration. We suppose the loop indices and parameters to be integers – otherwise they are not suitable as parts of subscript expressions (languages like `Fortran` even disallow the usage of any other type as loop counters or array subscripts).

Usually, the dimensions of $\mathbb{Z}^{n_{src}+n_{blob}}$ are enumerated as follows:

- The first n_{src} dimensions correspond to the loop counters in the textual order in which the loops appear in the program text.
- The next $n_{blob} - 2$ dimensions correspond to structural parameters.
- The next dimension corresponds to m_∞.
- The last dimension of $\mathbb{Z}^{n_{src}+n_{blob}}$ is the one reserved for m_c.

In order to represent the index space as a polyhedron, it is necessary for the loop bounds to represent hyperplanes of $\mathbb{Z}^{n_{src}+n_{blob}}$, i.e., they have to be linear expressions in surrounding loop variables, parameters and the additional dimension for m_c. Thus, in general, it is not possible to analyze a complete program. But intraprocedural program fragments can be analyzed in the polyhedron model if they

- contain only statements without side effects, i.e., only assignments and calls to `PURE` functions and subroutines as defined in `HPF` [Hig93, Hig97] and `Fortran` [Int97, MR98], and
- contain only accesses to scalars or arrays subscripted by linear expressions, and
- contain only loops as control structures, and
- loop bounds are linear expressions and strides are integer constants, or lower bounds are minima, upper bounds maxima, of linear expressions.

The last two conditions can be lifted somewhat [Gri96]. Nonetheless, in this work, we assume that all these restrictions hold, if not stated otherwise.

Strides in loops create holes in the index space. For example, the program fragment

```
DO i1=l1,u1,s1
  DO i2=l2,u2,s2
    A(i2,i1)=A(i2-1,i1)+A(i2,i1-1)
  END DO
END DO
```

corresponds to the polyhedron shown in Figure 2.8 for the values $l_1 = 0$, $l_2 = $ i1, $u_1 = u_2 = 10$, $s_1 = 2$, and $s_2 = 3$. Such polyhedra can be represented by a combination of

1. their borders (so-called faces), defined by the hyperplanes that tightly surround the index space, and
2. an integer lattice defining the exact integer points of the index space inside the rational polyhedron.

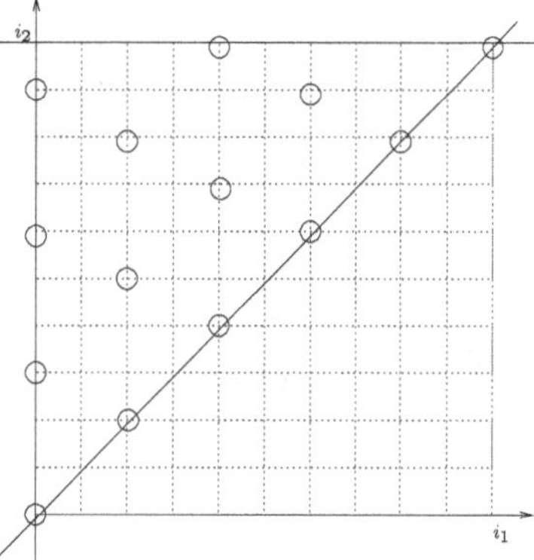

Figure 2.8: Polyhedron with holes representing non-unit strides in a loop program.

In Figure 2.8, the boundaries of the polyhedron are given by continuous lines. With unit strides, the lattice mentioned above is just the one built from all integer vectors (of appropriate dimension), represented by the dotted lines. This lattice can be represented in implementations by an implicit restriction to only integer values inside a rational polyhedron. With non-unit strides, we have to use a lattice that depends on these strides. The integer polyhedron is then also called a **$\mathbb{Z}$-polyhedron**. The lattice may be a bit more complicated, as shown by the actual index space points in the figure, where we have holes between the actual index points that correspond to integer coordinates but not to any index vector. These holes can be represented by a polyhedron without holes, if we introduce additional parameter dimensions to the index space (and adhere to the implicit rule of just considering the integer points in any given polyhedron). Restricting the original index vectors to multiples of these so-called **pseudo parameters** then results in holes in the projection of the polyhedron onto the corresponding original dimension. In contrast to normal index vector dimensions, these additional dimensions represent existentially quantified variables that do not have to be enumerated, but only tested for the existence of a solution. Section 3.7 describes the code corresponding to those dimensions in more detail. For example, the index space corresponding to Figure 2.8 above can be defined by the following restrictions (with b_1, b_2 as newly introduced existentially quantified variables as described above):

$$\begin{array}{rcl} i_1 & \geq & 0 \\ -i_1 + 10 & \geq & 0 \\ -i_1 + i_2 & \geq & 0 \\ -i_2 + 10 & \geq & 0 \\ i_1 - 2 * b_1 & = & 0 \\ i_2 - i_1 - 3 * b_2 & = & 0 \end{array}$$

Note that the maximal number of these artificially introduced, existentially quantified parameters is given by the maximal depth of loop nests in the considered program fragment.

The usual case in the polyhedron model is that the order in which points in a polyhedron are to be enumerated (according to the modelled program fragment) is the **lexicographic order** on $\mathbb{Z}^n$ (for an n-dimensional polyhedron). We denote this order with $\prec$. This is guaranteed by associating the i-th loop in the program text with the i-th dimension of the polyhedron.

In order to model the flow of data – and thus the computation performed – dependence analysis algorithms consider memory accesses during program execution. The basic construct is the **textual access**: a textual access is a reference t to a memory cell in the program. As mentioned before, these accesses are array accesses of the form

$\mathtt{A}(\Psi(i_1, i_2, m_1, m_2, m_\infty, m_c))$

with Ψ being a linear function in the indices (loop counter variables) i_1, i_2, (loop independent) parameters m_1, m_2, m_∞, and m_c. Note that it is thus possible to write down expressions that are linear in ∞ – in this case, the usual rules apply and, for any rational q, we have:

1. $q \cdot m_\infty = \operatorname{sgn}(q) \cdot m_\infty$.
2. $q \cdot (-m_\infty) = -\operatorname{sgn}(q) \cdot m_\infty$.
3. $q + m_\infty = m_\infty$.
4. $q - m_\infty = -m_\infty$.
5. $-m_\infty < q < m_\infty$.

However, these rules are relevant only when simplifying polyhedra: there we can assume a polyhedron that implies a finite m_∞ to be empty.

For an instancewise approach, we distinguish between different run time *instances*, i.e., different executions of the same access for different values of surrounding loop indices. Thus, an **access instance** $\langle \nu, \mathtt{t} \rangle$ of a textual access t in a statement S is defined by the index vector $\nu \in \mathrm{IS}(S)$ and the textual access t. It is the atomic item of the dependence analysis algorithms employed here.

An access instance of a program can either be a **read** or a **write** access (i.e., load from memory or store into memory). A dummy argument of a subroutine that may be written as well as read is, for this purpose, replaced by a read access followed by a write access.

Transformations or dependence relations between polyhedra are usually expressed by linear mappings or – more generally – by linear relations.

2.3.1 Going Fine Grain

In this thesis, we aim at constructing a program fragment that can be executed as efficiently as possible. For this purpose, we eliminate as many redundant calculations as possible and then try to produce efficient code from the description of the remaining calculations to be performed. In order to obtain an accurate description of the calulations in the program fragment, we aim at a precise model of the program execution; i.e., our goal is to model the complete execution of a program fragment by polyhedra and (dependence) relations between these polyhedra. In our refined model, we want to keep the property that the sequential program execution corresponds to the enumeration of the polyhedron in lexicographic order.

The structures we use for this purpose are generalizations of the conventional polyhedron model. We will therefore introduce some extensions to the model in this section.

Linear Expressions Linear expressions are considered atomic, i.e., they are computable at essentially no cost, and represent a well defined value that only depends on the place in the polyhedron where they are evaluated. Only the top level operator of a linear expression has a representation in our graphs – there is no reason to represent the complete text of the linear expression. Note, however, that even this top level operator does not actually *need* to be represented, if we can encode the value represented by this linear expression in a dependence relation. The question is rather: do we want to be able process linear expressions in the same way as non-linear expressions, e.g., do we want to extract a loop independent linear expression term `2*n+m` from a loop iterating through an index i? If the answer is *no*, we do not actually need to represent linear expressions at all.

We work on expressions which are compositions of several textual accesses, including operators and function calls. Thus, we generalize the concept of single textual accesses to the execution of operators.

Definition 2.17 *Let $\mathcal{F}$ be a set of* **function symbols**. *In addition to constants and user defined function names, the following symbols are elements of $\mathcal{F}$:*

`+,-,*,/,max,min,CALL,...`*:*
Intrinsic functions.

`LaExpr`*:* *Linear expression – considered atomic.*

`read_A`*:* *Read access to an element of array A.*

`write_A`*:* *Write access to an element of array A.*

`Assign`*:* *Assignment operator.*

Each function symbol $\odot \in \mathcal{F}$ represents an operator that can be executed by the processor. The function symbols are associated with an input arity by a function $\text{ArityIn} : \mathcal{F} \to \mathbb{N}$ *and an output arity by a function* $\text{ArityOut} : \mathcal{F} \to \mathbb{N}$ *indicating the number of input and output arguments, respectively. Thus the arity of a function symbol $\odot$ is* $\text{ArityIn}(\odot) + \text{ArityOut}(\odot)$. *Without loss of generality, we assume all input arguments of $\odot$ to be the first* $\text{ArityIn}(\odot)$ *arguments and the output arguments to be arguments number* $\text{ArityIn}(\odot) + 1$ *to* $\text{ArityIn}(\odot) + \text{ArityOut}(\odot)$. *In addition to output arguments as described above, every function is expected to return some return value (which might be ignored for some functions). A write access thus has an output arity of* 0 *(*$\text{ArityOut}(\texttt{write\_}\cdot)=0$*).*

In order to identify the operation the processor has to execute, it does not suffice to identify the values of the loop indices, since the order of execution also depends on the sequence of the operators in the program text. Beside the value of indices, the order of execution is defined by:

Statement order:
The statements are executed from top to bottom of the program text.

Argument evaluation:
Before the processor can perform a computation, the input must be present; in general, an operation is executed in three steps:

1. Evaluate the memory locations and stored values of input arguments, and the memory locations of output arguments.
2. Compute the operation itself.
3. Write the calculated values of output arguments into memory for later use.

In correspondence to this definition of execution order – which holds for most strict programming languages – we define two more dimensions of the polyhedra that model the program execution: *occurrences* and *operand numbers*. These definitions represent the central building blocks for our transformation framework.

Definition 2.18 (Occurrence) *With the operator symbols $\mathcal{F}$ defined in Definition 2.17 and the special operator symbols* `;`*,* `if` *and* `loop` *representing control statements, a program fragment can be viewed as a* **term** *on* $\mathcal{F} \cup \{\texttt{;}, \texttt{if}, \texttt{loop}\}$. *In contrast to the usual definition of a term [BCL82], the arguments of a function symbol can be input or output arguments. Each point $\mathfrak{t}$ in the program text is assigned a unique integer number* $\text{Occ}(\mathfrak{t})$*, its* **occurrence**, *such that:*

- *If* $\mathfrak{t} = \,;(\mathfrak{t}_1, \mathfrak{t}_2)$, *then* $\text{Occ}(\mathfrak{t}_1) < \text{Occ}(\mathfrak{t}_2)$.
- *If* $\mathfrak{t} = \odot(\mathfrak{t}_1, \ldots, \mathfrak{t}_j, \mathfrak{s}_1, \ldots, \mathfrak{s}_k)$ *with* $\odot \in \mathcal{F}$, $\mathfrak{t}_{j'}$ *input arguments and* $\mathfrak{s}_{k'}$ *output arguments, then the occurrences of these arguments are ordered in the following way:*
  - $\text{Occ}(\mathfrak{t}_{j'}) < \text{Occ}(\mathfrak{t}) < \text{Occ}(\mathfrak{s}_{k'})$.
  - *All* $\mathfrak{s}_{k'}$ *have the form* $\mathfrak{s}_{k'} = \texttt{write\_A}(\mathfrak{s}_{k',1}, \ldots, \mathfrak{s}_{k',l})$*; the* $\mathfrak{s}_{k',l'}$ *are enumerated such that:* $\text{Occ}(\mathfrak{s}_{k',l'}) < \text{Occ}(\mathfrak{t})$.

This definition guarantees that occurrences are enumerated in the order of execution within a single loop iteration.

In addition, the memory location of an output parameter is determined before execution of an operator; this property is actually only relevant if we admit non-linear subscript expressions – which we will not consider in this work.

In order to re-establish the calculation to be performed, we use the inverse mapping of Occ, Occ^{-1}, and apply a function called head to the result which yields the corresponding function symbol, as sketched in Figure 2.10.

Note the differences of the definitions presented here with the usual definitions in computer algebra:

- A term can have *several* output arguments. We also consider all operators to have *at least* one output argument, which stands for the execution of the operation and can be identified with its return value. This return value may be `void` – which is the case for the `CALL` and the `Assign` operator. The return value is the single output argument that can be put directly into expressions (such as $F(i) + 1$) in order to use a result of the performed calculation. The only reason for such a distinguished output argument is the fact that most existing programming languages allow up to one output argument to be treated in this way.
- According to the above definition, an occurrence is given by a single integer. In computer algebra, it is usually a sequence of integers defined by the argument positions of the subtrees containing the one subtree that is to be identified by the given occurrence. Our approach corresponds to a linearization of this sequence.

For the determination of occurrence numbers, we use a **syntax tree**, in which control structures govern their bodies as well as the arguments that control the execution, while function and subroutine calls govern their arguments [ASU86]. By traversing this tree following an appropriate pattern, we can assign occurrence numbers in increasing order to each node so that the execution order of a single loop iteration is defined by the order on the occurrence numbers.

Example 2.19 *The following simple program fragment contains five occurrences whose executions depend on each other:*

```
A=2*n-A
```

The five occurrences are:

Occurrence 1:
: `read_A` *is an operator without input arguments (in general, there may be input arguments – the array subscripts that specify the exact location from where to read); the operator returns the current value of the 0-dimensional array A in its single output argument (its return value).*

Occurrence 2:
: `n*2` *is a linear expression composed of the structural parameter* `n` *and the integer constant 2; since linear expressions are atomic, this corresponds to a single occurrence with no input arguments and one output argument.*

Occurrence 3:
: *The addition* `-(2*n,read_A)` *takes the output arguments of Occurrences 1 and 2 as input arguments.*

Occurrence 4:
: *The assignment operation* `Assign(-(*(2,n),read_A),write_A)` *takes the output of Occurrence 3 as input argument and passes this value through to its single output argument.*

Occurrence 5:
: `write_A` *is an operator without input arguments (just as* `read_A`*) that writes the currently calculated value to memory.*

Note that, although Occurrence 1 does not *take any input arguments, the result of its execution will* depend *on the* current value of A. *This dependence is discovered by the usual dependence analysis in the polyhedron model. Likewise, the calculation (writing to memory) to be performed depends on the currently calculated value, which may be some i-th output argument of an operator execution.*

In order to model the complete program text, we still have to have a way of telling the argument position of an expression in a function call: in Example 2.19, it is important that we want to compute the expression $A - 2 \cdot n$, and not $2 \cdot n - A$, i.e., we have to be able to distinguish between the different argument positions in the model. Since our model is based on points in $\mathbb{Z}^n$ and relations between these points, this difference should be expressed as different positions in $\mathbb{Z}^n$. For this purpose, we introduce one more integer number:

Definition 2.20 *With each function symbol* $\odot$*, we associate the set of* **operand numbers** *via a function* $\mathrm{OpSet} : \mathcal{F} \rightarrow \mathcal{P}(\mathbb{Z})$*:*

$$\mathrm{OpSet}(\odot) = \{-\mathrm{ArityIn}(\odot), \ldots, \mathrm{ArityOut}(\odot)\}$$

The execution of a single occurrence corresponds to the enumeration of its operand numbers in increasing order:

1. *A negative operand number $-j$ corresponds to loading the j-th input argument of the operator someOp to be executed onto the stack - or into a register, if appropriate.*

2. *Operand number* 0 *represents the execution of the operator itself (by calling a function or a processor instruction), and storing back the return value.*

3. *A positive operand number j represents the storing of the $(j+1)$-st output value of the operator.*[3]

Now we can model all the operations the processor has to perform as elements of a polyhedron. Therefore, we build polyhedra of *occurrence instances.*

Definition 2.21 (Occurrence instance) *Let o be an occurrence within a program fragment containing n_{blob} parameters and n_{src} loops. Let $\mathfrak{P}$ be the $(n_{src} + n_{blob})$-dimensional polyhedron that represents the index space of the loop nest. An* **occurrence instance** $\alpha \in \mathbb{Z}^{n_{src}+n_{blob}+2}$ *is a vector*

$$\alpha = (\alpha_1, \ldots, \alpha_{n_{src}+n_{blob}+2})^T$$

such that

$$\begin{array}{rcl} (\alpha_1, \ldots, \alpha_{n_{src}+n_{blob}})^T & \in & \mathfrak{P} \\ \alpha_{n_{src}+n_{blob}+1} & = & o \\ \alpha_{n_{src}+n_{blob}+2} & \in & \mathrm{OpSet}(\mathrm{head}(\mathrm{Occ}^{-1}(o))) \end{array}$$

We denote the mapping of an occurrence instance α to its occurrence number (o) and its operand number with $\mathrm{OccId}(\alpha)$ *and* $\mathrm{OpId}(\alpha)$*, respectively. I.e., in the source program* $\mathrm{OccId}(\alpha) = \alpha[occdim_{src}]$, $\mathrm{OpId}(\alpha) = \alpha[opdim_{src}]$*, and in the target program* $\mathrm{OccId}(\alpha) = \alpha[occdim_{tgt}]$, $\mathrm{OpId}(\alpha) = \alpha[opdim_{tgt}]$.
Let us now consider an occurrence instance α representing the application of an operator $\mathrm{head}(\mathrm{Occ}^{-1}(\mathrm{OccId}(\alpha))) = \odot$*. In order to execute α, we first have to load the corresponding operands in registers or on the stack (depending on the machine architecture). Somehow, the*

[3]The first output value is the one with operand number 0, which is the return value of the expression.

operand have to be identified and associated to an argument position. For determining the occurrence instance representing operand i of α, we define an **operand selector**

$$OpndSel : \{-\text{ArityIn}(\odot), \ldots, \text{ArityOut}(\odot)\} \times \mathfrak{OI} \to \mathfrak{OI} : (i, \alpha) \mapsto \begin{pmatrix} \alpha\texttt{[1]} \\ \vdots \\ \alpha\texttt{[}opdim - 1\texttt{]} \\ i \end{pmatrix}$$

(where $opdim = opdim_{src}$ if α is taken from the source index space and $opdim = opdim_{tgt}$ if α is taken from the target index space).

We define the set $\mathfrak{OI}$ above as the set of occurrence instances of the program fragment under consideration. The index vector of an occurrence instance α (its position in the index space) is denoted with $\text{Idx}(\alpha)$ (i.e., the first n_{src} dimensions of $\mathbb{Z}^{n_{src}+n_{blob}+2}$ as $\text{Idx}(\alpha) = (\alpha\texttt{[1]}, \ldots, \alpha\texttt{[}n_{src}\texttt{]})$. Note that we will usually argue about a subset $\mathfrak{O}_i \subseteq \mathfrak{OI}$. If $\mathfrak{O}_i$ is a polyhedron, we usually assume that all its elements exhibit the same occurrence number – i.e., $\#(\text{OccId}(\mathfrak{O}_i)) = 1$. In this way, we can easily identify $\mathfrak{O}_i$ with a single mathematical operator that is computed in this set.

Example 2.22 *Figure 2.9 shows the set of occurrence instances of the following program fragment*

```
DO i1=0,1
  B(i1)=2*n-A
END DO
```

The parameter dimensions are ignored in this figure, the occurrence dimension reaches from bottom to top, the operand dimension from left to right, and the index dimension for i_1 together with the operand dimension, forms the ground plane. The arrows indicate dependence relations between a term and its subterms. Linear expressions are atomic, write accesses are executed after computations, and memory locations (defined by the subscript i_1 of $B(i_1)$) are calculated before the complete expression. Note that, since it is clear whether an array access is a read or write here, we represent the term `Assign(-(read_A,*(2,n)),write_B(i))` *just as* `Assign(-(A,*(2,n)),B(i))`. *We will do so where appropriate to save space. Note further that, since* `write_B(i)` *is the output argument of the assignment operator, it is placed* above *the complete term representing the assignment. Correspondigly, the dependence emerging from* `i` *goes to* `write_B(i)` *but not to the complete term* `Assign(-(A,*(2,n)),B(i))`.

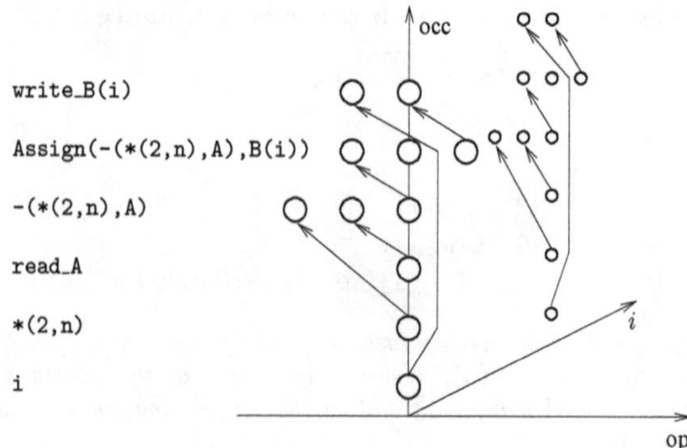

Figure 2.9: Occurrence instances of the code fragment of Example 2.22.

Since a single term t is now associated with several values from OpSet(head(t)), we can identify different operand positions for the function execution represented by an occurrence instance α.

The mappings between occurrences, occurrence instances, and points in the program text, are sketched in Figure 2.10.

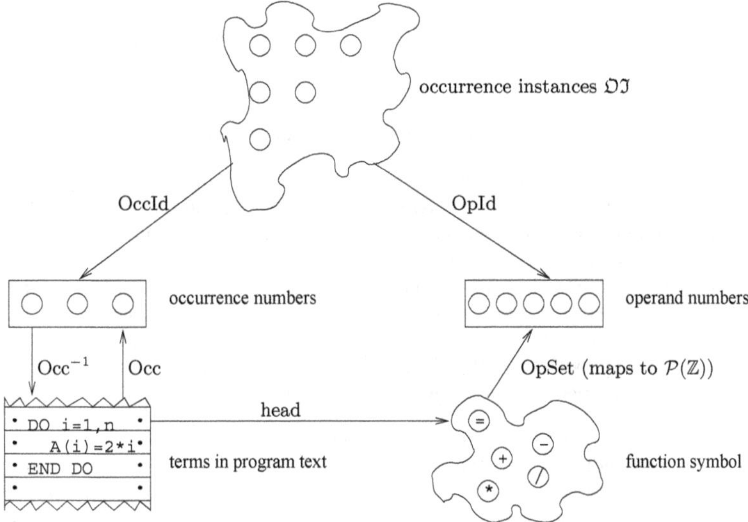

Figure 2.10: Mappings between occurrence numbers, operand numbers, occurrence instances, points in the program text, and operator symbols.

As described above, we assume that there is a distinct parameter m_∞ representing infinity. This is a convention suggested by Feautrier for his PIP tool [Fea03] – where this is called a *big parameter*. We use this parameter in order to represent any occurrence instance α of a program fragment containing n_{src} loops and n_{blob} parameters as an $(n_{src} + n_{blob})$-dimensional vector, no matter how many loops actually surround $\text{Occ}^{-1}(\text{OccId}(\alpha))$. If the term of an occurrence instance α is not surrounded by the i-th loop of the program fragment, we define the i-th component of α as

$$\alpha_i = \begin{cases} -m_\infty & \text{if } \alpha \text{ is textually placed above the } i\text{-th loop} \\ m_\infty & \text{if } \alpha \text{ is textually placed below the } i\text{-th loop} \end{cases}$$

This encoding of the occurrence instances ensures that the lexicographic order on occurrence instances corresponds exactly to the execution order of the occurrence instances in the (sequential) program:

Theorem 2.23 *Let α, β be occurrence instances of a loop program. Then $\alpha \prec \beta$ iff α is executed before β in the program.*

Proof:

Let us first consider occurrence instances $\alpha \prec \beta$; let i be minimal with $\alpha[i] \neq \beta[i]$. We discern the following cases:

1. Dimension i belongs to a loop surrounding both α and β: then, the index of that loop is smaller for α, i.e., it is executed *before* β.

2. Dimension i belongs to a loop surrounding α, but not β: in this case, $\alpha[i] > -m_\infty$, since it is enumerated in the corresponding loop; since β is not surrounded by the loop, $\beta[i] \in \{-m_\infty, m_\infty\}$. Therefore $\beta[i] = m_\infty$, i.e., β is placed below the corresponding loop in the program and will therefore be executed later than α.

3. Dimension i belongs to a loop surrounding β, but not α: just as above, but with the order the other way round ($\beta\,[i] > -m_\infty \Rightarrow \alpha\,[i] = -m_\infty$), therefore α is placed textually before the loop).

4. Dimension i belongs to a loop surrounding neither α nor β: in this case, $\alpha\,[i] = -m_\infty, \beta\,[i] = +m_\infty$. Therefore α is placed before the loop in question, β below the loop, and α is executed before β.

On the other hand, if α is executed before β, there may be two alternative reasons:

1. α is textually ordered before β, but the indices of the loops embracing both α and β have the same values in both α and β. For a non-surrounding loop, α and β may be placed on the same side of the loop (either before the loop or after), in which case the corresponding coordinate of α and β are equal, or α may be placed before, and β after the loop, in which case α is executed before β. For all loops textually between α and β corresponding to the i-th index space dimension, we have $-m_\infty = \alpha\,[i] < \beta\,[i] = m_\infty$; therefore (since the coordinates of all non-surrouding loops for which α and β are on the same side are equal), we have $\alpha \prec \beta$. If α and β are nested within exactly the same nest of loops, $\text{OccId}(\alpha) < \text{OccId}(\beta)$ (remember: occurrences are represented as integers), and therefore $\alpha \prec \beta$.

2. α executes before β, and some loops surrounding both α and β first enumerate α. For all loops not surrounding α or β that appear textually before a loop that surrounds α and β, the coordinates of α and β must be $-m_\infty$ (both appear textually below that loop). The first loop that surrounds both α and β and whose coordinate differs between α and β must therefore make the difference in the lexicographic ordering. Since α is executed before β, its index value for that coordinate is smaller, and thus $\alpha \prec \beta$.

✓

2.3.2 Dependences in the Polyhedron Model

Dependences between memory accesses dictate the order in which the occurrence instances of a program fragment may be executed. If there is a dependence from occurrence instance α to β, $\alpha \prec \beta$, we write $\alpha\,\Delta\,\beta$ (for $(\alpha,\beta) \in \Delta$). Note that the lexicographic order on occurrence instances corresponds exactly to the execution order, even if α and β only differ in their textual order. This dependence relation may fall into one of four classes:

1. Input Dependence ($\alpha\,\Delta^i\,\beta$): α is a read access, β is a read access.

2. Anti Dependence ($\alpha\,\Delta^a\,\beta$): α is a read access, β is a write access.

3. Flow Dependence ($\alpha\,\Delta^f\,\beta$): α is a write access, β is a read access.

4. Output Dependence ($\alpha\,\Delta^o\,\beta$): α is a write access, β is a write access.

Dependences between read and write accesses can be calculated by a variety of dependence analysis algorithms with differing accuracy [Kei97]. Dependence analysis algorithms such as the array dataflow analysis (ADA) of Feautrier [Fea91] or fuzzy array dataflow analysis (FADA) [BCF97] and the improved version CfFADA [CG99] represent dependences between access instances as linear functions – so-called *h-transformations.*

An **h-transformation** maps the index vector of a target access instance to the index vector of the unique source access instance. Since we view access instances as special cases of occurrence instances, we view an h-transformation as a mapping from a target occurrence instance to its source occurrence instance (which means that the places in the program text are also defined in the h-transformation). In the settings defined above (linear array accesses and loop bounds), such an h-transformation that describes the dependence relation between different iterations of the same statements can be defined by a linear mapping

$$H : \mathfrak{D} \rightarrow \text{im}(H) : \alpha \mapsto M_H \cdot \alpha$$

with $\mathfrak{D} \subseteq \mathfrak{DI} \subseteq \mathbb{Z}^n$ and $M_H \in \mathbb{Q}^{n \times n}$ for a suitable $\mathfrak{D}$. Otherwise it is defined by a piecewise linear function (as we will see in Section 2.3.2 below). For the analysis of the source program, we have $n = n_{src} + n_{blob} + 2$. Note that, for such a representation of a dependence relation, M_H has to be rational, since otherwise we cannot represent dependence relations like the flow dependence in the following code fragment:

```
DO i1=1,n
  A(2*i1)=i1
END DO
DO i2=1,n
  B(i1)=A(i1)
END DO
```

Here, iteration i_2 of the second loop depends on $\frac{i_2}{2}$. In addition, since both source and target space of a dependence are not rational but integer, i.e., $\mathfrak{DI} \subseteq \mathbb{Z}^n$, this dependence relation does not hold for the whole index space $\{\begin{pmatrix} \infty \\ 1 \end{pmatrix}, \ldots, \begin{pmatrix} \infty \\ n \end{pmatrix}\}$. The dependence holds only for even values of i_2 (the targets of odd values for i_2 are not integer vectors and, thus, this relation would not be a subset of $\mathbb{Z}^n \times \mathbb{Z}^n$). Therefore, matrix M_H does not suffice to define mapping H. We also need to add the index space in which the dependence holds. In the above case, this is $\{\begin{pmatrix} \infty \\ 2 \cdot i \end{pmatrix} \Big| \, i \in \{1, \ldots, \lfloor \frac{n}{2} \rfloor\}\}$.

Note further that a single linear mapping does not suffice to represent all dependences between array accesses in a program fragment. If we provide an additional loop initializing array A in the above example, we obtain code like the following:

```
      DO i1=1,2*n
!       Statement S1:
        A(i1)=0
      END DO
      DO i1=1,n
!       Statement S2:
        A(2*i1)=i1
      END DO
      DO i2=1,n
!       Statement S3:
        B(i1)=A(i1)
      END DO
```

In this case, Statement S_3 depends on both Statement S_1 and S_2, in turn. Although each dependence relation itself can be described by a linear mapping, this does not hold for the complete dependence relation! The h-transformation is therefore defined by a piecewise linear function – or, equivalently, by several scoped linear functions whose domains are mutually disjoint and cover the complete domain of the target statement. In this thesis, we assume the latter representation, i.e., a single h-transformation is always defined as a scoped linear function (a linear function defined on some restricted domain), possibly representing part of a dependence relation that is completely defined by a family of h-transformations.

A dependence $(\alpha, \beta) \in \Delta$ is direct if there is no access executed between α and β that conflicts with α (and β).

In order to model the flow of data into and out of the program fragment considered, we suppose that, for each array defined as `A(m:n)`, **dummy loops** of the following form are inserted:

```
DO i=m,n
  A(i)=A(i)
END DO
```

With these loops, it is actually not necessary to calculate output dependences directly using a dependence analysis algorithm; it is also possible, to calculate output dependences from flow dependences. Correspondingly, input dependences can be deduced from flow dependences that originate from the dummy loops at the beginning of the program. Note that it actually suffices to create two imperfectly nested loops (one at the beginning of the program fragment, and one at the end) with each dimension enumerated from $-m_\infty$ to m_∞. The occurrence instances representing the write accesses enumerated by dummy loops can be viewed as a way to represent the position in memory occupied by the corresponding array element: the index values represent the value to be added to the base address of the array, and the base address is represented by the occurrence of the write access.

Example 2.24 *Consider the following program fragment:*

```
   DO i1 = 1, min(10, m2)
     DO i2 = i1,i1+m1+5
!    [13]    [1]     [2]  [12] [7]      [3]      [4]     [9] [8]     [5]       [6]      [11] [10]
       a  ( i1    , i2  )   =   a  ( i1-1 , i2  )  +   a  ( i1-2 , i2-2 )   +    c
     END DO
   END DO
```

Occurrences are written above the corresponding positions in the program text in brackets. We obtain the following flow dependences:

$$\begin{aligned}
\Delta^f &= \{((i_1,i_2,13,0,m_1,m_2,m_\infty,m_c),(i_1+1,i_2,7,0,m_1,m_2,m_\infty,m_c)) \mid \\
&\qquad 1 \leq i_1 \leq \min(9,m_2-1), i_1 \leq i_2 \leq i_1+m_1+5\} \\
&\cup \{((i_1,i_2,13,0,m_1,m_2,m_\infty,m_c),(i_1+2,i_2+2,8,0,m_1,m_2,m_\infty,m_c)) \mid \\
&\qquad 1 \leq i_1 \leq \min(8,m_2-2), i_1 \leq i_2 \leq i_1+m_2+3\}
\end{aligned}$$

These are represented by h-transformations (we omit the scope):

$$\begin{aligned}
H_1 &: (i_1,i_2,7,0,m_1,m_2,m_\infty,m_c)^T \mapsto (i_1-1,i_2,13,0,m_1,m_2,m_\infty,m_c)^T \\
H_2 &: (i_1,i_2,8,0,m_1,m_2,m_\infty,m_c)^T \mapsto (i_1-2,i_2-2,13,0,m_1,m_2,m_\infty,m_c)^T
\end{aligned}$$

In addition, we insert special occurrence instances that correspond to a loop of the form

```
DO i=-m∞ ,m∞
  num(i)=i
END DO
```

The instances of this loop correspond to integer numbers. In this way, the execution of a read of an integer number can be represented by an occurrence instance whose value depends on the corresponding instance of the write access `num(i)`. This is necessary in order to deduce the equivalence of terms like

`A(i)*2` and `A(j)*2`

(which holds for $i = j$) while asserting that terms like

`A(i)*2` and `A(i)*3`

are *not* equivalent.

Note that the restriction to integer values is not really necessary. It stems only from the representation of the original write access (`num(i)`) by a point in $\mathbb{Z}^n$. In order to include real numbers in our representation, we could for example represent a real number x by the instance of a different occurrence, where the binary representation of the instance corresponds to the binary representation of x. Note further that, with a dimension of $\mathbb{Z}^n$ reserved for m_c, whose unit value represents the integer number 1, it is not really necessary to reserve also a dimension for representing integer numbers. However, we feel this strategy to be the conceptually cleanest one. The reason for this is a bit complicated: In order to ensure that every computation of the program fragment depends on some other computation, we need a special computation that stands for an integer value – just as we need artificial read and write operations that represent the actual memory addresses with which read and write operations in the considered program fragment interact. We could work around this representation by not assuming that every n-ary operator has to be dependent on all its n operands. This would have the side effect that we would not be able to argue about linear expressions, for example if we wanted to extract the calculation of linear expressions from loops. But for the sake of a cleaner representation, we prescribe that an n-ary operation depends on all n operands. Since this representation requires a special dummy loop, we also need an index bound by this dummy loop. Conceptually, the variable bound by this dummy loop actually has to be an index, because *any* integer can be referenced in the considered program fragment. Only counting dimensions, one might think it a good idea to use the dimension reserved for m_c as the one enumerated by the index of this dummy loop. But in order not to mix up indices with parameters, we neglect to use this reserved dimension, so that m_c is a parameter that *only* takes the value 1 throughout the whole program execution (enabling simplification of polyhedra that might be empty, if that component can only take the value 1). Thus, independently of the polyhedron given, we only have two parameters with a special meaning – m_c is always guaranteed to be 1 and m_∞ is guaranteed to be ∞.

We view dependences as relations between occurrence instances. So far, we have only considered dependences between access instances. However, since the value of any term

$$\odot(\mathfrak{t}_1, \ldots, \mathfrak{t}_j, \mathfrak{s}_1, \ldots, \mathfrak{s}_k)$$

depends on the values of its subterms $\mathfrak{t}_1 \ldots, \mathfrak{t}_j$, we have to introduce non-loop-carried flow dependences between the occurrences representing the subterms and those representing the term constructed by these subterms. Figure 2.9 shows an example of these dependences.

Definition 2.25 (Structural Dependences) *Let* $\alpha = (\alpha_1, \ldots, \alpha_{n_{src}+n_{blob}+2})^T$ *be an occurrence instance with* $\mathrm{OpId}(\alpha) = 0$, $\mathrm{OccId}(\alpha) = o$, $\mathrm{Occ}^{-1}(o) = \odot(\ldots, \mathfrak{t}, \ldots)$, $\odot \in \mathcal{F}$. *Let* β *be the occurrence instance with* $\mathrm{OpId}(\beta) = 0$, $\mathrm{Occ}^{-1}(\mathrm{OccId}(\beta)) = \mathfrak{t}$ *and* $\mathrm{Idx}(\beta) = \mathrm{Idx}(\alpha)$. *For each* $i \in \mathrm{OpSet}(\mathrm{head}(\mathfrak{t}))$, *we define a* **structural dependence**

$$\begin{array}{ll}(\beta, (\alpha_1, \ldots, \alpha_n, o, i)) \in \Delta^s & \text{if } i < 0, \text{ and} \\ ((\alpha_1, \ldots, \alpha_n, o, i), \beta) \in \Delta^s & \text{if } i > 0\end{array}$$

A structural dependence represents the fact that a term $\mathfrak{t} = \odot(\mathfrak{t}_1, \ldots, \mathfrak{t}_n)$ can only be evaluated after all subterms $\mathfrak{t}_1, \ldots, \mathfrak{t}_n$ have been evaluated. This holds for all programming languages that follow an eager evaluation scheme. As discussed above, an operator may have several operands, some of which may be input arguments and therefore correspond to read accesses, others may be output arguments corresponding to write accesses. So, a single operation $\odot$ is performed as an interplay of three different parts.

1. A loading stage that transfers all data needed for the calculation to a place at which it can be used by the operation (occurrence instances α with negative operand numbers, $\mathrm{OpId}(\alpha) < 0$).

2. A computation stage that does the actual algorithmic operation – we identify this stage with the result value of the operation (occurrence instances α with operand number 0, $\mathrm{OpId}(\alpha) = 0$).

3. A storing stage that transfers all additional result values – if there are any explicit output arguments – to their actual destination (occurrence instances α with positive operand numbers, $\mathrm{OpId}(\alpha) > 0$).

Of course, these three stages have to be executed in this order (i.e., again, the lexicographical order on occurrence instances). Therefore, we always assume implicit structural dependences $\Delta^{s}_{\mathrm{impl}} \not\subseteq \Delta^{s}$ between occurrence instances that differ only in their operand position ($opdim_{src}$ in the original program, $opdim_{tgt}$ in the target program):

$$\left(\forall \alpha : \alpha \in \mathfrak{OI} : \begin{array}{l} \mathrm{OpId}(\alpha) < \max(\mathrm{OpSet}(\mathrm{head}(\mathrm{Occ}^{-1}(\mathrm{OccId}(\alpha))))) \Rightarrow \\ ((\alpha\texttt{[1]},\ldots,\mathrm{OpId}(\alpha)),(\alpha\texttt{[1]},\ldots,\mathrm{OpId}(\alpha)+1)) \in \Delta^{s}_{\mathrm{impl}} \end{array}\right).$$

However, although we have to ascertain that every execution of an operation obeys the order defined by $\Delta^{s}_{\mathrm{impl}}$, we will not handle these dependences explicitly. Moreover, there are two special cases to the execution in three stages:

Write accesses:
: These represent only the output of a certain value to a place in memory. Therefore, they are represented only by the execution occurrence instance (formally, $(\forall \alpha : \alpha \in \mathfrak{OI} : \mathrm{head}(\mathrm{Occ}^{-1}(\mathrm{OccId}(\alpha))) = \texttt{write\_}\cdot \Rightarrow \mathrm{OpId}(\alpha) = 0)$).

Read accesses:
: These represent exactly the opposite action – namely the input of a certain value. Again, these operations are represented only by the execution occurrence instance (formally, $(\forall \alpha : \alpha \in \mathfrak{OI} : \mathrm{head}(\mathrm{Occ}^{-1}(\mathrm{OccId}(\alpha))) = \texttt{read\_}\cdot \Rightarrow \mathrm{OpId}(\alpha) = 0)$).

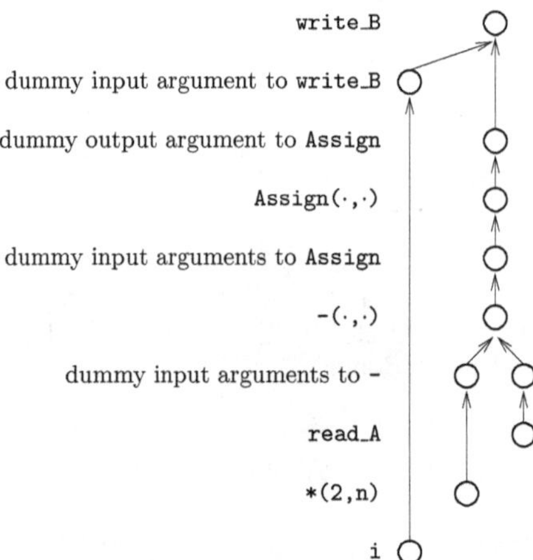

Figure 2.11: Structural dependences of Example 2.22 for a single iteration, including implicit dependences.

Consequently, we always assume implicit dependences between the different occurrence instances performing any given operation. Figure 2.11 shows the structural dependences of the first iteration of Example 2.22, including implicit dependences between occurrence instances that together represent a single operation. The coordinates of the occurrence instances are *not* related to their positions in the figure. Rather, the positions represent the partial order on occurrence instances induced by the structural dependences. Each operator that takes input arguments is represented by several occurrence instances on which the actual operator execution depends; if an operator has (explicit) output arguments – such as the assignment operator `Assign` – these are represented by additional occurrence instances. As noted earlier, these occurrence instances do *not* represent the actual arguments (as `*(2,n)` and `read_A` for the subtraction operator in the example), but the dummy (or formal) arguments. I.e., the occurrence instances that differ only in the operand dimension are needed in a dependence graph representation in order to discern, e.g., the first from the second output argument of a procedure with several output arguments, or the first from the second operand of an operation with arity > 1, such as the subtraction in the example. Otherwise, a dependence graph representation of the calculation `2*n-A` would not differ from `A-2*n`. This use of several occurrence instances representing a single operation is only hinted at in the figure by different positions from left to right. The dependences represented by the arrows reflect the fact that *all* input arguments have to be evaluated before the execution of an operator, and the operator before writing to the output arguments. Read and write accesses do not take arguments besides the subscript expressions, because they only represent the respective memory access.

Note that one may also choose a representation as three-address code instead of the one supporting a varying number of arguments that we choose here. Although three-address code is the state-of-the-art representation for intermediate code today, it does not hold any benefits for our purpose here. Quite to the contrary, our representation with varying argument number is easier to extend to the analysis of associative and commutative operators and also allows an easier code generation in a source-to-source compiler such as LooPo. When using three-address code, the graph presented in Figure 2.11 does not get any simpler either, because we still need to represent the complete flow of data: there are only more kinds dependences to consider for the graph.

Structural dependences generally fall into one of two dependence classes:

Structural flow dependences:
: A dependence between a storage stage and a read stage model the transfer of a value from a subterm to a larger one, and thus is a **structural flow dependence**. We will denote the set of structural flow dependences of the program fragment given by $\Delta^{(s,f)}$.

Structural output dependences:
: A dependence between two storage stages is a **structural output dependence**. These are exactly those dependences that exist between the output operand position of an operator and the actual write access to a memory cell (e.g., the dependence between the rightmost occurrence instance for the `Assign` operator in Figure 2.9 and the occurrence instance representing the execution of `write_B`). We denote the set of structural output dependences of the program fragment given by $\Delta^{(s,o)}$.

Note that there are no structural anti or input dependences. This is because structural dependences are not defined for the ; operator. Otherwise, one might be inclined to define such dependences, e.g. for occurrence instances writing to the same array element $A(1)$. Flow dependences that are loop-carried or that hold between terms that are not related via a term-subterm-relation hold between occurrence instances with operand coordinate 0 (i.e., between the read or write accesses themselves).

Example 2.26 *In the following code fragment, an assignment is appended to the loop of Example 2.22.*

```
DO i1=0,1
  B(i1)=2*n-A
END DO
C=2*B(0)+2*B(1)
```

In this example, we can see both structural and non-structural flow dependences. Figures 2.12(a) through 2.14 show dependence graphs for this example as used in the conventional polyhedron model and in our more finely grained model. Non-structural dependences are here marked with bold arrows.

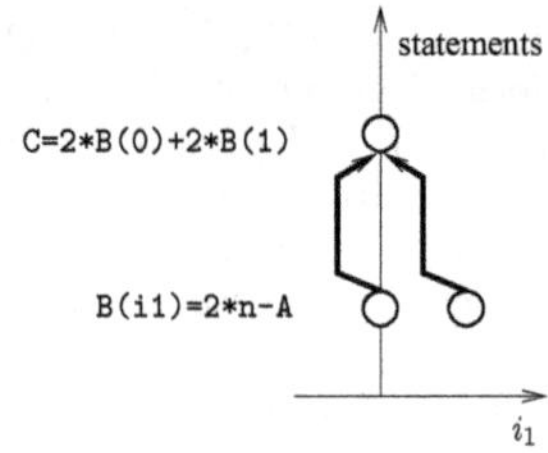

(a) Dependence graph on operations.

C=2*B(0)+2*B(1) $H_1 : \mathbb{Z}^0 \rightarrow \mathbb{Z} : 0 \mapsto 1$

$H_2 : \mathbb{Z}^0 \rightarrow \mathbb{Z} : 0 \mapsto 0$

B(i1)=2*n-A, $i_1 \in \{0, \ldots, 1\}$

(b) Reduced form of the dependence graph in Figure 2.12(a).

Figure 2.12: Dependence graphs of Example 2.26 at the granularity of statements.

Combining non-structural and structural flow dependences enables an accurate model of the flow of data in the code fragment considered. As usual in the polyhedron model, place and time of the execution of a given occurrence instance are determined by a **space-time mapping**, a piecewise linear mapping from the index space into a target space representing sequential and parallel loops. The usual granularity of the polyhedron model is the statement. Therefore, the space-time mapping is usually calculated on the grounds of a dependence graph based on statement instances, as depicted in Figure 2.12(a) for the code in Example 2.24 – each iteration of the loop creates a new instance of the statement in the loop body. Figure 2.12(a) shows the dependence graph with the loop laid out in one dimension and the different statements in the other: the calculation of C dependens on the values – and thus the calculations – of $B(0)$ and $B(1)$. Since such a graph can get infinitely large, actual algorithms represent it in a reduced form. Figure 2.12(b) shows this reduced dependence graph in which a set of statement instances (whose index space can be described as a polyhedron) is represented as a single vertex in the graph. The edges represent again dependences; however the h-transformations H_j with which the edges are labelled, represent a mapping from the target index space of the dependence to its source index space (thus, edges point in the opposite direction as h-transformations). Note that the index space of a statement not enclosed in any loop in the usual polyhedron model is actually $\mathbb{Z}^0 = \{0\}$.[4] Figures 2.13 and 2.14 show the finer granularity of occurrence instances. The calculations and dependences between them in the finely grained model based on occurrence instances use the *Occurrence Instance Graph* (OIG) and its reduced form – shown in Figures 2.13 and 2.14, respectively. Note that in Figure 2.14, the different iterations of the loop in Example 2.24 are represented by the same axis as the different iterations of the dummy loop representing integer values. That is, they are implicitly

[4] Any vector space must contain the neutral element of vector addition, and therefore be non-empty. This has a correspondence in the space of array elements of a 0-dimensional array (a scalar), which *does* have exactly one element (and not – as one might think – none) and in the index space of a statement that is not enclosed in any loop (it is executed *once*).

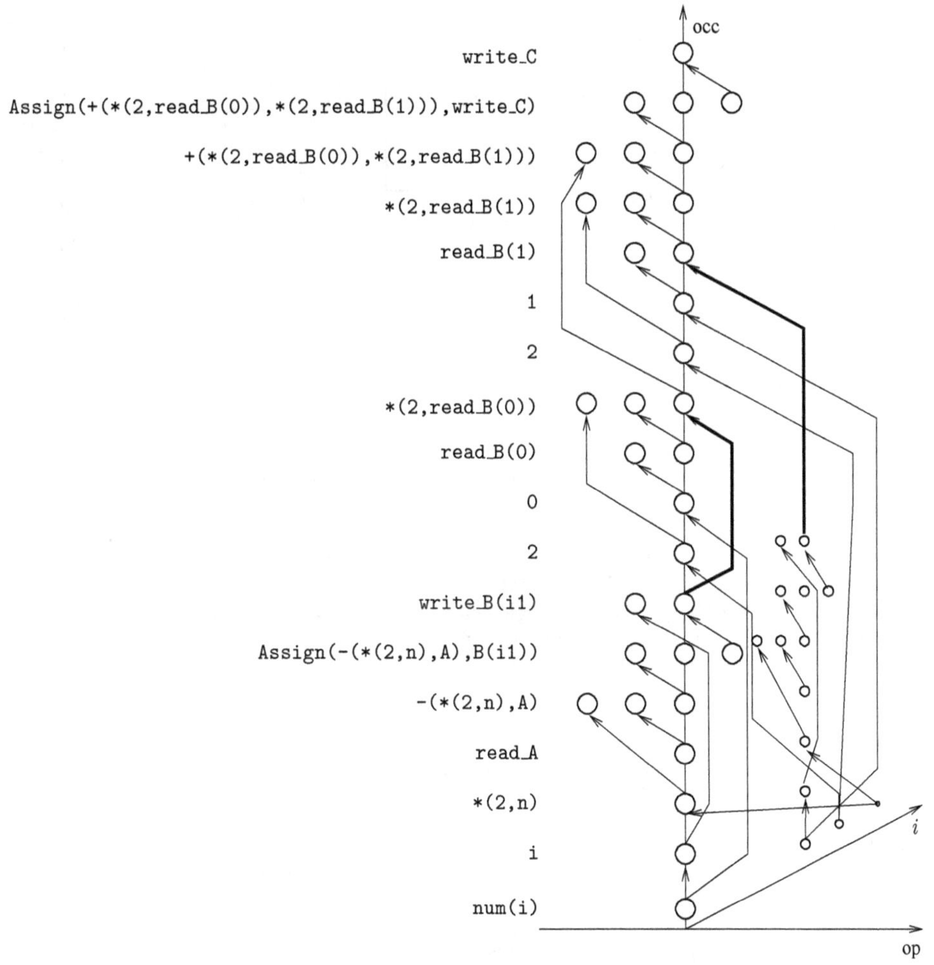

Figure 2.13: Fine grain dependence graph at the granularity of occurrences.

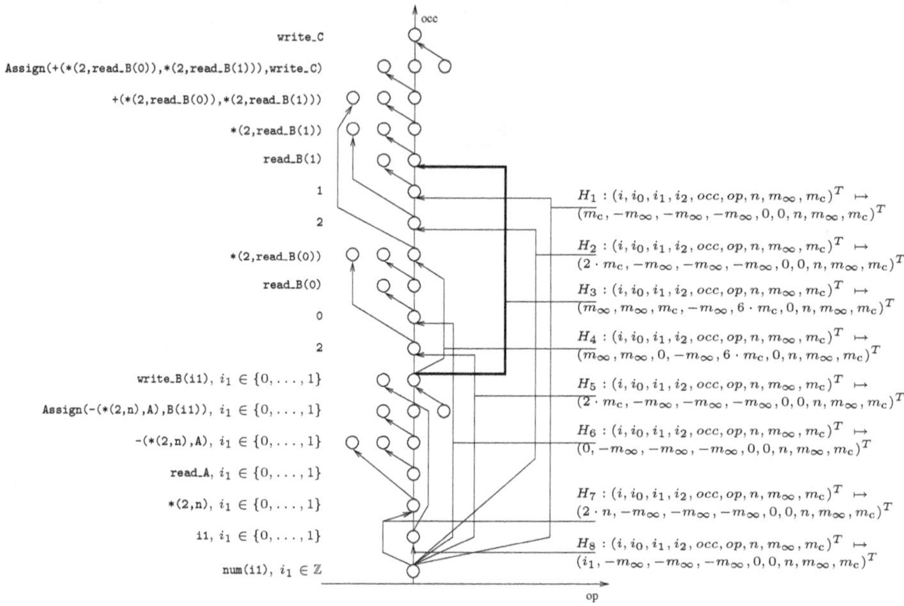

Figure 2.14: Reduced form of the dependence graph in Figure 2.13.

prepended to each program fragment, as discussed in Section 2.3.2. On the other hand, the dummy loops for the arrays of the program fragment are completely dropped in the figure. The implicit dependences between occurrence instances representing different operands of an operation are omitted in the figures to avoid clutter; these dependences would otherwise always point from the left to the right and either start or end at operand position 0.

Definition 2.27 *An* **Occurrence Instance Graph** *is a graph* $OIG = (\mathfrak{OI}, \Delta)$*, where* $\mathfrak{OI}$ *is the set of occurrence instances of the program fragment and* Δ *is a set of dependences between occurrence instances,* $\Delta \subseteq \mathfrak{OI} \times \mathfrak{OI}$*.*

Note that the calculations performed by a given program fragment are completely described by its set $\mathfrak{OI}$ of occurrence instances and the mapping Occ^{-1} which associates an operator to each occurrence. The occurrence instance graph, i.e., the set of occurrence instances together with a dependence relation on these occurrence instances, however, is a convenient way of representing all constraints that actually hold with respect to the execution order of occurrence instances. Depending on which task we want to accomplish, it may be helpful to view different occurrence instance graphs with different edge relations – Δ^f, Δ^a, Δ^i, Δ^o, Δ^s, or any combination thereof. Since the type of a dependence only depends on the operators of its source and target occurrence instances, all these subgraphs can be obtained from a single graph $(\mathfrak{OI}, \Delta^f \cup \Delta^a \cup \Delta^i \cup \Delta^o \cup \Delta^s)$.

As Example 2.24 clearly shows, an OIG is obviously by far larger and more complicated than a graph based on statement instances. Even the reduced form of the graph is rather complicated. However, this fine grained structure is only necessary for the *loop-carried code placement* technique in Chapter 3. The computation of space-time mappings in Chapter 4 is done on a more compact form.

2.4 Relation Representations

In the polyhedron model, the complete source code of a loop program is represented by polyhedra. Dependences are defined as relations between different polyhedra, and in the same way, we may also define a space-time mapping as a relation between a polyhedron which defines the occurrence instances executed by the original program and another polyhedron whose axes represent space and time, and which defines the occurrence instances of a transformed, parallel program. In Chapter 4.1, we will take a closer look at the placement part of such a space-time mapping.

Sometimes it is necessary to restrict a polyhedron to the exact index space in which its elements are executed – for example, if the corresponding index space is to be enumerated. However, in some cases, these restrictions are not necessary. In this section, we consider different representations of relations between polyhedra.

Let us start with an example:

Example 2.28 *Consider the following data distribution in an* HPF *program:*

```
!HPF$ TEMPLATE T(0:3,0:3,0:3)
!HPF$ DISTRIBUTE T(BLOCK,BLOCK,BLOCK)
!HPF$ ALIGN A(i1,i2) WITH T(0,i1,*)
```

The asterisk in the WITH*-clause tells the compiler to store each row* A(i,:) *of array A on different coordinates of the three-dimensional processor array, but to keep copies of each of these rows on the complete third dimension of the processor array. Only processors whose first coordinate is 0 store any data at all.*

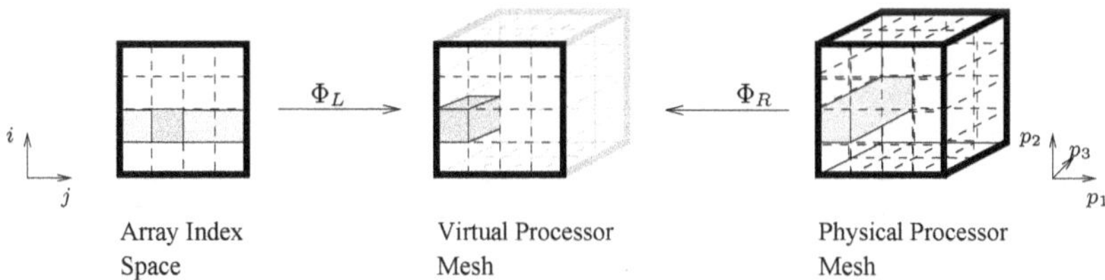

Figure 2.15: Representation of a placement relation by a pair of mappings Φ_L and Φ_R..

The distribution of Example 2.28 is sketched in Figure 2.15. It represents a linear relation between the processors that have to store a given element $A(i, j)$ and its array subscript. This relation can easily be represented by a pair of linear mappings $\Phi = (\Phi_L, \Phi_R)$ that define a data distribution: a given virtual processor (p_1, p_2) has to store a given element (i_1, i_2), iff

$$\begin{array}{rcl} 0 & = & p_1 \\ & \text{and} & \\ 1 \cdot i_1 & = & 1 \cdot p_2 \end{array}$$

without any condition for p_3, i.e., the condition is

$$(p_1, p_2, p_3) \in \Phi_R^{-1} \circ \Phi_L(i_1, i_2)$$

with

$$\begin{array}{rl} \Phi_R: \ \mathbb{Z}^3 \to \mathbb{Z}^2: & (p_1, p_2, p_3) \mapsto (1 \cdot p_1, 1 \cdot p_2) \\ & \text{and} \\ \Phi_L: \ \mathbb{Z}^2 \to \mathbb{Z}^2: & (i_1, i_2) \mapsto (0, 1 \cdot i_2) \end{array}$$

Although the distribution cannot be described by a linear function, it is possible to describe it by a pair of mappings that map into a common subspace. The information we left out of our description here are the ranges of the processor array and the array to be distributed – we only supposed "given coordinates". This information may not even be available until run time and is not necessary for expressing the alignment of different data elements with respect to each other.

Note that the fact that function images – i.e., the finite bounds of the polyhedra representing array and processor index space – are ignored effectively eliminates the differences between considering linear relations on $\mathbb{Q}$ or on $\mathbb{Z}$, since a solution in $\mathbb{Z}$ is always also a solution in $\mathbb{Q}$, and a solution in $\mathbb{Q}$ also defines a solution in $\mathbb{Z}$, if only the nominator is considered (considering the denominator, on the other hand, is only of interest if there is some boundary value for the solution that must not be exceeded).

The virtual processor mesh in Figure 2.15 is determined by the number of equalities that can be established between the index space of the array to be distributed and the processor array. This connection will be made clear in Section 2.4.1.

Actually, the data distributions allowed by HPF are quite restrictive; i.e., many linear relations are *not* legal HPF distributions, as we have seen in Section 2.1.1. However, this is of minor interest at this point.

2.4.1 Higher-Dimensional Polyhedra

The most general way to represent linear relations between integer vector spaces is to build a new polyhedron that – possibly in addition to domain restrictions – expresses the relation between them. This is the standard representation of, e.g., CLooG [Bas03, Bas04]. The advantage of this representation is that it is able to capture restrictions on relations between (lower-dimensional) polyhedra, i.e., subsets of vector spaces, rather than whole vector spaces, so that the complete description of a loop program can be stored.

Given two polyhedra $\mathfrak{P}_1 = \{\nu_1 \in \mathbb{Z}^{r_1} \mid M_{\mathfrak{P}_1} \cdot \nu_1 \geq 0\}$ and $\mathfrak{P}_2 = \{\nu_2 \in \mathbb{Z}^{r_2} \mid M_{\mathfrak{P}_2} \cdot \nu_2 \geq 0\}$ with defining matrices $M_{\mathfrak{P}_1} \in \mathbb{Z}^{q \times r_1}$ and $M_{\mathfrak{P}_2} \in \mathbb{Z}^{q_2 \times r_2}$, the cross product of these polyhedra is given by:

$$\mathfrak{P}_1 \times \mathfrak{P}_2 = \left\{ \begin{pmatrix} \nu_1 \\ \nu_2 \end{pmatrix} \in \mathbb{Z}^{r_1 + r_2} \middle| \begin{pmatrix} M_{\mathfrak{P}_1} & 0 \\ 0 & M_{\mathfrak{P}_2} \end{pmatrix} \cdot \begin{pmatrix} \nu_1 \\ \nu_2 \end{pmatrix} \geq 0 \right\}$$

A relation between these polyhedra is then expressed as a subset of $\mathfrak{P}_1 \times \mathfrak{P}_2$, i.e., in addition to the constraints given by $M_{\mathfrak{P}_1}$ and $M_{\mathfrak{P}_2}$, further constraints have to hold: for a relation

$$P = \left\{ (\nu_1, \nu_2) \in \mathfrak{P}_1 \times \mathfrak{P}_2 \middle| M_{\mathfrak{P}_1, \mathfrak{P}_2} \cdot \begin{pmatrix} \nu_1 \\ \nu_2 \end{pmatrix} = 0 \right\}$$

as in Definition 2.15, constraints of the form

$$\begin{pmatrix} M_{\mathfrak{P}_1, \mathfrak{P}_2} \\ -M_{\mathfrak{P}_1, \mathfrak{P}_2} \end{pmatrix} \cdot \begin{pmatrix} \nu_1 \\ \nu_2 \\ \nu_1 \\ \nu_2 \end{pmatrix} \geq 0$$

are added so that the complete relation can be expressed as a polyhedron.[5]

[5] Of course the equations do not really have to be rewritten into inequations – this is just to show that the

2.4.2 Pairs of Mappings

Section 3.3 will focus on the generation of equivalence classes of occurrence instances. In order to find candidate occurrence instances that may be equivalent to other occurrence instances, it is not useful to take the exact bounds of the index space of these occurrence instances into account: we may well want to view some occurrence instances as equivalent if they were executed by the program, even if they are actually not executed due to restrictions in their loop bounds. When we are only interested in the relation of points with respect to each other, but not in the respective domains, it suffices to view a linear relation as a composition of two mappings as in Example 2.28. The set $\mathfrak{N}$ of points to be associated with a vector ν is the pre-image Λ_R^{-1} of an image point $\Lambda_L(\nu)$ of the other mapping:

$$\mathfrak{N} := (\Lambda_R^{-1} \circ \Lambda_L)(\nu)$$

(In the example, $\Lambda_L = \Phi_L$ and $\Lambda_R = \Phi_R$.) This pre-image can be described by a specific solution and a linear subspace:

$$\mathfrak{N} \quad = \quad (\Lambda_R^g \circ \Lambda_L(\nu)) + \ker(\Lambda_R)$$

Λ_R^g yields a specific solution υ for a pre-image of a point μ under Λ_R (i.e., it yields an υ with $\Lambda_R(\upsilon) = \mu$), provided that μ indeed is the image of Λ_R. All other points that are mapped to μ by Λ_R differ from υ only by the addition of vectors from $\ker(\Lambda_R)$.

As can be gleaned from Figure 2.15, the advantage of this approach lies in the symmetric description of relations: we can choose whether we are only interested in a point (as existence proof) satisfying the relation, whether we need the whole set of points within the relation, or whether we want to view the "image" or the "pre-image" of our relation.

The vector space into which both Λ_L and Λ_R map here depends on the number of equations that have to hold. In our context, the sets $\mathfrak{U}$ and $\mathfrak{V}$ are actually $\mathbb{Z}$-polyhedra; however, since the domains are unimportant, we view $\mathbb{Z}$ as subset of $\mathbb{Q}$ and only consider vector spaces on the rationals.

For technical reasons, we want to be able to compare these representations; therefore we need a unique representation of a linear relation in this form.

The correspondence between the equalities defining a relation and the mappings that can be used to express these equalities is as follows:

Theorem 2.29 *Consider a linear relation* $\Lambda \subseteq \mathbb{Q}^{n_L} \times \mathbb{Q}^{n_R}$ *such that*

$$\Lambda = \left\{ (\nu, \upsilon) \in \mathfrak{U}_L \times \mathfrak{U}_R \middle| M_\Lambda \cdot \begin{pmatrix} \nu \\ \upsilon \end{pmatrix} = 0 \right\}$$

with $M_\Lambda \in \mathbb{Q}^{q \times (n_L + n_R)}$. *Then there are linear mappings* $\Lambda_L : \mathbb{Q}^{n_L} \to \mathbb{Q}^r : \nu \mapsto M_{\Lambda_L} \cdot \nu$, $\Lambda_R : \mathbb{Q}^{n_R} \to \mathbb{Q}^r : \upsilon \mapsto M_{\Lambda_R} \cdot \upsilon$ *such that*

$$(\nu, \upsilon) \in \Lambda \Leftrightarrow \Lambda_L(\nu) = \Lambda_R(\upsilon)$$

with $r = \mathrm{rk}(M_\Lambda)$.

Proof:

The proof is actually trivial, since the reduced echelon form of M_Λ yields exactly the desired result. Let $M_\Lambda' \in \mathbb{Q}^{q \times (n_L + n_R)}$ be the reduced echelon form of M_Λ; M_Λ' is unique and contains $r = \mathrm{rk}(M_\Lambda)$ non-zero rows. Since M_Λ' is obtained from M_Λ via basic row transformations, $M_\Lambda' \cdot \begin{pmatrix} \nu \\ \upsilon \end{pmatrix} = 0 \Leftrightarrow M_\Lambda \cdot \begin{pmatrix} \nu \\ \upsilon \end{pmatrix} = 0$.

general description using only inequations can be obtained from the added equation system.

Put differently, $M_\Lambda \cdot \begin{pmatrix} \nu \\ \upsilon \end{pmatrix} = 0$, iff for each row i of $M_\Lambda{}'$, we have

$$\begin{aligned} &(\textstyle\sum j : j \in \{1,\ldots,n_L\} :\ M_\Lambda{}'[i,j] \cdot \nu[j]) \\ +&(\textstyle\sum j : j \in \{n_L+1,\ldots,n_L+n_R\} :\ M_\Lambda{}'[i,j] \cdot \upsilon[j-n_L]) \ = \ 0, \end{aligned}$$

i.e., we can split the matrix $M_\Lambda{}'$ in two parts,

$$M_{\Lambda_L} = \begin{pmatrix} -M_\Lambda{}'[1,1] & \ldots & -M_\Lambda{}'[1,n_L] \\ -M_\Lambda{}'[r,1] & \ldots & -M_\Lambda{}'[r,n_L] \end{pmatrix} \quad \text{and}$$

$$M_{\Lambda_R} = \begin{pmatrix} M_\Lambda{}'[1,n_L+1] & \ldots & M_\Lambda{}'[1,n_L+n_R] \\ M_\Lambda{}'[r,n_L+1] & \ldots & M_\Lambda{}'[r,n_L+n_R] \end{pmatrix}$$

Since only the first r rows of $M_\Lambda{}'$ are non-zero, M_{Λ_L} and M_{Λ_R} need only r rows. Thus, it is possible to create a pair of mappings that define our new representation and that only use $\mathrm{rk}(M_\Lambda)$ dimensions in the image of both Λ_L and Λ_R. ✓

Theorem 2.29 shows that we can derive a representation of a linear relation via the reduced echelon form. This representation can be used to compare linear relations syntactically, since the reduced echelon form of a matrix is unique.

Note that the model works with rational coordinates. This means that we may have rational valued occurrence instances that are in relation with each other. Of course, we will ultimately create these occurrence instances by counting through loops that enumerate integer values; however, with these integer values representing the nominator of the fraction of the occurrence instance, we can represent the rational occurrence instance by simply dividing by the denominator.

Linear Function and Subspace

Let us suppose a linear relation $\Lambda \subseteq \mathbb{Q}^r \times \mathbb{Q}^s$ is given by a mapping T and a linear subspace $\mathfrak{W} \subseteq \mathbb{Q}^s$ such that

$$(\nu, \upsilon) \in \Lambda \Leftrightarrow (\exists \omega : \omega \in \mathfrak{W} :\ \nu = T(\upsilon) + \omega)$$

Using a pair of mappings as sketched in Figure 2.16 instead of a single mapping and a subspace is a more symmetric approach (and thus better suited for non-scoped relations, especially when building their inverses), and is equivalent to this representation, as we will see now.

Actually, this problem is closely related to the homomorphism theorem of linear algebra: the idea is to partition $\mathbb{Q}^s$ into the quotient space $\mathbb{Q}^s/\mathfrak{W}$ and its complement in $\mathbb{Q}^s$. Then we can map a vector ν to some representative of $T(\nu) + \mathfrak{W}$ by a function Λ_L. Correspondingly, a second mapping Λ_R maps all elements of $\mathfrak{W}$ to 0 – so that the pre-image of Λ_R is given by a specific solution for obtaining the above representative of $T(\nu) + \mathfrak{W}$ in addition to the kernel of Λ_R, i.e., $\mathfrak{W}$.

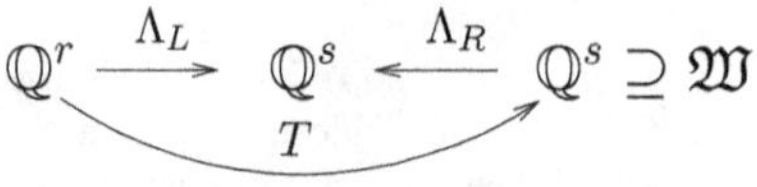

$$\mathbb{Q}^r \xrightarrow{\Lambda_L} \mathbb{Q}^s \xleftarrow{\Lambda_R} \mathbb{Q}^s \supseteq \mathfrak{W} \qquad (T : \mathbb{Q}^r \to \mathbb{Q}^s)$$

Figure 2.16: Representation of a relation $T + \mathfrak{W}$ by two mappings Λ_L and Λ_R.

Theorem 2.30 *Let $T : \mathbb{Q}^r \rightarrow \mathbb{Q}^s$ be a linear mapping, $\mathfrak{W} \subseteq \mathbb{Q}^s$ a q-dimensional $\mathbb{Q}$-vector space. Then, there exist linear mappings $\Lambda_L : \mathbb{Q}^r \rightarrow \mathbb{Q}^s$ and $\Lambda_R : \mathbb{Q}^s \rightarrow \mathbb{Q}^s$ such that*

$$\mathrm{im}(\Lambda_L) \subseteq \mathrm{im}(\Lambda_R) \tag{2.5}$$

and

$$(\forall \nu : \nu \in \mathbb{Q}^r : \ (\Lambda_R^{-1} \circ \Lambda_L)(\nu) = T(\nu) + \mathfrak{W}) \tag{2.6}$$

Proof:

The second condition (2.6) can be written as:

$$\left(\forall \nu : \nu \in \mathbb{Q}^r : \left(\exists \{a_1, \ldots, a_q\} : a_1, \ldots, a_q \in \mathbb{Q} : \begin{array}{l}(-(\Lambda_R^g \circ \Lambda_L))(\nu) + \\ (\sum j : j \in \{1, \ldots, q\} : \ a_j \cdot \omega_j) = T(v)\end{array}\right)\right) \tag{2.7}$$

and

$$(\forall \nu : \nu \in \mathbb{Q}^r : (\forall v : v \in (\Lambda_R^{-1} \circ \Lambda_L)(\nu) : (\exists \omega : \omega \in \mathfrak{W} : T(\nu) + \omega = v))) \tag{2.8}$$

The subspace $\mathfrak{W}$ is generated by some spanning vectors $\{\omega_1, \ldots, \omega_q\}$. Therefore, the kernel of Λ_R must be created by exactly the same vectors (the only way to define a relation between some point ν and a whole set of points in $\mathbb{Q}^s$ using this approach is that the pre-image of Λ_R is a set instead of a single point).

Therefore, we partition the source space of Λ_R into its designated kernel $\mathfrak{W} = \mathrm{Span}(\omega_1, \ldots, \omega_q)$ and an extension to the full space $\mathbb{Q}^s$ consisting of unit vectors:

1. Let $\mathfrak{B} = \{\omega_1, \ldots, \omega_q\}$.
2. Let $W = \begin{pmatrix} \omega_1^T \\ \vdots \\ \omega_q^T \end{pmatrix}$ be the matrix consisting of the vectors in $\mathfrak{B}$ as row vectors and W' the echelon form of W.
3. For each row i of W', let j_i be the column of the first non-zero entry of this row; we define $j_{q+1} = s + 1$.
4. For each row i, define the set of completing unit vectors $\mathfrak{C}_i = \left\{\iota_{(1+j_i)}, \ldots, \iota_{(j_{(i+1)}-1)}\right\}$.
5. Define $\mathfrak{C} = \bigcup_{i \in \{1,\ldots,q\}} \mathfrak{C}_i$.

Note that although $\mathfrak{C}_i$, as defined in Step 4, always exists and is well defined, the definition at this point is rather arbitrary – we could as well define $\mathfrak{C}_i = \left\{\iota_{(j_i)}, \ldots, \iota_{(j_{(i+1)}-2)}\right\}$ or something similar – this is just the definition that comes naturally with the echelon form. In this setting, we define Λ_R as the unique linear mapping that maps $\mathbb{Q}^s$ as follows:

$$\Lambda_R : \mathbb{Q}^s \rightarrow \mathbb{Q}^s : v \mapsto \begin{cases} 0 & \text{if } v \in \mathfrak{B} \\ v & \text{if } v \in \mathfrak{C} \end{cases} \tag{2.9}$$

Since Λ_R is defined as an endomorphism, this mapping always exists. Also note that Λ_R is defined in dependence on $\mathfrak{C}$. Therefore, the choice for $\mathfrak{C}_i$ does have impact on the mapping Λ_R – depending on that choice, one or the other row of the defining matrix (and thus one or the other image dimension of the mapping) will be 0.

In addition, $\mathbb{Q}^s$ is thus divided into the subsets $\mathbb{Q}^s = \mathfrak{W} \oplus \mathrm{im}(\Lambda_R)$. Therefore, we have

$$(\forall v : v \in \mathbb{Q}^s : (\exists c_1, \ldots, c_q : c_j \in \mathbb{Q} : \Lambda_R^g \circ \Lambda_R(v) = v + (\textstyle\sum j : j \in \{1, \ldots, q\} : \ c_j \cdot \omega_j))) \tag{2.10}$$

This holds, since otherwise there are υ, $\upsilon_0 \in \mathrm{im}(\Lambda_R)$ so that $\Lambda_R^g \circ \Lambda_R(\upsilon_0) = \upsilon_0 + \upsilon + (\sum j : j \in \{1,\ldots,q\} : c_j \cdot \omega_j)$ with $\upsilon \neq 0$, which implies that $\Lambda_R \circ \Lambda_R^g \circ \Lambda_R(\upsilon_0) = \Lambda_R(\upsilon_0 + \upsilon + (\sum j : j \in \{1,\ldots,q\} : c_j \cdot \omega_j)) = \Lambda_R(\upsilon_0) + \Lambda_R(\upsilon)$, with $\Lambda_R(\upsilon) \neq 0$, and thus $\Lambda_R \circ \Lambda_R^g \circ \Lambda_R \neq \Lambda_R$, which contradicts the definition of Λ_R^g.

We now go on to define Λ_L. The purpose of Λ_L is to "act like" T but with the restriction $\mathrm{im}(\Lambda_L) \subseteq \mathrm{im}(\Lambda_R)$. This can be asserted by chosing

$$\Lambda_L := \Lambda_R \circ T$$

Thereby, we guarantee proposition (2.5), and it is now straightforward to see proposition (2.6): $\mathbb{Q}^s$ is partitioned into $\mathfrak{W}$ and another subspace that can be spanned using unit vectors; i.e., we can represent any $T(\nu)$ as linear combination of vectors from those two vector spaces:

$$T(\nu) \quad = \quad (\textstyle\sum i : i \in \{1,\ldots,s-q\} : b_i \cdot v_i) + (\sum j : j \in \{1,\ldots,q\} : a_j \cdot \omega_j)$$

Note that this decomposition does not have to be orthogonal (in general it will not be orthogonal); it suffices to use linearly independent vectors. On the other hand, Λ_L is defined as $\Lambda_R \circ T$:

$$\begin{aligned} \Lambda_L(\nu) &= \Lambda_R \circ T(\nu) = \Lambda_R((\textstyle\sum i : i \in \{1,\ldots,s-q\} : b_i \cdot v_i) + (\sum j : j \in \{1,\ldots,q\} : a_j \cdot \omega_j)) \\ &= \Lambda_R((\textstyle\sum i : i \in \{1,\ldots,s-q\} : b_i \cdot v_i)) \text{ since } \mathfrak{W} = \ker(\Lambda_R) \end{aligned}$$

Therefore, with Equation (2.10) and suitable c_j, we have

$$\begin{aligned} \Lambda_R^g \circ \Lambda_L(\nu) &= \Lambda_R^g \circ \Lambda_R((\textstyle\sum i : i \in \{1,\ldots,s-q\} : b_i \cdot v_i)) \\ &= (\textstyle\sum i : i \in \{1,\ldots,s-q\} : b_i \cdot v_i) + (\sum j : j \in \{1,\ldots,q\} : c_j \cdot \omega_j) \end{aligned}$$

and thus condition (2.7). Correspondingly, condition (2.8) is met, since $\Lambda_R^{-1}(0) = \mathfrak{W}$ and thus $\Lambda_R^{-1} \circ \Lambda_L(\nu) \subseteq T(\nu) + \mathfrak{W}$. ✓

The above theorem states that we can represent any (rational) linear relation of the form

$$\Lambda = \{(\nu, T(\nu) + \omega) \mid \nu \in \mathbb{Q}^r, \omega \in \mathfrak{W} = \mathrm{Span}(\omega_1, \ldots, \omega_q), T(\nu) + \omega \in \mathbb{Q}^s\}$$

with two linear mappings, Λ_L and Λ_R, as $\Lambda_R^{-1} \circ \Lambda_L$. Thus, using the polyhedra $\mathfrak{P}$ and $\mathfrak{Q}$ as additional restrictions for the source spaces of our two mappings Λ_L and Λ_R, we can describe the complete relation by these mappings in combination with their source space.

A representation "in-between" polyhedra and this representation using a pair of mappings can be obtained by leaving out the dimensions in the target space of Λ_L and Λ_R whose coordinates are always 0 in both $\mathrm{im}(\Lambda_L)$ and $\mathrm{im}(\Lambda_R)$. This is stated by the following corollary.

Corollary 2.31 *Let $\Psi : \mathbb{Q}^r \rightarrow \mathbb{Q}^s$ be a linear mapping, $\mathfrak{W} \subseteq \mathbb{Q}^s$ a q-dimensional $\mathbb{Q}$-vector space. Then, for each $q_0 \in \{1,\ldots,q\}$, there exist linear mappings $\Lambda_L : \mathbb{Q}^r \rightarrow \mathbb{Q}^{s-q_0}$ and $\Lambda_R : \mathbb{Q}^s \rightarrow \mathbb{Q}^{s-q_0}$ such that*

$$\mathrm{im}(\Lambda_L) \subseteq \mathrm{im}(\Lambda_R) \tag{2.11}$$

and

$$(\forall \nu : \nu \in \mathfrak{U} : \Lambda_R^{-1} \circ \Lambda_L(\nu) = \Psi(\nu) + \mathfrak{W}) \tag{2.12}$$

Proof:

The only difference to the situation in Theorem 2.30 is that Λ_R is no longer restricted to being an endomorphism. In Theorem 2.30, Λ_R is constructed as the unique mapping satisfying Equation (2.9). Since we chose unit vectors for completing $\mathfrak{W}$ to an s-dimensional vector space, it is easy to see that, for each dimension i of $\mathfrak{W}$, Λ_R will exhibit a dimension j_i so that

$(\forall \upsilon : \upsilon \in \mathbb{Q}^s : \mathit{\Lambda}_R(\upsilon)[j_i] = 0)$, since, for each i, $\omega_i \notin \mathfrak{C}$ and therefore $(\forall \upsilon : \upsilon \in \mathbb{Q}^s : \omega_i^T \cdot \upsilon = 0)$. And, since Λ_L is defined as a composition with $\Lambda_R \circ T$, the dimensions i_j are also set to 0 in the image of Λ_L. Therefore, the restrictions $0 = \mathit{\Lambda}_L(\nu)[j_i] = \mathit{\Lambda}_R(\upsilon)[j_i] = 0$ are always true. They do not define any further restriction on the relation. Thus, these dimensions can be disregarded✓

This means that we can represent any linear relation by a pair of a homomorphism and an endomorphism – or, equivalently, by a pair of two homomorphisms. This is the representation chosen for array placement relations in Section 4.1.2.

Operations on Pairs of Mappings

When representing a relation by pairs of mappings instead of complete equation systems, we will need some operations on this new representation in order to do calculations. These basic operations are presented here.

Inversion Inversion is the simplest operation: since the relation represented by (Λ_L, Λ_R) is actually $\Lambda_R^{-1} \circ \Lambda_L$, the inverse relation is simply given by (Λ_R, Λ_L) which represents $\Lambda_L^{-1} \circ \Lambda_R$.

Composition The relation representing the composition of two linear relations $(\Lambda_{1L}, \Lambda_{1R})$ and $(\Lambda_{2L}, \Lambda_{2R})$ is given by

$$\Lambda_{2R}^{-1} \circ \Lambda_{2L} \circ \Lambda_{1R}^{-1} \circ \Lambda_{1L}$$

Since Λ_{iL} and Λ_{iR} are linear functions, the sets $\Lambda_{iL}^{-1}(\alpha)$ and $\Lambda_{iR}^{-1}(\beta)$ (for some α and some β) can be computed using arbitrary generalized inverses Λ_{iL}^g and Λ_{iR}^g, together with the appropriate kernel – $\ker(\Lambda_{iL})$ and $\ker(\Lambda_{iR})$, respectively:

$$\Lambda_{2R}^{-1} \circ \Lambda_{2L} \circ \Lambda_{1R}^{-1} \circ \Lambda_{1L} = \Lambda_{2R}^g \circ \Lambda_{2L} \circ \Lambda_{1R}^g \circ \Lambda_{1L} + \ker(\Lambda_{2R}) + \Lambda_{2R}^g \circ \Lambda_{2L}(\ker(\Lambda_{1R})) \quad (2.13)$$

This composition can be expressed by two mappings as defined in Theorem 2.30 above.

We observe that there are two terms in this expression, where sets come into play:

1. Λ_1 may define a set that is associated with any given point due to the term $\Lambda_{2R}^g \circ \Lambda_{2L}(\ker(\Lambda_{1R}))$.
2. Λ_2 may define a set that is associated with any given point due to the term $\ker(\Lambda_{2R})$.

In general, there are at least two ways to handle the resulting new sets: either we unite them, or we build the difference. For each of these options, we define an individual operator. Let us first examine the case of building the union:

Definition 2.32 (Operator $\circ_+$) *The composition as shown in Equation 2.13 above – using the union of the two sets in question – defines the composition operator $\circ_+$:*
Let $\Lambda_1 = (\Lambda_{1L}, \Lambda_{1R}), \Lambda_2 = (\Lambda_{2L}, \Lambda_{2R})$ be two linear relations. The operator $\circ_+$ defines a result relation $\Lambda_2 \circ_+ \Lambda_1$ as:

$$\begin{aligned}(\Lambda_{2L}, \Lambda_{2R}) \circ_+ (\Lambda_{1L}, \Lambda_{1R}) &= (\Lambda_{3L}, \Lambda_{3R}) \text{ such that} \\ \Lambda_{3R}^{-1} \circ \Lambda_{3L} &= \Lambda_{2R}^g \circ \Lambda_{2L} \circ \Lambda_{1R}^g \circ \Lambda_{1L} + \ker(\Lambda_{2R}) + \Lambda_{2R}^g \circ \Lambda_{2L}(\ker(\Lambda_{1R}))\end{aligned}$$

An implementation may compute $(\Lambda_{3L}, \Lambda_{3R}) = \Lambda_2 \circ_+ \Lambda_1$ by defining Λ_{3L} and Λ_{3R} as the mappings Λ_L and Λ_R of Theorem 2.30 that represent the above relation.

Thus, the union is quite easily described, and this is the usual case found: all points for which a relation exists have to be considered. And yet, sometimes it may be appropriate to take the difference of the sets above: $\Lambda_1 = (\Lambda_{1L}, \Lambda_{1R}) \subseteq \mathbb{Q}^{r \times s}$ specifies a set that has to be considered for each point υ of $\mathbb{Q}^r$, while for each point in $\mathbb{Q}^s$, $\Lambda_2 = (\Lambda_{2L}, \Lambda_{2R}) \subseteq \mathbb{Q}^{s \times t}$ defines a subset of $\mathbb{Q}^t$, from which we may *choose* any point. This will be the case in Section 4.1, when some data element is stored in a replicated fashion. There, the processor reading some data may choose the specific copy that it actually reads. For this case, we now need another composition operation $\circ_-$ that picks a single point υ from the possibilities supplied by Λ_2 (for some given ν, $\upsilon \in ({\Lambda_2}_R^{-1} \circ \Lambda_{2L} \circ {\Lambda_1}_R^{-1} \circ \Lambda_{1L})(\nu)$ should hold as discussed above). Since we now have the freedom to choose some point, we need a notion of what may be a good point to choose. The criterion we apply here is as follows: with ultimate source and target space being identical, the resulting mapping is such that, given some input vector ν, the output vector υ features the same coordinates as ν for as many dimensions as possible. Moreover, the result of the $\circ_-$ operator is again a linear relation. Theorem 2.34 will show how to select a vector υ that features the same coordinates as ν in as many dimensions as possible: the resulting linear relation that picks the appropriate value for υ can be derived from the relation obtained from the $\circ_+$ operator by adding appropriate base vectors to the defining matrix. But, in order to prove this, Lemma 2.33 first tells us how to obtain a linear mapping Ψ' from an original mapping Ψ and a subspace $\mathfrak{V}$ so that we have $\Psi'(\nu) \in \Psi(\nu) + \mathfrak{V}$ for all vectors ν.

Lemma 2.33 *Let $\Psi : \mathbb{Q}^r \to \mathbb{Q}^u : \mu \mapsto M_\Psi \cdot \mu$ and $V \in \mathbb{Q}^{u \times s}$, $C \in \mathbb{Q}^{s \times r}$ arbitrarily, then the mapping $\Psi' : \mathbb{Q}^r \to \mathbb{Q}^u : \mu \mapsto M_{\Psi'} \cdot \mu$ with*

$$M_{\Psi'} = M_\Psi + V \cdot C$$

maps any point $\mu \in \mathbb{Q}^r$ to an element of $\Psi(\mu) + \mathrm{Span}(V[\cdot,1], \ldots, V[\cdot,s])$.

Proof:

This property follows directly from the linearity of the matrix-vector product. Let $\upsilon = \Psi(\mu) = M_\Psi \cdot \mu$, $\upsilon' = \Psi'(\mu) = M_{\Psi'} \cdot \mu$. Then we have

$$\begin{aligned}
\upsilon[i] &= (\textstyle\sum k : k \in \{1,\ldots,r\} : M_\Psi[i,k] \cdot \mu[k]) \\
&= (\textstyle\sum k : k \in \{1,\ldots,r\} : (M_{\Psi'}[i,k] - (V \cdot C)[i,k]) \cdot \mu[k]) \\
&= (\textstyle\sum k : k \in \{1,\ldots,r\} : M_{\Psi'}[i,k] \cdot \mu[k]) - (\textstyle\sum k : k \in \{1,\ldots,r\} : (V \cdot C)[i,k] \cdot \mu[k]) \\
&= \upsilon'[i] - (\textstyle\sum k : k \in \{1,\ldots,r\} : (\textstyle\sum p : p \in \{1,\ldots,s\} : V[i,p] \cdot C[p,k]) \cdot \mu[k]) \\
&= \upsilon'[i] - (\textstyle\sum k : k \in \{1,\ldots,r\} : (\textstyle\sum p : p \in \{1,\ldots,s\} : V[i,p] \cdot C[p,k] \cdot \mu[k])) \\
&= \upsilon'[i] - (\textstyle\sum p : p \in \{1,\ldots,s\} : V[i,p] \cdot (\textstyle\sum k : k \in \{1,\ldots,r\} : C[p,k] \cdot \mu[k])) \\
&= \upsilon'[i] - \left(\textstyle\sum p : p \in \{1,\ldots,s\} : V[i,p] \cdot \underbrace{(C[p,\cdot] \cdot \mu)}_{=:x_p \in \mathbb{Q}} \right)
\end{aligned}$$

I.e., there is a vector $\chi = C \cdot \mu \in \mathbb{Q}^s$ with components $\chi[p] = x_p$ so that using x_p as the coefficient of vector $V[\cdot,p]$ in the sum $(\sum p : p \in \{1,\ldots,s\} : x_p \cdot V[\cdot,p])$ guarantees that $\upsilon = \upsilon' - (\sum p : p \in \{1,\ldots,s\} : x_p \cdot V[\cdot,p])$, i.e., $\upsilon' = \upsilon + (\sum p : p \in \{1,\ldots,s\} : x_p \cdot V[\cdot,p])$. Thus, the resulting image point $\upsilon' = \Psi'(\mu)$ lies within $\upsilon + \mathrm{Span}(V[\cdot,1], \ldots, V[\cdot,s])$. ✓

Now, given a linear subspace $\mathfrak{V} \subseteq \mathbb{Q}^u$ and a linear function $\Psi : \mathbb{Q}^r \to \mathbb{Q}^u$, we want to obtain a linear function Ψ' that maps each element to an element of the same equivalence class wrt. $\mathbb{Q}^u/\mathfrak{V}$ as Ψ but otherwise leaves as many dimensions as possible constant – in the case that $r = u$.[6]

Theorem 2.34 *Let $\Psi : \mathbb{Q}^r \to \mathbb{Q}^u : \mu \mapsto M_\Psi \cdot \mu$ be a linear function, defined by matrix M_Ψ, and $\mathfrak{V} = \mathrm{Span}(\nu_1, \ldots, \nu_n)$ be a subspace of $\mathbb{Q}^u$. Let $j_1 < \cdots < j_m \in \{1, \ldots, r\}, m \le \min(n, r)$, be a sequence of different integers. Then there is a sequence of integers $(j'_1, \ldots, j'_m)$ and a linear*

[6] If $r \neq u$, we want to pick at least a set of corresponding dimensions in the source and target space of Ψ' whose coordinates are left unchanged by the application of Ψ'.

function $\Psi' : \mathbb{Q}^r \to \mathbb{Q}^u : \mu \mapsto M_{\Psi'} \cdot \mu$ *such that*

$$(\forall \mu : \mu \in \mathbb{Q}^r : \Psi'(\mu) \in \Psi(\mu) + \mathfrak{V}) \tag{2.14}$$

and for each $i \in \{1, \ldots, m\}$*:*

$$(\forall \mu : \mu \in \mathbb{Q}^r : \Psi'(\mu)[j'_i] = \mu[j_i]) \tag{2.15}$$

Proof:

Let $N = \begin{pmatrix} \nu_1^T \\ \vdots \\ \nu_n^T \end{pmatrix}$ be the matrix whose rows consist of the base vectors of $\mathfrak{V}$. Let N' be the reduced echelon form of N, and $V = N'^T$. Then we define the sequence $(j'_1, \ldots, j'_m)$ as:

$$j'_i = \min(\{j \in \{1, \ldots, u\} \mid V[i,j] \neq 0\})$$

Let us further define the matrix $C \in \mathbb{Q}^{n \times r}$ as follows:

$$C[i,j] = \begin{cases} 1 - M_\Psi[i,j] & \text{if } i \leq m, j = j_i \\ -M_\Psi[i,j] & \text{if } i \leq m, j \neq j_i \\ 0 & \text{if } i > m \end{cases} \tag{2.16}$$

With these definitions, we obtain the new mapping Ψ' as $\Psi' : \mu \to M_{\Psi'} \cdot \mu$ with

$$M_{\Psi'} = M_\Psi + V \cdot C \tag{2.17}$$

Note that, for $m < n$, we define the rows of C below row m to be zero – this results in changing the mapping Ψ in only m dimensions.

Equation 2.14 follows immediately from Lemma 2.33, since $\text{Span}(V[\cdot,1], \ldots, V[\cdot,m]) \subseteq \text{Span}(\nu_1, \ldots, \nu_m)$.

For Equation 2.15, we examine dimension j'_i in the result vector for some given i:

Since, according to the definition in Equation 2.17,

$$M_{\Psi'}[i,j] = M_\Psi[i,j] + (\textstyle\sum k : k \in \{1, \ldots, n\} : V[i,k] \cdot C[k,j])$$

we have

$$\begin{aligned} M_{\Psi'}[j'_i,j] &= M_\Psi[j'_i,j] + (\textstyle\sum k : k \in \{1, \ldots, n\} : V[j'_i,k] \cdot C[k,j]) \\ &\underset{V^T \text{ in reduced echelon}}{=} M_\Psi[j'_i,j] + \left(\sum k : k \in \{1, \ldots, n\} : \underbrace{V[j'_i,k]}_{= \begin{cases} 1 & \text{if } k = i \\ 0 & \text{otherwise} \end{cases}} \cdot C[k,j] \right) \\ &\underset{\text{Equation 2.16}}{=} M_\Psi[j'_i,j] + \underbrace{C[i,j]}_{= \begin{cases} 1 - M_\Psi[j'_i,j] & \text{if } j = j_i \\ -M_\Psi[j'_i,j] & \text{otherwise} \end{cases}} \\ &= \begin{cases} 1 & \text{if } j = j_i \\ 0 & \text{otherwise} \end{cases} \end{aligned}$$

And therefore,

$$\begin{aligned} M_{\Psi'}[j'_i, \cdot] &= \iota^T_{r,j_i} \\ \text{and thus} & \\ (\Psi'(\mu))[j'_i] &= (M_{\Psi'} \cdot \mu)[j'_i] \\ &= (M_{\Psi'}[j'_i, \cdot]) \cdot \mu = \iota_{r,j_i} \cdot \mu = \mu[j_i] \end{aligned}$$

✓

With the help of this theorem, we build a mapping Ψ' that acts like Ψ up to the addition of vectors from a subspace $\mathfrak{V}$. Note that, in general, the choice of Ψ' is actually quite arbitrary at first, since the choice of sequences $(j_i)_{i\in\{1,\dots,m\}}$ and $(j'_i)_{i\in\{1,\dots,m\}}$ is somewhat arbitrary. However, the interesting case for us is the case $r = u$, in which we can pick $j_i := j'_i$ (with j'_i being not the only choice, but a natural one, for dimensions in the target space of Ψ that we may arrange ourselves). In this case, our aim is to create a mapping with as many unit vectors as possible as eigenvectors. However, even then, it is not immediately clear which Ψ' to use – choosing Ψ' as in the proof of Theorem 2.34 (and later by Algorithm 2.4.1) may even be counterproductive, as the following example shows.

Example 2.35 *Let* $M_\Psi = \begin{pmatrix} 2 & 0 & 0 \\ 0 & 2 & 0 \\ 0 & 1 & 1 \end{pmatrix}$. *Let* $\mathfrak{V} = \text{Span}(\nu_1)$, $\nu_1 = \begin{pmatrix} 1 \\ 1 \\ 1 \end{pmatrix}$. *We may now obtain matrix* $M_{\Psi'}$ *from* M_Ψ *by adding some product of a rational number with* ν_1 *for each column separately, i.e., we build*

$$M_{\Psi'} = \begin{pmatrix} 2 & 0 & 0 \\ 0 & 2 & 0 \\ 0 & 1 & 1 \end{pmatrix} + \begin{pmatrix} 1 \\ 1 \\ 1 \end{pmatrix} \cdot \begin{pmatrix} c_{1,1} & c_{1,2} & c_{1,3} \end{pmatrix}$$

Since $\begin{pmatrix} 1 & 1 & 1 \end{pmatrix}$ *is already in reduced echelon form, our procedure in the proof of Theorem 2.34 chooses to map the first dimension by the identity, i.e., we obtain:*

$$\begin{aligned} M_{\Psi'} &= \begin{pmatrix} 2 & 0 & 0 \\ 0 & 2 & 0 \\ 0 & 1 & 1 \end{pmatrix} + \begin{pmatrix} 1 \\ 1 \\ 1 \end{pmatrix} \cdot \begin{pmatrix} -1 & 0 & 0 \end{pmatrix} \\ &= \begin{pmatrix} 1 & 0 & 0 \\ -1 & 2 & 0 \\ -1 & 1 & 1 \end{pmatrix} \end{aligned}$$

Thus, we have sacrificed the possibility of obtaining a unit vector in rows two or three for a unit vector in the first row. But we may also choose different coefficients $c_{1,1}, c_{1,2}, c_{1,3}$. *The mapping*

$$\begin{aligned} M_{\Psi'} &= \begin{pmatrix} 2 & 0 & 0 \\ 0 & 2 & 0 \\ 0 & 1 & 1 \end{pmatrix} + \begin{pmatrix} 1 \\ 1 \\ 1 \end{pmatrix} \cdot \begin{pmatrix} 0 & -1 & 0 \end{pmatrix} \\ &= \begin{pmatrix} 1 & -1 & 0 \\ 0 & 1 & 0 \\ 0 & 0 & 1 \end{pmatrix} \end{aligned}$$

maps two dimensions according to the identity – not only one.

In general, we can obtain a unique solution by introducing a larger linear equation system and removing iteratively different combinations of equations in order to make the system satisfiable again – a procedure that is reliable, but complex. However, we are mainly interested in the case that $\mathfrak{V} = \text{Span}(\iota_{F(1)}, \ldots, \iota_{F(n)})$ (for some function F). And in this case, defining $j_i = F(i)$ *is* a straightforward way to obtain a unique Ψ' that maps as many dimensions as possible to the identity – since we may only add unit vectors to M_Ψ in order to satisfy Equation 2.14, and adding unit vectors has an impact in only one image dimension. I.e., adding such a vector to the different columns of a given matrix can only take effect on one row of the matrix (for which we already compute the corresponding coefficients in order to obtain the desried result) – unlike the situation in Example 2.35, other rows of the matrix cannot be influenced.

We have introduced the $\circ_+$ operator that computes the usual composition of linear relations. And yet, sometimes it may be appropriate to take the difference of the sets above. In this thesis, this is the case in Chapter 4. When some data element is stored in a replicated fashion – which is defined by a linear relation – one has to choose which of the copies to read when this data element should be processed. Therefore, we define the $\circ_-$ operator. In contrast to the $\circ_+$ operator, a composition with the $\circ_-$ operator will be used where relation Λ_2 actually allows us to pick some point of its source space, while relation Λ_1 determines a set of target points that is to be considered. This means that, for the composition of these relations, we actually want to build the difference instead of a union as with the $\circ_+$ operator. In this case, we will not necessarily have to represent the complete set $\Lambda_2{}^g_R \circ \Lambda_{2L}(\ker(\Lambda_{1R}))$. The formal definition of the $\circ_-$ operator that implements this strategy is as follows.

Definition 2.36 (Operator $\circ_-$) *Let $\Lambda_1 = (\Lambda_{1L}, \Lambda_{1R}) \subseteq \mathbb{Q}^r \times \mathbb{Q}^s$ and $\Lambda_2 = (\Lambda_{2L}, \Lambda_{2R}) \subseteq \mathbb{Q}^s \times \mathbb{Q}^r$ be two relations. We define the operator $\circ_-$ as follows:*

$$(\Lambda_{2L}, \Lambda_{2R}) \circ_- (\Lambda_{1L}, \Lambda_{1R}) = (\Lambda_{3L}, \Lambda_{3R})$$

where Λ_{3R} is the endomorphism on $\mathbb{Q}^r$ with kernel $(\Lambda_2{}^{go}_R \circ \Lambda_{2L})(\ker(\Lambda_{1R}))$ from Theorem 2.30, where the subspace $\mathfrak{W}$ of the theorem is $\mathfrak{W} = (\Lambda_2{}^{go}_R \circ \Lambda_{2L})(\ker(\Lambda_{1R}))$. The proof of Theorem 2.30 constructs a unique mapping with this property. Let $\mathfrak{V} = \Lambda_{3R}(\ker(\Lambda_{2R})) = \text{Span}(\nu_1, \ldots, \nu_n)$, $V = \begin{pmatrix} \nu_1^T \\ \vdots \\ \nu_n^T \end{pmatrix}$, and let V' be the reduced echelon form of V. Then we define Λ_{3L} as the function Ψ' from Theorem 2.34, defining Ψ and the j_i of Theorem 2.34 as: $\Psi = \Lambda_{3R} \circ \Lambda_2{}^{go}_R \circ \Lambda_{2L} \circ \Lambda_1{}^{go}_R \circ \Lambda_{1L}$ and, for $i \in \{1, \ldots, n\}$, $j_i = \min(\{j \in \{1, \ldots, r\} \mid V'[i, j] \neq 0\})$.

Note that, in contrast to $\circ_+$, $\circ_-$ is only defined for the case that $\text{dom}(\Lambda_{2R}) = \text{dom}(\Lambda_{1L})$, i.e., $\Lambda_2{}_R^{-1}$ maps back to the domain of Λ_{1L}. Therefore, the result of the $\circ_-$ operator is always a relation between elements of the same vector space: $\Lambda_2 \circ_- \Lambda_1 \subseteq \mathbb{Q}^r$. This enables us to introduce a notion of maximum simplicity: having as many unit vectors as possible as eigenvectors.

It is clear how to compute $\circ_+$. Obtaining $\circ_-$ algorithmically, however, may not be completely straightforward. Algorithm 2.4.1 computes $\Lambda_2 \circ_- \Lambda_1$ with the help of Theorem 2.34.

Algorithm 2.4.1 [*CompositionSubset*]:
Input:

Λ_1 : linear relation $\Lambda_1 = (\Lambda_{1L}, \Lambda_{1R}) \subseteq \mathbb{Q}^r \times \mathbb{Q}^s$.

Λ_2 : linear relation $\Lambda_2 = (\Lambda_{2L}, \Lambda_{2R}) \subseteq \mathbb{Q}^s \times \mathbb{Q}^r$.

Output:

$\Lambda_3 = (\Lambda_{3L}, \Lambda_{3R})$:
subset composition of $\Lambda_3 = \Lambda_2 \circ_- \Lambda_1$.

Procedure:

/* STEP 1: create the base vectors for the set to be represented */
Let $M_{\Lambda_{1L}}$, $M_{\Lambda_{2L}}$ be the defining matrices so that $\Lambda_{1L} : \mu \mapsto M_{\Lambda_{1L}}\mu$, $\Lambda_{2L} : \mu \mapsto M_{\Lambda_{2L}}\mu$;
Let $M_{\Lambda_{1R}}$, $M_{\Lambda_{2R}}$ be the defining matrices so that $\Lambda_{1R} : \mu \mapsto M_{\Lambda_{1R}}\mu$, $\Lambda_{2R} : \mu \mapsto M_{\Lambda_{2R}}\mu$;
Let $\mathfrak{W} = \text{Span}(\kappa_1, \ldots, \kappa_m) = \Lambda_2{}^{go}_R \circ \Lambda_{1L}(\ker(\Lambda_{1R}))$;

```
/* STEP 2: build an equation system A · X = R whose solution ensures that exactly the set 𝔎 */
/* is represented by ker(X^T) */
Let A, R ∈ ℚ^{r×r} arrays of rationals;
/* S will now be filled with the expressions that have to evaluate to 0 in the first n rows */
/* and with unit vectors (that are mapped to themselves) in the rest of the row; */
Let K = (κ_1^T ; ... ; κ_m^T), and K' be the reduced echelon form of K;
Let FirstNonZero be an array of m + 1 integers;
for i = 1 to m
  /* Let FirstNonZero[i] be the column of the first non-zero entry of K[i,·] */
  FirstNonZero[i] := min({j | K[i,j] ≠ 0});
endfor
FirstNonZero[m + 1] := r + 1;
/* Those dimensions of our output space ℚ^r that are not specified */
/* in FirstNonZero will be mapped to themselves; */
/* the others have to be mapped to 0 */
k := 0;
for i = 1 to m
  k := k + 1;
  A[k,·] := K'[i,·];
  R[k,·] := 0;
  for j = FirstNonZero[i] to FirstNonZero[i + 1]
    k := k + 1;
    A[k,·] := ι_{r,j};
    R[k,·] := ι_{r,j};
  endfor
endfor
/* STEP 3: compute mappings Λ_{3L}, Λ_{3R} from the condition */
/* X^T · A^T = R^T with X^T defining Λ_{3R} */
For each column j of R, solve the system A · X[·,j] = R[·,j] to obtain a column j of X;
Let Λ_{3R} be the function Λ_{3R} : ℚ^r → ℚ^r : μ ↦ X^T · μ;
/* We now build Λ_{3L} according to Theorem 2.34 */
L := X^T · M_{Λ_{2R}}^{go} · M_{Λ_{2L}} · M_{Λ_{1R}}^{go} · M_{Λ_{1L}};
/* We may choose the mapping we use for Λ_{1L} according to Lemma 2.33: */
Let Span(ν_1, ..., ν_n) = Λ_{3R}(ker(Λ_{2R}));
Let V = (ν_1^T ; ... ; ν_n^T);
Let V' the reduced echelon form of V;
Let FirstNonZeroV be an array of n integers;
for i = 1 to n
  /* Let FirstNonZeroV[i] be the column of the first non-zero entry of V' */
  FirstNonZeroV[i] := min({j | V'[i,j] ≠ 0});
endfor
Let C ∈ ℚ^{n×r}, C = 0;
for i = 1 to n
  for j = 1 to r
    C[i,j] := { 1 − L[FirstNonZeroV[i], j]   if j = FirstNonZeroV[i]
              {   − L[FirstNonZeroV[i], j]   else                     ;
  endfor
endfor
```

```
/* Compute M_{Λ_{3L}} */
M_{Λ_{3L}} := L + V^T · C;
Let Λ_{3L} be the function Λ_{3L} : Q^r → Q^r : μ ↦ M_{Λ_{3L}} · μ;
return (Λ_{3L}, Λ_{3R});
```

Algorithm 2.4.1 works as follows:

1. Get the defining matrices. In the case of W, the non-zero columns of the reduced echelon form of $M_{\Lambda_{2R}^{90}} \cdot M_{\Lambda_{1L}} \cdot X$, with X solving $M_{\Lambda_{1R}} \cdot X = 0$, build the base vectors $\kappa_1, \ldots, \kappa_m$.

2. Compute $M_{\Lambda_{3R}}$. This is the unique solution for the equation system that asserts $M_{\Lambda_{3R}} \cdot \kappa_i = 0$ and for which the following holds: Going from one row to the next in the reduced echelon form $K' = \begin{pmatrix} \kappa_1^T \\ \vdots \\ \kappa_m^T \end{pmatrix}$ of the base vectors for $\mathfrak{W}$, if there is a step from column j to j' larger than 1 ($j' > j+1$), then we have $M_{\Lambda_{3R}} \cdot \iota_{r,j+1} = \iota_{r,j+1}, \ldots, M_{\Lambda_{3R}} \cdot \iota_{r,j'-1} = \iota_{r,j'-1}$

3. M_{Λ_R} is then easily computed using Theorem 2.34. An initial version of M_{Λ_L}, called $L = \begin{pmatrix} \lambda_1 & \ldots & \lambda_r \end{pmatrix}$, is also easily obtained; however, we may still modify M_{Λ_L} by adding the transposed rows of $V' = \begin{pmatrix} \nu_1'^T \\ \vdots \\ \nu_n'^T \end{pmatrix}$ (with V' as in Definition 2.36). Figure 2.17 sketches how we proceed here: we accumulate the coefficients needed for a sweep of each row ν_i' of V' after transposition over the columns of L so that we can guarantee that the row of $L + V'^T \cdot C$ which corresponds to the first non-zero coordinate of ν_i' has the desired form (the unit vector corresponding to that first non-zero coordinate).

Combination A placement in Chapter 4 can be defined by a linear relation. The basic procedure in that chapter is to examine different placement relations for a given set of set of occurrence instances. Placement candidates may define completely different placements for a given data element, such as Φ_1 and Φ_2 in

$$\Phi_1(\alpha) = \{0\} \qquad \Phi_2(\alpha) = \{1\}$$

(we only view dimensions interesting for us at the moment in this representation). Nevertheless it is also possible that two placement candidates select placements that do not contradict each other, for example if one placement defines a superset of the other one, as in the following example:

$$\Phi_1'(\alpha) = \{0\} \qquad \Phi_2'(\alpha) = \{1, \ldots, n\}$$

The idea is now to combine placement candidates into a new placement that represents the linear hull of the union of the two. In the example above this yields a placement

$$\Phi_3'(\alpha) = \Phi_2'(\alpha) = \{1, \ldots, n\}$$

Although we will not pursue this idea further in this thesis, this combination into the linear hull of the union may lead to new placement relations that possibly should also be considered in a compiler.

In order to compute the linear hull of the union, we introduce the **combination** of relations: the combination of two linear relations Λ_1, Λ_2 is is a superset of both Λ_1 and Λ_2, and therefore

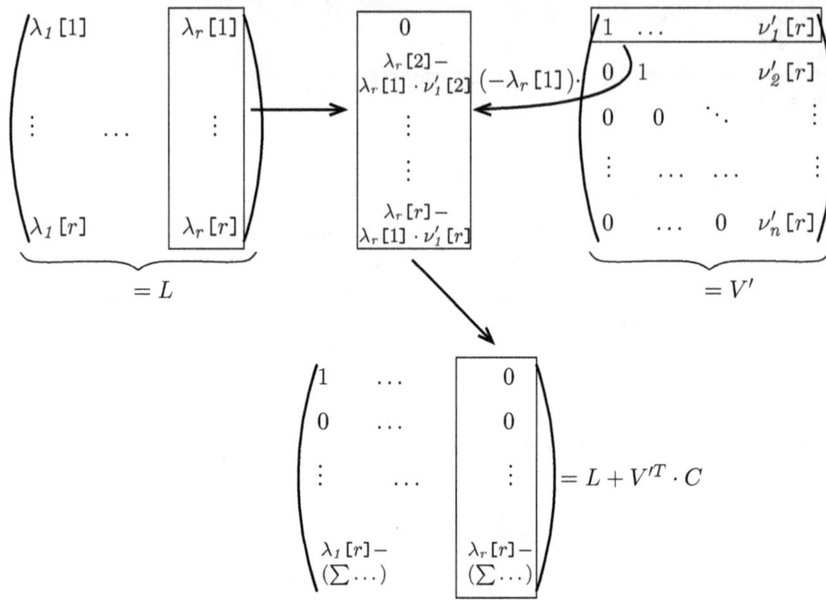

Figure 2.17: Addition of a vector $\nu_1'^T$ of V' to a set of vectors $L = (\lambda_1 \ldots \lambda_r)$, modifying the image of L.

does not contradict either, Λ_1 or Λ_2. However, the combination operator will occur only briefly as an interesting idea in Section 4.1.

Definition 2.37 *Let* $\Lambda_1 = (\Lambda_{1L}, \Lambda_{1R}), \Lambda_{1R}^{-1} \circ \Lambda_{1L} \subseteq \mathbb{Q}^k \times \mathbb{Q}^l$ *and* $\Lambda_2 = (\Lambda_{2L}, \Lambda_{2R}), \Lambda_{2R}^{-1} \circ \Lambda_{2L} \subseteq \mathbb{Q}^k \times \mathbb{Q}^l$ *be linear relations. The combination of* Λ_1 *and* Λ_2*,* $\mathrm{Comb}(\Lambda_1, \Lambda_2)$*, is the linear hull of* $\Lambda_1 \cup \Lambda_2$*.*

Algorithm 2.4.2 [*Combination*]:
Input:
Λ_1 : linear relation $\Lambda_1 = (\Lambda_{1L}, \Lambda_{1R})$

Λ_2 : linear relation $\Lambda_2 = (\Lambda_{2L}, \Lambda_{2R})$
Output:
$\Lambda_3 = (\Lambda_{3L}, \Lambda_{3R})$:
the combination of Λ_1 with Λ_2

Procedure:
Let $\ker(\Lambda_{1R}) = Span(v_1, \ldots, v_k)$;
Let $\ker(\Lambda_{2R}) = Span(v_{k+1}, \ldots, v_{k+l})$;
Let $\mathrm{im}(((\Lambda_{1R}^{go} \circ \Lambda_{1L}) - (\Lambda_{2R}^{go} \circ \Lambda_{2L}))) = Span(v_{k+l+1}, \ldots, v_{k+l+m})$;
Let $\mathfrak{K} = \ker(\Lambda_{1R}) \cup \ker(\Lambda_{2R}) \cup \mathrm{im}(((\Lambda_{1R}^{go} \circ \Lambda_{1L}) - (\Lambda_{2R}^{go} \circ \Lambda_{2L})))$;

$$M_{\mathfrak{K}} := \begin{pmatrix} v_1^T \\ \vdots \\ v_{k+l+m}^T \end{pmatrix};$$

/* i.e., matrix $M_{\mathfrak{K}}$ consists of base vectors of $\mathfrak{K}$ as row vectors */
Let $M_{\mathfrak{K}^\perp}$ be a matrix consisting of a base for $\mathfrak{K}^\perp$ as row vectors;
Let $M_{\Lambda_{3R}} \in \mathbb{Q}^{l \times l}$ be the unique solution to $\left(\frac{M_{\mathfrak{K}}}{M_{\mathfrak{K}^\perp}}\right) \cdot {M_{\Lambda_{3R}}}^T = \left(\frac{0}{M_{\mathfrak{K}^\perp}}\right)$;

```
let Λ3_R be the linear mapping defined by M_Λ3_R;
Λ3_L := Λ3_R ∘ Λ1_L;
return Λ3 = (Λ3_L, Λ3_R);
```

Algorithm 2.4.2 computes the combination of two relations. It defines first the mapping Λ_{3R} as the unique endomorphism with kernel $\mathfrak{K}$ that maps all elements of the orthogonal complement $\mathfrak{K}^{\perp}$ of $\mathfrak{K}$ to themselves. This mapping is obtained as the solution of a linear equation system (the uniqueness follows from the fact that $M_{\mathfrak{K}}$ together with $M_{\mathfrak{K}^{\perp}}$ in the definition has full row rank). Then Λ_{3L} is defined according to Theorem 2.30 so that the resulting relation corresponds to Λ_1 and Λ_2 where these relations do not differ. Λ_{3L} can be defined via either Λ_{1L} or Λ_{2L}, since the difference between the two mappings will be non-existent in the resulting relation $\Lambda_{3R}^{-1} \circ \Lambda_{3L}$ due to the definition of Λ_{3R}.

Example 2.38 *Let* $\Phi_1 = (\Phi_{1L}, \Phi_{1R})$ *with* $\Phi_{1L} : \nu \mapsto \begin{pmatrix} 1 & 0 \\ 0 & 1 \end{pmatrix} \cdot \nu$, $\Phi_{1R} : \nu \mapsto \begin{pmatrix} 1 & 0 \\ 0 & 1 \end{pmatrix} \cdot \nu$, *and* $\Phi_2 = (\Phi_{2L}, \Phi_{2R})$ *with* $\Phi_{2L} : \nu \mapsto \begin{pmatrix} 1 & 0 \\ 0 & 0 \end{pmatrix} \cdot \nu$, $\Phi_{2R} : \nu \mapsto \begin{pmatrix} 1 & 0 \\ 0 & 0 \end{pmatrix} \cdot \nu$. Φ_1 *represents the identity on* $\mathbb{Q}^2$, *while* Φ_2 *represents a relation between all elements of* $\mathbb{Q}^2$ *whose first coordinates are equal. Since* $\Phi_2 \supseteq \Phi_1$, *the smallest relation that is compatible with both* Φ_1 *and* Φ_2, *is* Φ_2. *In this case, Algorithm 2.4.2 iterates through the following steps:*

1. $\ker(\Lambda_{1R}) = \emptyset \Rightarrow k = 0.$

2. $\ker(\Lambda_{1R}) = \begin{pmatrix} 0 \\ 1 \end{pmatrix} \cdot \mathbb{Q} \Rightarrow l = 1, v_1 = \begin{pmatrix} 0 \\ 1 \end{pmatrix}.$

3. $\mathrm{im}((\Lambda_{1R}^{go} \circ \Lambda_{1L}) - (\Lambda_{1R}^{go} \circ \Lambda_{1L})) = \left(\begin{pmatrix} 1 & 0 \\ 0 & 1 \end{pmatrix} - \begin{pmatrix} 1 & 0 \\ 0 & 0 \end{pmatrix} \right) \cdot \mathbb{Q}^2 = \begin{pmatrix} 0 & 0 \\ 0 & 1 \end{pmatrix} \cdot \mathbb{Q}^2 = \begin{pmatrix} 0 \\ 1 \end{pmatrix} \cdot \mathbb{Q}$, *i.e.,* $m = 1, v_2 = \begin{pmatrix} 0 \\ 1 \end{pmatrix}.$

4. $\mathfrak{K} = \begin{pmatrix} 0 \\ 1 \end{pmatrix} \cdot \mathbb{Q} + \begin{pmatrix} 0 \\ 1 \end{pmatrix} \cdot \mathbb{Q}.$

5. $M_{\mathfrak{K}} = \begin{pmatrix} 0 & 1 \\ 0 & 1 \end{pmatrix}.$

6. *There are several ways to obtain a matrix for* $M_{\mathfrak{K}^{\perp}}$. *Schmidt's orthogonalization [Usm87, p. 143] yields* $M_{\mathfrak{K}^{\perp}} = \begin{pmatrix} 1 & 0 \end{pmatrix}.$

7. $M_{\Lambda_{3R}}$ *is the unique solution to* $\begin{pmatrix} 0 & 1 \\ 0 & 1 \\ 1 & 0 \end{pmatrix} \cdot M_{\Lambda_{3R}} = \begin{pmatrix} 0 & 0 \\ 0 & 0 \\ 1 & 0 \end{pmatrix}$, *i.e.,* $M_{\Lambda_{3R}} = \begin{pmatrix} 1 & 0 \\ 0 & 0 \end{pmatrix}.$

8. $M_{\Lambda_{3R}}$ *defines* Λ_{3R}.

9. $M_{\Lambda_{3L}} = \begin{pmatrix} 1 & 0 \\ 0 & 0 \end{pmatrix} \cdot \begin{pmatrix} 1 & 0 \\ 0 & 1 \end{pmatrix} = \begin{pmatrix} 1 & 0 \\ 0 & 0 \end{pmatrix}$; $M_{\Lambda_{3L}}$ *defines* Λ_{3L}.

We see that the result is exactly Φ_2 *– the smallest relation containing both* Φ_1 *and* Φ_2.

Testing Relations for Equality In order to argue about different relations between polyhedra, it may be necessary to determine whether two representations indeed represent the same relation. This question will crop up in Section 4.1. For a single set of points that we want to place on the processor array, there may be several placement relations whose quality we want to evaluate for their. In order to evaluate as few placement relations as possible, duplicates should be eliminated.

We have established that we can represent a relation Λ between two polyhedra $\mathfrak{P}$ and $\mathfrak{Q}$ of the form

$$\Lambda = \{(\nu, T(\nu) + \mathfrak{W} \mid \nu \in \mathfrak{P} \subseteq U, \mathfrak{W} \subseteq \mathfrak{V}, \mathfrak{W} = \mathrm{Span}(\omega_1, \ldots, \omega_n), \mathfrak{Q} \subseteq \mathfrak{V}\}$$

by two linear mappings, $\Lambda_R : \mathfrak{U} \to \mathfrak{V}$ and $\Lambda_L : \mathfrak{V} \to \mathfrak{V}$ and descriptions of source and target space of these mappings using inequations ($\mathfrak{P} = \{\nu \in \mathfrak{U} \mid M_{\mathfrak{P}} \cdot \nu \geq 0\}$ and $\mathfrak{Q} = \{\upsilon \in \mathfrak{V} \mid M_{\mathfrak{Q}} \cdot \upsilon \geq 0\}$).

For the purpose of deciding, whether two representations $(\Lambda_{1L}, \Lambda_{1R}, \mathfrak{P}_1, \mathfrak{Q}_1)$ and $(\Lambda_{2L}, \Lambda_{2R}, \mathfrak{P}_2, \mathfrak{Q}_2)$ represent the same relation, we generate a normal form from the relation representations as described below. These normal forms consist of series of matrices that are unique for a given relation. Note that unions of linear relations are *not* transformed into a unique form in this way. We only consider the transformation of a single – convex – linear relation. Such a representation is produced as follows.[7]

1. Create a matrix M_i defining $\mathfrak{P}_i$ and $\mathfrak{Q}_i$: M_i has a column for each dimension of $\mathfrak{P}_i$, $\mathfrak{Q}_i$, and each target dimension of Λ_{iL} (or Λ_{iR}), as described in Section 2.4.1.

2. Simplify the inequations by computing the dual representation of the defined polyhedron as vertices, rays and cones employing Chernikova's algorithm [Che64, Che65, Che68], and create the corresponding equations and inequations again – this step creates a unique representation of the polyhedron up to the order of (in)equations, the use of equalities, and the value of parameters.

3. Separate the resulting relations into equations and inequations.

4. Disambiguate equality relations (since equality relations cannot have been deleted by the previous step, we have at least one equality relation for each target dimension of Λ_{iL}). Compute the reduced echelon form of the matrix defining the equalities from above. As noted earlier, this form is unique.

5. Regenerate the representation as pair of mappings as discussed in Section 2.4.1.

Actually, our only application of this comparison of linear relations is to reduce memory and time requirements of the implementation of the placement method described in Chapter 4 (which we will not discuss in any greater detail). In this case we do not care about the scope of our mappings, which is represented by the inequations of $\mathfrak{P}_i$ and $\mathfrak{Q}_i$. Therefore, in our case it suffices to view the equations that are defined by Λ_{iL} and Λ_{iR} and compare the reduced echelon forms of the corresponding matrices.

[7] In order to get a usable description of our mappings, the bases have to be ordered as above, i.e. first the target dimensions of the mapping, then the source dimensions, then parametric dimensions, and finally the dimension for the constant part in homogenous coordinates.

Chapter 3

Loop-Carried Code Placement

Optimization techniques like *dead code elimination*, *strength reduction* [ACK81], *value numbering* [RL77], and (loop invariant) *code placement* [MR79, ASU86, KRS94, Muc97, Mor98] are well understood and are widely and successfully applied to increase in performance in the calculation of scalar expressions. However, the basic ideas behind these optimizations apply to arrays just as well as to scalars and, with the advent of precise dependence analysis algorithms for loop-carried dependences [Fea92a, Fea92b, CG97, CC96, BCF97, CG99], it is now possible to lift these optimization techniques to the case of array references in loop nests.

In this thesis, we focus on the generalization of code placement to the case of loop-carried dependences [FGL01c, FGL01a, FGL04]. The basic idea behind **loop-carried code placement** – or LCCP, for short – is to create a simple, "minimized" representation of the computation to be performed: if we can prove a certain value to be computed several times during the execution of a loop nest, we generate a new loop nest with as few loops as possible that computes each value only once. To this end, we take advantage of term equivalences as much as possible. Since each occurrence instance in our model represents an action the CPU has to take, reducing the number of occurrence instances also reduces the number of actions to be performed. Therefore, it is a natural choice to search for occurrence instances that represent the same value as others and eliminate them, so that recomputation of a given value does not occur.

Specifically, we aim at a representation that sorts the polyhedra representing computations according to their dimensionality. On a parallel machine, this procedure has the advantage that we can choose either to make use of the available processors in an $\#pdim$-dimensional layout for a k-dimensional expression (i.e., an expression that actually depends on k indices), and possibly calculate the same value several times on different processors, or to compute each value on only one processor of a k-dimensional processor arrangement: using the same number of available processors in a lower-dimensional configuration results in more available processors in a given dimension and thus in better processor utilization, albeit at the possible cost of increased communication needs.

In sequential programs, these communication needs translate into increased costs for memory references in sequential, since the above expressions have to be stored in auxiliary variables. These communication needs come as increased costs for memory references in sequential programs, since the above expressions have to be stored in auxiliary variables. Since these auxiliary variables usually *have* to be arrays, their handling is quite expensive compared to the original program, where intermediate results are usually stored in registers. There are exceptions to the rule that intermediate results are stored in extremely fast register memory: There are also memory-to-memory architectures that do not use registers at all with explicit load and store commands or allow for a very flexible management of the memory area containing registers [Bal98, SM98]. However, even with these architectures, one may encounter increased costs due to the fact that the auxiliary arrays used do not necessarily fit in a single memory page, or due to (possibly false) data sharing between threads in a parallel execution scheme. So, in general, we may assume that storing these auxiliary arrays results in more expensive memory references. In the parallel setting with distributed memory, these references may even refer to memory on different compute nodes that do have to be realized by explicit communication between those nodes. Therefore, especially in the parallel setting, we need a data placement strategy that may reduce these communication costs again by (partially) undoing the effects of LCCP, if this seems appropriate. This case will be examined in Chapter 4.

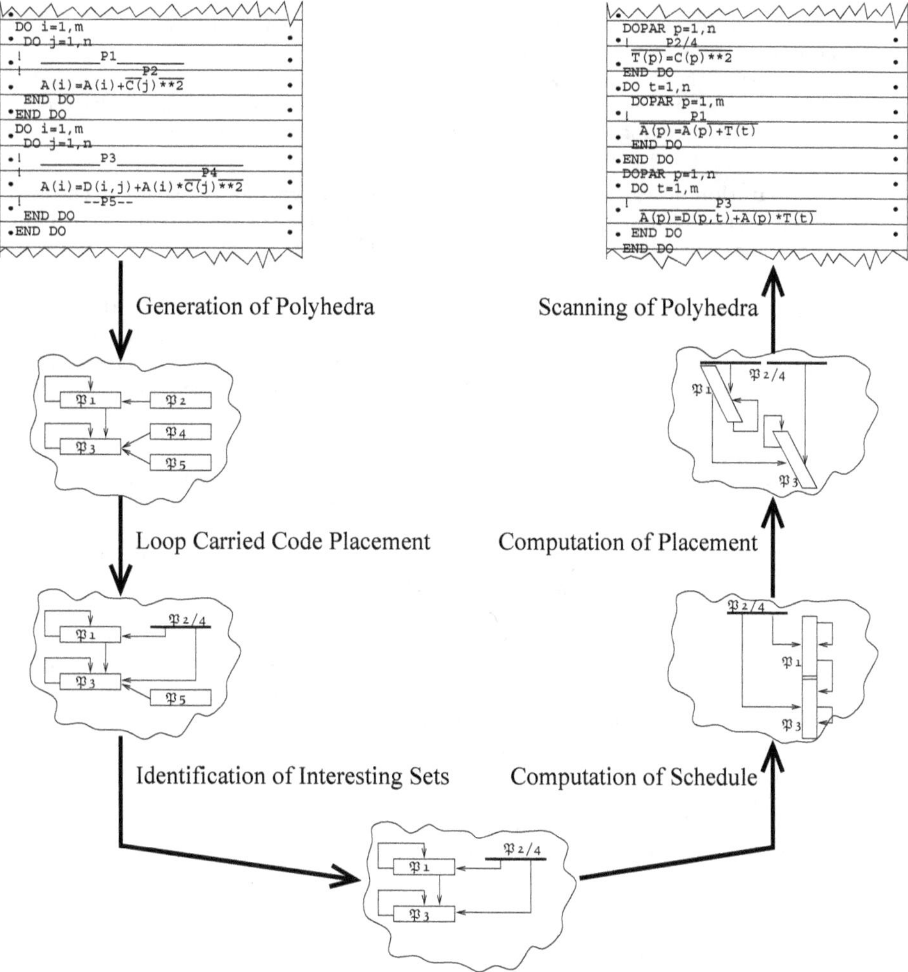

Figure 3.1: Overview of the sequence of code transformations.

Figure 3.1 sketches this approach. The program transformation is done in six steps:

1. **Generation of Polyhedra:** Convert the syntactic program representation into a representation by polyhedra and dependence relations between these polyhedra. Figure 3.1 shows polyhedra $\mathfrak{P}1$ to $\mathfrak{P}5$ with the corresponding code fragments marked with `P1` to `P5` in the comments above or below the program text, respectively. Note that the program is represented by even more polyhedra that represent non-linear subexpressions. This is because *every* place in the code is represented by an occurrence, whether the corresponding subterm is a linear expression or not. For the sake of clarity, we leave out these additional polyhedra representing non-linear subexpressions in the model.

2. **Loop-Carried Code Placement:** LCCP takes the description of the polyhedra generated in Step 1 and creates new polyhedra with appropriate dimensionality for the given subexpression. In the figure, the polyhedra $\mathfrak{P}2$ and $\mathfrak{P}4$ can be combined into a single, one-dimensional polyhedron $\mathfrak{P}2/4$.

3. **Identification of *Interesting Sets*:** In order to obtain a legal execution order for the occurrence instances of a code fragment, we need not consider *all* occurrences, but only certain sets. These so-called interesting sets are polyhedra that represent values that may be reused later and therefore have to be stored in a variable. Polyhedron $\mathfrak{P}5$ does not have to be distinguished from polyhedron $\mathfrak{P}3$ for the purpose of computing a space-time mapping.

4. **Computation of a Schedule:** Find a legal execution order for the points enumerated in the representation as polyhedra. The operations represented by both $\mathfrak{P}1$ and $\mathfrak{P}3$ have to be executed sequentially. The operations represented by $\mathfrak{P}2/4$ can all be executed in parallel.

5. **Computation of a Placement:** Find a processor dependent placement for the polyhedra. In order to minimize communication, it may pay off to replicate the computation of an, e.g., 1-dimensional expression along the second dimension of a 2-dimensional processor array. Note that, in this case, the effect of reducing the number of dimensions of a polyhedron in Step 2 may be undone.

6. **Scanning of Polyhedra:** Placement and schedule represent together define a new basis in the coordinate system. Use the polyhedra in their new bases to regenerate loop code.

Note that – although, in this procedure, we compute space-time mappings for the polyhedra representing the program – we do *not* necessarily aim at producing a parallel program. The only important step is the computation of a schedule that guarantees a legal execution order. Reducing the dimensions of polyhedra, and thus the number of iterations executed by the target program, represents an optimization for a sequential program as well as for a parallel one. Depending on the way we produce polyhedra of reduced dimensionality in Step 2 above, we may or may not have to determine a legal execution order of the points in the polyhedra; if we do have to create a new execution order, and if this execution order does not already specify different time steps for each point in the generated polyhedra, we may execute some of these points in parallel (for which we then have to create a placement). However, we do not really *have* to execute those points in parallel – we just have to execute them at all (otherwise we would change the semantics of the program). Let us consider the following example.

Example 3.1 *The program fragment on the left enables loop invariant code placement for the term* C*D, *yielding the code on the right:*

```
!HPF$ INDEPENDENT, NEW(i2)
      DO i1=1,n
!HPF$ INDEPENDENT
        DO i2=1,n
          A(i2,i1)=B(i2,i1)+C*D
        END DO
      END DO
```

```
      T=C*D
!HPF$ INDEPENDENT, NEW(i2)
      DO i1=1,n
!HPF$ INDEPENDENT
        DO i2=1,n
          A(i2,i1)=B(i2,i1)+T
        END DO
      END DO
```

These two program fragments may lead to different communication patterns. Let us suppose that C, D, *and* T *are aligned with* A(1,1). *Then, in the first case, we need to broadcast the values of* C *and* D *while, in the second case, only communication for* T *is necessary. The code generated by ADAPTOR-7.1 for these two program fragments looks as follows (the code on the left corresponds to the left program fragment):*

```
! --------  P A R T   I     --------
! COMMUNICATION FOR READ ACCESS TO C
 call DALIB_home_set_var              &
&(A_DSP,2,2,1,1,1)
 call DALIB_home_set_all ()
 call DALIB_home_assign (TMP0,C,8)

! COMMUNICATION FOR READ ACCESS TO D
 call DALIB_home_set_var              &
&(A_DSP,2,2,1,1,1)
 call DALIB_home_set_all ()
 call DALIB_home_assign (TMP1,D,8)

! --------  P A R T   I I   --------
!ACTUAL COMPUTATION
 call DALIB_array_my_slice            &
&(A_DSP,2,1,N,A_START2,A_STOP2)
 do I=A_START2,A_STOP2
    call DALIB_array_my_slice         &
&(A_DSP,1,1,N,A_START1,A_STOP1)
    do J=A_START1,A_STOP1
       A(A_ZERO+I*A_DIM1+J) =         &
&B(B_ZERO+I*B_DIM1+J)+TMP0*TMP1
    end do
 end do
```

```
! --------  P A R T   I     --------
! COMPUTATION FOR C*D
 if (DALIB_is_local(A_DSP,2,1)) then
   if (DALIB_is_local(A_DSP,1,1))    &
&then
      T = C*D
   end if
 end if

! COMMUNICATION FOR T
 call DALIB_home_set_var              &
&(A_DSP,2,2,1,1,1)
 call DALIB_home_set_all ()
 call DALIB_home_assign (TMP0,T,8)

! --------  P A R T   I I   --------
! ACTUAL COMPUTATION
 call DALIB_array_my_slice            &
&(A_DSP,2,1,N,A_START2,A_STOP2)
 call DALIB_array_my_slice            &
&(A_DSP,1,1,N,A_START1,A_STOP1)
 do I=A_START2,A_STOP2
    do J=A_START1,A_STOP1
       A(A_ZERO+I*A_DIM1+J) =         &
&B(B_ZERO+I*B_DIM1+J)+TMP0
    end do
 end do
```

For the sake of a clearer representation, we have augmented the ADAPTOR output is with comments indicating the action performed by the respective part of the code. Although the right program fragment is a bit longer, we may expect higher performance from it: in Part I, the program fragment on the left broadcasts both C and D; however, in Part I of the second code fragment, these variables are both locally available for the computation of $C \cdot D$, and whether this computation has to be done is guarded by two relatively inexpensive `if`*-statements, so that only* `T` *has to be broadcast in the following code snippet. In Part II, we can see the effect of loop invariant code motion: in the program fragment on the left, the value $C \cdot D$ has to be calculated for each iteration of the loop nest (which is done using the term* `TMP0*TMP1`*), while this calculation is replaced by a simple lookup to TMP0 in the program fragment on the right.*

As Example 3.1 shows, even conventional optimization techniques as loop invariant code motion may have an impact on the communication structure of a parallel program. This communication is often managed by subroutine calls that are inserted into the program in order to mimic the communication pattern on a distributed memory machine and may even represent the dominating cost factor of a parallel program. Therefore, it is vital for distributed memory architectures to apply optimization techniques that change the communication pattern of the program before the insertion of communication code.

3.1 Input Dependences and Equivalence Relations

The first step of LCCP consists of identifying redundant computations. Conventional code motion techniques use data flow analysis based on syntactic structures such as basic blocks [MR79, ASU86, KRS94, Mor98] or statements [KKS98]. Data flow analysis in the polyhedron model supplies us with the information as to which occurrence instance uses input of which other occurrence instance. An occurrence instance α of a read access (i.e., $\text{head}(\text{Occ}^{-1}(\text{OccId}(\alpha))) = \texttt{read\_A}$) represents the same value as another occurrence instance β (head(Occ^{-1}(OccId(β))) = `read_A`) if there is a direct input dependence between α and β ($(\alpha, \beta) \in \Delta^i$ or $(\beta, \alpha) \in \Delta^i$). In other words, equality of read accesses can be defined as the equivalence relation induced by the input dependences Δ^i – i.e., its symmetric, reflexive, and transitive closure $(\Delta^{i^-})^*$.

Let us briefly recall how dependence analysis in the polyhedron model works. Consider a dependence between occurrence instances α, $\mathrm{Occ}^{-1}(\mathrm{OccId}(\alpha)) = \mathtt{A}(\Psi_1(\alpha))$, and β, $\mathrm{Occ}^{-1}(\mathrm{OccId}(\beta)) = \mathtt{A}(\Psi_2(\beta))$, that is to be discovered by a dependence analysis algorithm. Then, the analysis algorithm usually performs three steps:

1. Conflict equation system: there has to be a conflicting memory access, i.e., the dependence relation $\Delta = \{(\alpha, \beta) \mid \beta \text{ depends on } \alpha\}$ holds only if condition $\Psi_1(\alpha) = \Psi_2(\beta)$ is satisfied.

2. Existence inequation system: both source and destination of the dependence have to exist, i.e., Δ is restricted to the index space of the occurrences of α, β ($\Delta \subseteq \mathfrak{OI} \times \mathfrak{OI}$). Actually, we restrict Δ even further, depending on what kind of dependence is to be computed (anti, flow, input, output), to occurrences that represent the corresponding array accesses.

3. Optimization: in order to find direct dependences, such that β depends on α, we have to find the lexicographically maximal α, for which the above conditions hold and for which $\alpha \prec \beta$. Here, even if we compute input dependences, α has to be larger than any occurrence instance representing a write access.

Up to the third step, our conditions represent a reflexive, symmetric, and transitive relation, in other words, an equivalence relation. However, this is not the reflexive, symmetric, and transitive closure of the input dependences (which we actually need): there could still be a conflicting write access executed between two read accesses that are related to each other in the above. In such a case, these read accesses do not refer to the same value and should not be considered equivalent. Therefore, we have to compute the actual dependence relation, and the equivalence we want to represent is the reflexive, symmetric, and transitive closure of this relation.

Whether two terms represent the same value depend not only on the environment (i.e., the values of variables in the term) but also on the underlying algebra. The method described here can be adapted to the case in which operators may have certain properties, such as associativity and commutativity. However, we do not restrict ourselves to these special cases in this thesis. Instead we consider the the general case of Herbrand equivalences. I.e., the interpretation of a function symbol is that function symbol itself; it does not stand for any given algebraic function on, say, the integers. We will not go into further details of algebra and logic. Instead, the interested reader is referred to the literature [HK91, BN98].

Limits of Equivalence Detection Overall, we want to achieve performance improvements by detecting equivalent occurrence instances. Note that the program fragments we consider here – with the restrictions presented in Section 2.3 – can be represented by **systems of affine recurrence equations** (SAREs). The only necessary step apart from the reformulation of the loops is a conversion into static single assignment form (SSA form) [Fea91, Coh99]. The equivalence of two SAREs is undecidable, as proven by Barthou, Feautrier and Redon [BFR01]. Such statements depend on the underlying algebra in which the recurrence equations are formed. However, this undecidability is true even in the simplest case in which the recurrence equations are formed in the Herbrand algebra. Therefore, we cannot expect to be able to devise an algorithm that can identify *all* equivalent occurrence instances, even if we restrict ourselves to purely syntactic equivalence (up to affine expressions). We can only hope to identify "many" equivalences. This will be our aim in the rest of this thesis.

3.1.1 Deducing Equivalences of Larger Terms

Once we have established an equivalence relation on read accesses, the question arises how to deduce the equivalence of larger terms. We will extend the equivalence relation to larger terms using a set of *equivalence propagating dependences.* The following operator is used to propagate a relation from some operands to the combined term of operator and operands.

Definition 3.2 *Let α and β be occurrence instances. Let Λ be a relation on occurrence instances. Let Δ be a set of flow dependences (represented relations with source occurrence instances as first elements and target occurrence instances as second elements), $\alpha, \beta \in \mathrm{im}(\Delta)$, and $\mathrm{head}(\mathrm{Occ}^{-1}(\mathrm{OccId}(\alpha))) = \odot$. Then we define $\mathit{LiftRel}_{\Delta}(\Lambda)$:*

$$(\alpha,\beta) \in \mathit{LiftRel}_{\Delta}(\Lambda) :\Leftrightarrow \mathrm{OpId}(\alpha) = 0 = \mathrm{OpId}(\beta) \wedge \mathrm{head}(\mathrm{Occ}^{-1}(\mathrm{OccId}(\beta))) = \odot \wedge \left(\begin{array}{l} \forall i : i \in \{s,\dots,t\} : \\ \left(\exists \gamma_i, \delta_i : \gamma_i, \delta_i \in \mathfrak{OI} : \begin{array}{l} (\gamma_i,\delta_i) \in \Lambda \wedge \\ (\gamma_i, \mathit{OpndSel}(-i,\alpha)) \in \Delta \wedge (\delta_i, \mathit{OpndSel}(-i,\beta)) \in \Delta \end{array} \right) \end{array} \right)$$

with

$$(s,t) = \begin{cases} (0,0) & \text{if } \Delta^{-1}(\mathit{OpndSel}(0,\alpha)) \cup \Delta^{-1}(\mathit{OpndSel}(0,\beta)) \neq \emptyset \\ (1, \mathrm{ArityIn}(\odot)) & \text{otherwise} \end{cases}$$

The flow dependences Δ here are called **equivalence propagating dependences**. *Equivalence propagating dependences are usually denoted with Δ^e.*

The operator *LiftRel* lifts the relation Λ one level higher in direction of Δ. It formally defines the equality of two occurrence instances that represent the execution on the same operator $\odot$ on the same input data ($(\gamma_i, \delta_i) \in \Lambda$). If Δ represents the structural flow dependences of Definition 2.25 and Λ is the symmetric, reflexive, and transitive closure of input dependences on arrays, then we can use the operator *LiftRel* to deduce the equivalence on occurrence instances, given the equivalence of all occurrence instances that are related to each other by Λ. The case distinction in the definition of (s,t) plays the role of a special dependence on operand 0: if we know that an operator that depends on a certain input will produce a certain output regardless of the other operands, a relation between this input and operand position 0 can be built to express that dependence. This is actually only used for instances of read accesses, which are known to read a specific value due to the dependence analysis.

Example 3.3 *Let us again consider the code of Example 2.22:*

```
DO i1=0,1
  B(i1)=2*n-A
END DO
```

Figure 3.2 sketches the closure of input dependences $\Lambda := \left(\Delta^{i^-}\right)^*$ (dashed arrows) along with the (structural) flow dependences (solid arrows) of this code fragment. The closure of the input dependences (whose reflexive part is omitted) is lifted along the structural dependences $\Delta := \Delta^s$. The equivalence classes of occurrence instances that can be shown to represent the same value correspond to this lifted equivalence relation Λ_1 (dotted arrows). In this example, the two computations of $2 \cdot n$ are clear to represent the same value from the start; likewise, the two lookups of array A are already known to represent the same value. The application of LiftRel then results in Λ_1 showing the equivalence of the two computations of $2 \cdot n - A$.

LiftRel represents the usual equality for operators that states that any two applications of the same operator on the same (i.e., equivalent wrt. Λ) input data will, again, yield the same – equivalent – result. This also means that *LiftRel*, applied to an equivalence relation, has to be an equivalence relation, again.

Lemma 3.4 *If Λ is an equivalence relation, so is $\mathit{LiftRel}_{\Delta}(\Lambda)$.*

Proof:

In the following, let $I(\alpha) := \mathrm{ArityIn}(\mathrm{Occ}^{-1}(\mathrm{OccId}(\alpha)))$.

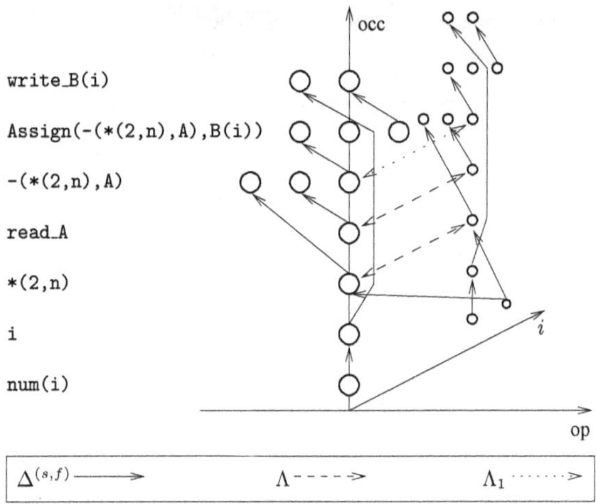

Figure 3.2: Occurrence instances of Example 3.3 with structural flow dependences ($\Delta^{(s,f)}$), originally given equivalences (Λ) and propagated equivalences (Λ_1).

Transitivity:

If $(\alpha, \beta) \in \mathit{LiftRel}_\Delta(\Lambda)$ and $(\beta, \gamma) \in \mathit{LiftRel}_\Delta(\Lambda)$, we also have predecessors (input arguments) α_j, β_j and γ_j $(j \in \{1, \ldots, I(\alpha)\})$, so that $(\alpha_j, \mathit{OpndSel}(-j, \alpha)) \in \Delta$, $(\beta_j, \mathit{OpndSel}(-j, \beta)) \in \Delta$, $(\gamma_j, \mathit{OpndSel}(-j, \gamma)) \in \Delta$ and $(\alpha_j, \beta_j) \in \Lambda$, $(\beta_j, \gamma_j) \in \Lambda$. Therefore, we also have $(\alpha_j, \gamma_j) \in \Lambda$ for all j and thus $(\alpha, \gamma) \in \mathit{LiftRel}_\Delta(\Lambda)$.

Symmetry:

In just the same way, if $(\alpha, \beta) \in \mathit{LiftRel}_\Delta(\Lambda)$, we also have predecessors (input arguments) α_j and β_j $(j \in \{1, \ldots, I(\alpha)\})$, so that $(\alpha_j, \mathit{OpndSel}(-j, \alpha)) \in \Delta$, $(\beta_j, \mathit{OpndSel}(-j, \beta)) \in \Delta$ and $(\alpha_j, \beta_j) \in \Lambda$. Again, it also follows that $(\beta_j, \alpha_j) \in \Lambda$ and therefore $(\beta_j, \alpha_j) \in \Lambda$ for all j and thus $(\alpha, \gamma) \in \mathit{LiftRel}_\Delta(\Lambda)$.

Reflexivity:

Reflexivity follows immediately from the reflexivity of Λ.

✓

The overall equivalence that tells us which occurrence instances actually represent the same value (in a Herbrand universe) is defined by all equivalences that may be deduced from the application of $\mathit{LiftRel}_{\Delta^{(s,f)}}((\Delta^{i^-})^*)$. Formally, it is defined via the corresponding functional given in the following Theorem 3.5. Not surprisingly, the equivalence we need is easily computed by repeatedly applying the *LiftRel* operator. Let us state this property formally.

Theorem 3.5 *Let Δ be the structural flow dependences of a program fragment and Λ an equivalence relation that defines an equivalence on some terms in the program fragment. Then the least fixed point (wrt. the lattice $(\mathcal{P}(\mathfrak{OI}) \times \mathcal{P}(\mathfrak{OI}), \subseteq)$) Λ_∞ of*

$$\mathit{LiftRel}_\Delta^{\Lambda \rightarrow} : \; (\mathcal{P}(\mathfrak{OI}) \times \mathcal{P}(\mathfrak{OI})) \rightarrow (\mathcal{P}(\mathfrak{OI}) \times \mathcal{P}(\mathfrak{OI})) : \; \mathfrak{O} \mapsto \begin{cases} \Lambda & \text{if } \mathfrak{O} = \emptyset \\ \mathfrak{O} \cup \mathit{LiftRel}_\Delta(\mathfrak{O}) & \text{otherwise} \end{cases}$$

as defined elementwise in Definition 3.2 exists and defines an equivalence on occurrence instances, iff the corresponding terms are Herbrand equivalent to each other for every environment that satisfies the equivalence Λ.

Proof:

- For all i, $\mathit{LiftRel}^i_\Delta(\Lambda)$ and thus $(\mathit{LiftRel}^{\Lambda\rightarrow}_\Delta)^i(\emptyset)$ is an equivalence relation according to Lemma 3.4.
- $\mathit{LiftRel}^{\Lambda\rightarrow}_\Delta$ is monotonically increasing in the complete lattice $(\mathcal{P}(\mathfrak{O}\mathfrak{I}) \times \mathcal{P}(\mathfrak{O}\mathfrak{I}), \subseteq)$, since the definition guarantees that, for $\Lambda_i \subseteq \Lambda'_i$, $\mathit{LiftRel}^{\Lambda\rightarrow}_\Delta(\Lambda'_i)$ defines a relation between *all* elements that are related via Λ_i, and possibly some others. Therefore, the least fixed point Λ_∞ of $\mathit{LiftRel}^{\Lambda\rightarrow}_\Delta$ exists according to the Knaster-Tarski Theorem [Kna28, Tar55] and can be computed as $\Lambda_\infty = (\mathit{LiftRel}^{\Lambda\rightarrow}_\Delta)^\infty(\emptyset)$.
- A term $\mathfrak{t}_0 = \odot(\mathfrak{t}_1, \ldots, \mathfrak{t}_n)$ with input arguments $\mathfrak{t}_i$ is equivalent to some other term $\mathfrak{u}_0 = \otimes(\mathfrak{u}_1, \ldots, \mathfrak{u}_m)$, if $\odot = \otimes$, $m = n$, and $\big(\forall i : i \in \{1, \ldots, n\} : \mathfrak{t}_i = \mathfrak{u}_i\big)$. The Herbrand equivalence holds if and only if this condition is true. In our case, equivalence Λ is enforced in addition to the above rule. However, a relation between occurrence instances α and β is established iff the equivalence of the input occurrence instances on which α and β depend is already established, i.e., the least fixed point represents exactly the equivalence of occurrence instances that is deducible from Λ.

✓

So, indeed, we may just repeatedly apply *LiftRel* in order to obtain the equivalence that tells us of the same value being computed several times. However, since equivalence is propagated along the levels of the operator tree of the program fragments (which is mimicked by the structural flow dependences), we can make life even easier by just traversing once through the operator tree from bottom to top.

Corollary 3.6 *If the transitive closure of Δ in Theorem 3.5, denoted Δ^+, is a strict partial order on $\mathfrak{O}\mathfrak{I}$, $(\mathfrak{O}_1, \ldots, \mathfrak{O}_n)$ is a topologically sorted sequence in $(\mathfrak{O}\mathfrak{I}, \Delta)$ (i.e. $\big(\forall i : i \in \{2, \ldots, n\} : \big(\forall \beta : \beta \in \mathfrak{O}_i : \big(\exists \alpha : \alpha \in \mathfrak{O}_{i-1} : (\alpha, \beta) \in \Delta\big)\big)\big)$) and Λ is a relation on $\mathfrak{O}\mathfrak{I}$ such that $\big(\forall \alpha, \beta : (\alpha, \beta) \in \Lambda : \alpha \in \mathfrak{O}_i \Rightarrow \beta \in \mathfrak{O}_i\big)$, then*

$$\begin{array}{rcl}
\Lambda_1 & = & \Lambda|_{\mathfrak{O}_1} \\
 & \vdots & \\
\Lambda_i & = & \mathit{LiftRel}_\Delta(\Lambda_{i-1}) \cup \Lambda|_{\left(\bigcup\limits_{j \in \{1,\ldots,i\}} \mathfrak{O}_j\right)} \\
 & \vdots & \\
\Lambda_n & = & \mathit{LiftRel}_\Delta(\Lambda_{n-1}) \cup \Lambda|_{\left(\bigcup\limits_{j \in \{1,\ldots,n-1\}} \mathfrak{O}_j\right)} \\
\Lambda_\infty & = & \Lambda_n
\end{array}$$

I.e., the fixed point Λ_∞ can be calculated by traversing the partitioning of occurrence instances only once *in the direction of the strict partial order Δ^+.*

Proof:

According to Definition 3.2, the operator *LiftRel* only adds pairs to a relation *in the direction of* Δ^+, i.e.,

$$(\alpha, \beta) \in \mathit{LiftRel}_\Delta(\Lambda) \Rightarrow \big(\exists \gamma, \delta : \gamma, \delta \in \mathfrak{O}\mathfrak{I} : (\gamma, \alpha) \in \Delta \wedge (\delta, \beta) \in \Delta \wedge (\gamma, \delta) \in \Lambda\big)$$

Since $(\gamma, \delta) \in \Lambda \Rightarrow \big(\exists i : i \in \{1, \ldots, n\} : \gamma \in \mathfrak{O}_i \ni \delta\big)$, i.e., γ and δ are not comparable, neither $(\alpha, \delta) \in \Delta^+$ nor $(\beta, \gamma) \in \Delta^+$ can hold – otherwise, we could compare γ to δ via $\gamma\, \Delta^+\, \alpha\, \Delta^+\, \delta$ or

$\delta\,\Delta^+\,\beta\,\Delta^+\,\gamma$, respectively. Therefore, the calculation of $\mathit{LiftRel}_\Delta(\Lambda_{i-1})$ can only introduce pairs (ϵ, ζ) with $\epsilon \in \mathfrak{O}_i \ni \zeta$, i.e., $\mathit{LiftRel}_\Delta(\Lambda_{i-1}) \subseteq \mathfrak{O}_i \times \mathfrak{O}_i$. And, since no new relation between occurrence instances in $\mathfrak{O}_j, j < i$ is introduced in any step i, and new relations are always introduced between occurrence instances that are greater wrt. Δ than the occurrence instances they are computed from, we always have $\mathit{LiftRel}_\Delta(\Lambda_{i-1}) = \mathit{LiftRel}_\Delta(\mathit{LiftRel}_\Delta(\Lambda_{i-1})) = \Lambda_i$. In particular, this holds for Λ_n and thus Λ_∞. ✓

In our case, Corollary 3.6 tells us that Λ_∞ can be computed by a single bottom-up traversal of the operator tree represented by the structural dependences $\Delta = \Delta^s$, since Δ^{s+} is obviously a strict partial order. But our case is still simpler: we do not even need to take into account equivalences originally defined by the original equivalence relation Λ, if they do not hold between occurrence instances on the lowest level of the operator tree (which is essentially just the read accesses). Indeed, we will choose not to take any additional equivalences into account. It should be noted, however, that even if we ignore such dependences here, our method can be trivially extended to support additional equivalences. For example, additional equivalences could be specified by the user. And this extension does not need to be restricted to array elements and memory locations: operations representing the same calculation could also be identified as equivalent by the user. Thus, the `Fortran` `EQUIVALENCE` statement could also be integrated into our approach as a way to introduce equivalences that cannot be detected by the compiler.[1]

The following lemma is actually just a reformulation of the definition of the *LiftRel* operator, Definition 3.2, that shows how to implement this operator. Here, we use the composition of linear relations and the notation $\mathit{OpndSel}_{0,i}$ for a mapping taking an occurrence instance α that represents the i-th input argument of an operation and returning an occurrence instance that represents the execution of that operation:

Lemma 3.7 *Let* $\text{head}(\text{Occ}^{-1}(\text{OccId}(\mathfrak{O}_1))) = \{\odot\} = \text{head}(\text{Occ}^{-1}(\text{OccId}(\mathfrak{O}_2)))$, $\Lambda \subseteq \mathfrak{O}_3 \times \mathfrak{O}_4$, $\Delta \subseteq (\mathfrak{O}_3 \times \mathfrak{O}_1) \cup (\mathfrak{O}_4 \times \mathfrak{O}_2)$ *relations, and for* $i \in \{0, \ldots, \text{ArityIn}(\odot)\}$ $\mathit{OpndSel}_{0,i} : \{\alpha_i \in \mathfrak{OI} \mid \text{OpId}(\alpha_i) = -i\} \to \mathfrak{OI} : \alpha_i \mapsto \mathit{OpndSel}(0, \alpha_i)$. *Then*

$$\begin{aligned}\mathit{LiftRel}_\Delta(\Lambda) &= (\mathit{OpndSel}_{0,0} \circ \Delta) \circ \Lambda \circ (\mathit{OpndSel}_{0,0} \circ \Delta)^{-1} \cup \\ &\quad \bigcap_{i \in \{1,\ldots,\text{ArityIn}(\odot)\}} (\mathit{OpndSel}_{0,i} \circ \Delta) \circ \Lambda \circ (\mathit{OpndSel}_{0,i} \circ \Delta)^{-1}\end{aligned}$$

Proof:

Note that $\mathit{OpndSel}_{0,i}$ is an injective mapping – i.e., $\alpha = \mathit{OpndSel}_{0,i}(\alpha_i) \Leftrightarrow (\alpha = \mathit{OpndSel}(0, \alpha_i) \wedge \text{OpId}(\alpha_i) = -i)$. For occurrence instances α, β, we then have:

$$\begin{aligned}&(\alpha, \beta) \in \bigcap_{i \in \{1,\ldots,\text{ArityIn}(\odot)\}} (\mathit{OpndSel}_{0,i} \circ \Delta) \circ \Lambda \circ (\mathit{OpndSel}_{0,i} \circ \Delta)^{-1} \\ \Leftrightarrow\ &(\forall i : i \in \{1, \ldots, \text{ArityIn}(\odot)\} : (\alpha, \beta) \in (\mathit{OpndSel}_{0,i} \circ \Delta) \circ \Lambda \circ (\mathit{OpndSel}_{0,i} \circ \Delta)^{-1}) \\ &\text{where} \\ &(\mathit{OpndSel}_{0,i} \circ \Delta) \circ \Lambda \circ (\mathit{OpndSel}_{0,i} \circ \Delta)^{-1} = (\mathit{OpndSel}_{0,i} \circ \Delta) \circ \Lambda \circ (\Delta^{-1} \circ \mathit{OpndSel}_{0,i}^{-1})\end{aligned}$$

[1]Note, however, that *only* adding equivalence relations here does not immediately suffice to handle `EQUIVALENCE` statements in the dependence analysis stage. Even if we extend our method to capture such equivalences, the dependence analysis stage also has to take them into account in order to enable later stages of the compiler to find a legal execution order for the program.

$$\Leftrightarrow \quad \left(\forall i : i \in \{1, \ldots, \text{ArityIn}(\odot)\} : \left(\exists \alpha_i, \beta_i : \begin{array}{l} \beta = \mathit{OpndSel}_{0,i}(\beta_i), \\ \alpha = \mathit{OpndSel}_{0,i}(\alpha_i) \end{array} : (\alpha_i, \beta_i) \in \Delta \circ \Lambda \circ \Delta^{-1}\right)\right)$$

$$\Leftrightarrow \quad \left(\forall i : i \in \{1, \ldots, \text{ArityIn}(\odot)\} : \left(\exists \alpha_i, \beta_i : \begin{array}{l} \mathit{OpndSel}(-i, \beta) = \beta_i, \\ \mathit{OpndSel}(-i, \alpha) = \alpha_i \end{array} : (\alpha_i, \beta_i) \in \Delta \circ \Lambda \circ \Delta^{-1}\right)\right)$$

Since *OpndSel* is well-defined for these α, β, α_i, β_i do exist. Therefore, the above relation holds, iff:

$$(\forall i : i \in \{1, \ldots, \text{ArityIn}(\odot)\} : (\mathit{OpndSel}(-i, \alpha), \mathit{OpndSel}(-i, \beta)) \in \Delta \circ \Lambda \circ \Delta^{-1})$$

$$\Leftrightarrow \quad \left(\forall i : i \in \{1, \ldots, \text{ArityIn}(\odot)\} : \left(\exists \gamma_i, \delta_i : \begin{array}{l} (\gamma_i, \mathit{OpndSel}(-i, \alpha)) \in \Delta, \\ (\delta_i, \mathit{OpndSel}(-i, \beta)) \in \Delta \end{array} : (\gamma_i, \delta_i) \in \Lambda\right)\right)$$

This represents exactly the case $(s, t) = (1, \text{ArityIn}(\odot))$ of Definition 3.2. However, Definition 3.2 also declares $(\alpha, \beta) \in \mathit{LiftRel}_\Delta(\Lambda)$, if $\Delta^{-1}(\mathit{OpndSel}(0, \alpha)) \cup \Delta^{-1}(\mathit{OpndSel}(0, \beta)) \neq \emptyset$ and $(\exists \gamma, \delta : \gamma, \delta \in \mathfrak{DI} : (\gamma, \delta) \in \Lambda \wedge (\gamma, \mathit{OpndSel}(0, \alpha)) \in \Delta \wedge (\delta, \mathit{OpndSel}(0, \beta)) \in \Delta)$. This is (correspondingly) the case iff $(\alpha, \beta) \in ((\mathit{OpndSel}_{0,0} \circ \Delta) \circ \Lambda \circ (\mathit{OpndSel}_{0,0} \circ \Delta)^{-1})$. ✓

Let us now state some basic properties of *LiftRel*. Since we will use it to follow equivalences from the reflexive, symmetric, transitive closure of input dependences by propagating these equivalences along structural dependences, with all dependence relations defined by h-transformations (i.e., linear functions), the result of applying *LiftRel* should not only be well-defined, but also in some sense well-behaved – i.e., we should be able to handle it within our framework of linear functions and polyhedra. The following corollary states that, indeed, we can handle the result as union of polyhedra.

Corollary 3.8 *In the situation of Lemma 3.7, the following holds:*

1. *If Δ and Λ are invertible mappings, so is $\mathit{LiftRel}_\Delta(\Lambda)$.*

2. *If Δ and Λ are linear mappings (linear relations), so is $\mathit{LiftRel}_\Delta(\Lambda)$.*

3. *If Δ and Λ are polyhedra representing linear relations, $\mathit{LiftRel}_\Delta(\Lambda)$ is a polyhedron representing a linear relation (that is possibly further restricted in its domain or image).*

4. *If Δ and Λ are unions of polyhedra representing linear relations, $\mathit{LiftRel}_\Delta(\Lambda)$ is a union of polyhedra representing linear relations (that are possibly further restricted in their domain or image).*

Proof:

1. For each i, $(\mathit{OpndSel}_{0,i} \circ \Delta) \circ \Lambda \circ (\mathit{OpndSel}_{0,i} \circ \Delta)^{-1} = \mathit{OpndSel}_{0,i} \circ \Delta \circ \Lambda \circ \Delta^{-1} \circ \mathit{OpndSel}_{0,i}^{-1}$ is an invertible mapping (since all mappings above are invertible). Since mappings always map each input element to exactly one output element, the intersection of mappings is always either the same as the first mapping – aside from possibly reducing its domain – or the empty set (which we will consider an invertible mapping for this purpose).

2. The composition of linear mappings (linear relations) is, again, a linear mapping (linear relation).

3. The difference to the above point is the restriction in the domain (or image) of the relation; however, it is clear that the composition of relations honours domain/image restrictions

of the original relations. For example, if $\Delta(\alpha) = \begin{cases} \Delta_1(\alpha) & \text{if } \alpha \in \mathfrak{A} \subseteq \mathfrak{OI} \\ \Delta_2(\alpha) & \text{if } \alpha \in \mathfrak{B} \subseteq \mathfrak{OI}, \mathfrak{A} \cap \mathfrak{B} = \emptyset \end{cases}$, the result is the union of the $\mathit{OpndSel}_{0,i} \circ \Delta \circ \Lambda' \circ \Delta^{-1} \circ \mathit{OpndSel}_{0,i}^{-1}$ with Λ' being one of

$$\Lambda'_1 : \mathit{OpndSel}_{0,i}(\Delta_1(\mathfrak{A})) \to \mathfrak{OI} : \quad \alpha \mapsto (\mathit{OpndSel}_{0,i} \circ \Delta_1 \circ \Lambda \circ \Delta_1^{-1} \circ \mathit{OpndSel}_{0,i}^{-1})(\alpha)$$
$$\Lambda'_2 : \mathit{OpndSel}_{0,i}(\Delta_2(\mathfrak{B})) \to \mathfrak{OI} : \quad \alpha \mapsto (\mathit{OpndSel}_{0,i} \circ \Delta_2 \circ \Lambda \circ \Delta_2^{-1} \circ \mathit{OpndSel}_{0,i}^{-1})(\alpha)$$

4. The composition of the unions of relations is the union of the compositions of these relations:

$$(R_1 \cup R_2) \circ (S_1 \cup S_2) = (R_1 \circ S_1) \cup (R_1 \circ S_2) \cup (R_2 \circ S_1) \cup (R_2 \circ S2)$$

✓

In particular, if Δ is a set of structural dependences, represented by corresponding h-transformations

$$\begin{aligned} H_{1,1} &: \mathit{OpndSel}(-1, \mathfrak{D}_1) \to \mathfrak{D}_3 \\ &\vdots \\ H_{1,\mathrm{ArityIn}(\odot)} &: \mathit{OpndSel}(-\mathrm{ArityIn}(\odot), \mathfrak{D}_1) \to \mathfrak{D}_3 \\ H_{2,1} &: \mathit{OpndSel}(-1, \mathfrak{D}_2) \to \mathfrak{D}_4 \\ &\vdots \\ H_{2,\mathrm{ArityIn}(\odot)} &: \mathit{OpndSel}(-\mathrm{ArityIn}(\odot), \mathfrak{D}_2) \to \mathfrak{D}_4 \end{aligned}$$

and Λ is a polyhedron representing a linear relation that is restricted in its scope

$$\Lambda : \mathfrak{D}_\Lambda \subseteq \mathfrak{D}_3 \times \mathfrak{D}_4$$

the resulting polyhedron again represents a linear relation that is restricted by its scope $\Lambda' = \mathit{LiftRel}_\Delta(\Lambda)$ is defined by

$$\Lambda' = \bigcap_{i \in \{1,\ldots,\mathrm{ArityIn}(\odot)\}} \mathit{OpndSel}_{0,i} \circ H_{2,i}^{-1} \circ \Lambda \circ H_{1,i} \circ \mathit{OpndSel}_{0,i}^{-1}$$

In our case, Δ is defined by a family of linear mappings – the h-transformations of equivalence propagating dependences. Due to the fact that the basic equivalences on read accesses and linear expressions are defined by linear equation systems as described on page 63 – only with possible additional restrictions representing conflicts with write accesses, the original equivalence relation Λ that we start with is defined by a union of polyhedra.

Let us now define our equivalence on occurrence instances formally:

Definition 3.9 *Let Δ^i be a set of input dependences on read accesses and $\Delta^{(s,f)}$ a set of structural flow dependences. We define the $(\Delta^{(s,f)}, \Delta^i)$-equivalence $\equiv_{(\Delta^{(s,f)},\, \Delta^i)}$ on occurrence instances as the equivalence that we can follow from the input dependences:*

$$\equiv_{(\Delta^{(s,f)},\, \Delta^i)} := (\mathit{LiftRel}_{\Delta^{(s,f)}}^{(\Delta^{i-})^* \to})^\infty(\emptyset)$$

It is plain to see that, for $\Delta^{(s,f)}$ comprising of all structural flow dependences and Δ^i containing all input dependences of the program fragment, Definition 3.9 is exactly the definition of Herbrand equivalence with only the basic equivalence on constants replaced by the equivalence relation $(\Delta^{i-})^*$.

Note, however, that, as was mentioned earlier, these equivalence classes by no means represent the equivalence of *all* occurrence instances one might actually find to represent the same value:

Example 3.10 *Let us reconsider the code fragment of Example 2.26:*

```
DO i1=0,1
  B(i1)=2*n-A
END DO
C=2*B(0)+2*B(1)
```

The execution of the loop results in both B(0) *and* B(1) *representing the same value* $(2 \cdot n - A)$*. However, a subsequent use of* B(0) *cannot be identified as equivalent to a use of* B(1)*. In the example, we could actually assign* 2*B(1)+2*B(1) *to* C *instead of* 2*B(0)+2*B(1)*. However, since there is no input dependence between the occurrence instances representing the read accesses* B(0) *and* B(1)*, respectively, these occurrence instances are* not $\equiv_{\left(\Delta^{(s,f)},\, \Delta^{i}\right)}$*-equivalent. Therefore, we will not consider the above alternate formulation for the result* C *a valid alternative. In other words, although the corresponding SAREs clearly define the same values for* B(0) *and* B(1)*, we cannot confirm this equivalence using the above scheme.*

The above example shows that the $\equiv_{\left(\Delta^{(s,f)},\, \Delta^{i}\right)}$-equivalence does *not* suffice to determine whether two code fragments actually compute the same value – despite Theorem 3.5. The problem is that $\equiv_{\left(\Delta^{(s,f)},\, \Delta^{i}\right)}$-equivalence does not hold between variable accesses that do represent the same value, as soon as this value has been stored in different memory cells.

The above situation can be improved by introducing equivalences also on write accesses (or introducing direct flow dependences between output occurrence instances and read access, eliminating the need for most write accesses in the program). Equivalences of terms can then be propagated beyond the sequence of a write and a read access using the following scheme:

1. Propagate input dependences Δ^{i} along the structural flow dependences $\Delta^{(s,f)}$ calculating the least fixed point $\equiv_{\left(\Delta^{(s,f)},\, \Delta^{i}\right)}$ of $\mathit{LiftRel}_{\Delta^{(s,f)}}^{\left(\Delta^{i-}\right)^{*}\rightarrow}$.
2. Define input dependences between write accesses by propagating along structural output dependences, i.e. calculating the least fixed point $\equiv_{\left(\Delta^{(s,f)},\, \Delta^{i}\right)}'$ of $\mathit{LiftRel}_{\Delta^{(s,o)}}^{\equiv_{\left(\Delta^{(s,f)},\, \Delta^{i}\right)}\rightarrow}$.

The propagation of input dependences along structural output dependences then defines input dependences between write accesses that do actually write the same value to memory. With the expanded equivalence induced by the resulting input dependences, situations as those in Example 3.10 can be solved: once the equality of write accesses is established, we can also deduce equality on the corresponding read accesses and so on. But what do we do if there is not a single write access between creation of a value and its use, but a whole chain as in the following Example 3.11?

Example 3.11 *Let us obfuscate the code of Example 3.3 even further: now, the value* $2 \cdot n - A$ *is not directly written to memory and then read again, but copied over and over again to different places in memory. Each of these places can then be used as a valid address to access this value:*

```
      DO i1=0,1
!       Statement S1:
        T00=2*n
!       Statement S2:
        T10=2*n-A
!       Statement S3:
        T11=T10
!       Statement S4:
        B(i1)=T11
      END DO
```

```
!       Statement S5:
        T12=B(0)
!       Statement S6:
        T13=B(1)
!       Statement S7:
        T14=T00-A
```

Not only `B(0)` *can be used to refer to value* $2 \cdot n - A$*, but also* `T10`*,* `T11`*,* `T12`*,* `T13`*, and* `T14`*. However, in order to discover this fact, e.g., for* `T12`*, one must not only trace back the assignment to* T_{12} *in Statement* S_5*, but also the assignments in Statements* S_2 *through* S_4*. This problem is not a simple trace. We cannot just call occurrence instances equivalent as long as a value is only copied from one occurrence instance to the other and no additional calculation is done: in Statement* S_7*, we compute* $2 \cdot n - A$ *from* T_{00} *and* A*. Therefore, it is always the complete equivalence relation – and not just some small piece of information – that has to be propagated through these assignments.*

Limits of the Method Again, we cannot determine the equivalence of all occurrence instances that represent the same value in Example 3.11. If we apply the above scheme using two steps – propagation along structural flow dependences and structural output dependences, respectively – not only once, but over and over again, the resulting equivalence relation represents all these equivalences. However, the termination of this repeated calculation of equivalence relations is *not* guaranteed. This is where the undecidability of the equivalence of two SAREs comes into play again: we cannot hope for an algorithm to compute an equivalence relation precisely. Of course, we can use some heuristics that decides when *all interesting* equivalences have been found or just iterate through the above scheme any given number of times – in this case an indirection through as many write and read accesses could be analyzed. In particular, the equivalence of A and B in the following example cannot be deduced with this scheme.

Example 3.12 *In the following code fragment,* A *and* B *always represent the same value. However, an equivalence cannot be deduced using our present procedure:*

```
A=0
B=0
DO i1=1,n
  A=A+1
  B=B+1
END DO
```

Although the first results of the calculations in the first iteration of the loop can be determined to be equivalent, if we allow propagation of equivalences over write accesses, the equivalence at the next iteration can only be deduced by propagating equivalence classes over write accesses once more, and so on: we can only deduce an equivalence of finitely many iterations.

Equivalence on Write Accesses The examples show that some semantic equivalences may only be discovered when equivalences on write accesses are allowed. Some of these equivalences may only be discovered with an infinite number of unrolling steps. Therefore, we consider only the simple case in this thesis: propagation is done only along structural flow dependences, and write accesses are *never* considered $\equiv_{(\Delta^{(s,f)},\,\Delta^{i})}$-equivalent to each other.

3.1.2 How to Obtain Equivalences on Read Accesses

Up to now, we just assumed that all dependence information – including input dependences that define an initial equivalence on input data – is supplied by the run of a dependence analysis algorithm. We have also seen that in order to determine, which occurrence instances can be removed from the original code dueto their equivalence, we need to consider the reflexive, symmetric and transitive closure of these input dependences. However, we did not really address the problem of computing this reflexive, symmetric and transitive closure that we need in order to define this $(\Delta^{(s,f)}, \Delta^i)$-equivalence. Computing the transitive closure is not necessarily possible – but as it turns out, we do not really need to compute any closure at all.

On the one hand, the transitive closure of an integer linear relation is not in general computable [KPRS94]. On the other hand, two occurrence instances α and β are (directly) input dependent on one another if they read from the same write access (and α immediately precedes β among the occurrence instances reading from the same write access). The $(\Delta^{(s,f)}, \Delta^i)$-equivalent occurrence instances of an occurrence instance α are those that read from the same write access as α, i.e., we can directly compute the reflexive, symmetric and transitive closure of the input dependences from direct flow dependences. Formally, this simplification is expressed by the following theorem.

Theorem 3.13 *Let Δ^f be the direct flow dependences of the program fragment considered and Δ^i its direct input dependences. Then*

$$\left(\Delta^{i^-}\right)^* = \mathit{LiftRel}_{\Delta^f}(\mathrm{id})$$

Proof:

1. Since id is an equivalence relation, $\mathit{LiftRel}_{\Delta^f}(\mathrm{id})$ is so, too (Lemma 3.4).

2. $\Delta^i \subseteq \mathit{LiftRel}_{\Delta^f}(\mathrm{id})$:

$$\begin{aligned} (\alpha,\beta) \in \Delta^i \quad &\Rightarrow \quad (\exists \gamma : \gamma \in \mathfrak{OI} : (\gamma,\alpha) \in \Delta^f \wedge (\gamma,\beta) \in \Delta^f) \\ &\Rightarrow \quad (\alpha,\beta) \in \mathit{LiftRel}_{\Delta^f}(\mathrm{id}) \\ & \quad \text{according to Defintion 3.2, since ArityIn}(\texttt{read\_}\cdot) = 0 \end{aligned}$$

3. $\left(\Delta^{i^-}\right)^* \supseteq \mathit{LiftRel}_{\Delta^f}(\mathrm{id})$:
Let $(\alpha,\beta) \in \mathit{LiftRel}_{\Delta^f}(\mathrm{id})$. Suppose $\alpha \neq \beta$ (otherwise, this is a trivial case). Without loss of generality, we assume $\alpha \prec \beta$ (we may do so, since $\mathit{LiftRel}_{\Delta^f}(\mathrm{id})$ is symmetric)

$$\Rightarrow \quad (\exists \gamma : \gamma \in \mathfrak{OI} : (\gamma,\alpha) \in \Delta^f \wedge (\gamma,\beta) \in \Delta^f)$$

In particular, α and β, both represent read accesses to the same memory cell. This means, we either have

(a) $(\alpha,\beta) \in \Delta^i$, or

(b) there is another memory access γ' to the same memory cell between these two accesses ($\alpha \prec \gamma' \prec \beta$). In the latter case, γ' has to be a read access, too, since otherwise $(\gamma,\beta) \in \Delta^f$ would not hold, i.e., by induction:

$$\begin{aligned} &\Rightarrow \quad \left(\exists \gamma' : \gamma' \in \mathfrak{OI} : (\alpha,\gamma') \in \Delta^i \wedge (\gamma',\beta) \in \Delta^{i^+}\right) \\ & \quad \text{i.e., in any case, we have:} \\ &\Rightarrow \quad (\alpha,\beta) \in \left(\Delta^{i^-}\right)^* \end{aligned}$$

✓

Corollary 3.14 *In the situation of Theorem 3.13 and with $\Delta^{(s,f)}$ being the structural flow dependences of the program fragment, the equivalence on occurrence instances defining the Herbrand equivalence with $\left(\Delta^{i^-}\right)^*$ as basic identity is given by computing the fixed point of*

$\mathit{LiftRel}^{\mathrm{id}\rightarrow}_{\Delta^{(s,f)} \cup \Delta^f}(\emptyset)$:

$$\equiv_{(\Delta^{(s,f)},\, \Delta^i)} \quad = \quad \equiv_{(\Delta^{(s,f)} \cup \Delta^f,\, \mathrm{id})}$$

Proof:
This follows directly from Theorem 3.13. ✓

In other words, we do not need to compute the reflexive, symmetric and transitive closure of Δ^i to obtain a basic equivalence relation for determining equivalent read accesses. Everything we need can already be computed directly from the flow dependences without even knowing the exact input dependences: we only need to take the structural flow dependences and the usual flow dependences from a write access to a read access into account. We then need one more step in the fixed point iteration for computing $\equiv_{(\Delta^{(s,f)} \cup \Delta^f,\, \mathrm{id})}$. Note that $\Delta^{(s,f)} \cup \Delta^{f^+}$ is a strict partial order on occurrence instances (and so is Δ^e together with the implicit dependences above): the transitivity is clear and, if $(\alpha, \beta) \in \Delta^{e+}$, (β, α) cannot be element of Δ^{e+} (i.e., we have $(\beta, \alpha) \notin \Delta^{e+}$), since targets of structural dependences always have greater occurrence numbers than their sources, and non-structural flow dependences always have write accesses as sources – which are excluded as targets of dependences in Δ^e. Thus, Corollary 3.8 still applies, and the least fixed point is still easily computable with only this single additional step.

3.2 Optimizing the Program

As already hinted on in the previous section, the next step in our method – after identifying equivalent occurrence instances – is to reduce the number of calculations to be executed. From now on, we do not aim to produce code for the calculations represented by $\mathfrak{OI}$, but for those represented by the different equivalence classes $\mathfrak{OI}/\equiv_{(\Delta^{(s,f)},\, \Delta^i)}$.

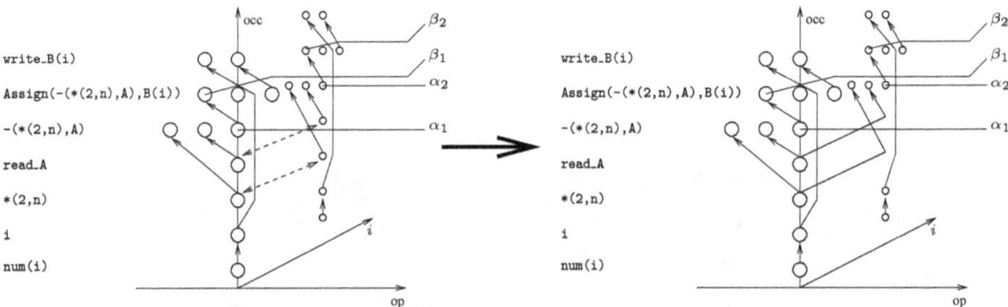

Figure 3.3: Left hand side: original OIG for Example 3.3; equivalences (dotted arrows) between basic expressions can be concluded. Right hand side: mixed graph with equivalent expressions condensed.

Example 3.15 *In Example 3.3 on page 64, the term* `2*n-A` *represents a loop independent value* $2 \cdot n - A$ *that is computed by occurrence instances* α_1 *and* α_2 *and assigned to* $B(i)$ *by occurrence instances* β_1 *and* β_2*. Figure 3.3 shows the OIG in a three-dimensional coordinate system. The two instances of* `2*n` *are equivalent, as indicated by the dotted arrows. Likewise, the two instances of the read access to A are equivalent. This can be used to form new equivalence classes, represented by the occurrence instances in the foreground. This results here in a mixed graph with rewritten dependences on the right hand side of Figure 3.3. From this representation, new equivalences between* α_1 *and* α_2 *can be deduced with the help of the LiftRel operator, depicted as dotted arrows on the left side of Figure 3.4. These equivalences lead to the transformed OIG on the right side.*

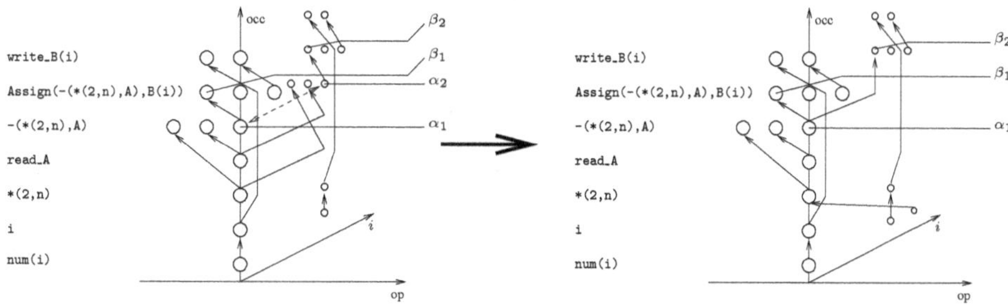

Figure 3.4: Left hand side: equivalences between the computations of `2*n-A` can be derived. Right hand side: possible transformed *OIG*with a single occurrence instance α for `2*n-A`.

The graph on the right represents a program that computes the loop invariant value $2 \cdot n - A$ *once and reuses it twice in the assignment in the loop. In the transformed OIG, this is represented by only one instance for all occurrences that are needed to build the term* `2*n-A`*. We obtain a single occurrence instance* α_1 *that represents this complete term. However, now, there are two occurrence instances* β_1 *and* β_2 *representing assignments that each depend on* α_1*.*

In particular, the dependence relation between α_2 *and* β_2 *does no longer exist and is replaced by a new dependence relation between* α_1 *and* β_2*.*

For the code generation later, this means that we will have to introduce new memory cells in which these values can be stored until all references to these values have been made.

The transformed OIG on the right hand side of Figure 3.4 is the goal we are aiming at. First, we will sketch the coarse structure of the method. Section 3.3 will then go into further detail about choosing representatives for equivalence classes, with two possibilities laid out in Sections 3.5.1 and 3.5.2.

But let us begin the description of this transformation to a graph based on equivalence classes with a formal definition of this new graph.

Definition 3.16 *The condensed graph of an occurrence instance graph OIG* $= (\mathfrak{OI}, \Delta)$ *relative to an equivalence relation* $\equiv$*, the* **condensed occurrence instance graph** *(COIG) of OIG, is the graph COIG* $=$ *OIG*$/\!\equiv\ = (\mathfrak{OI}/\!\equiv, \Delta/\!\equiv)$ *with* $\Delta/\!\equiv\ = \{(\mathfrak{A}, \mathfrak{B}) \mid (\exists \alpha, \beta : \alpha, \beta \in \mathfrak{OI} : (\alpha, \beta) \in \Delta \wedge \mathfrak{A} = [\alpha]_\equiv, \mathfrak{B} = [\beta]_\equiv)\}$ *(where* $[\alpha]_\equiv$ *is the equivalence class of* α *under* $\equiv$*).*

The simple strategy is then to generate code not for the OIG of a program fragment, but for its COIG. Since code is generated for vertices of the COIG – which may represent several vertices in the OIG, we enumerate fewer occurrence instances in the transformed program than in the original one. In particular the fact that we now deal with occurrence instance sets of reduced dimensionalities leads to a potential performance improvement (if occurrence instances along a complete dimension of the original index space are equivalent to each other, this dimension can be safely removed in the transforemd index space).

The COIG is prepared using *GroupOccurrenceInstances* and *LCCPHighLevel*. Algorithm 3.2.1 (*GroupOccurrenceInstances*) partitions the OIG so that we can compute the equivalence classes of occurrence instances in a single sweep through the partitioned graph. This computation of equivalence classes is then performed by Algorithm 3.2.2 (*LCCPHighLevel*), for which we will present a refined version using several helper algorithms in later sections.

As sketched in Figure 3.1, the first step in the target code generation is to compute a space-time mapping that determines when and where to execute the occurrence occurrence instances. It should be noted that it is not necessary to compute a space-time mapping for all sets of occurrence instances to produce the target code. As long as each occurrence instance of a set is a source for at most one target occurrence instance, it is not necessary to compute a space-time mapping: it may simply be inherited from the (unique) target occurrence instance. Moreover, this means that the code for the source occurrence instance may be produced as a subterm of the term corresponding to the target occurrence instance. For example, consider some read access `A(i)`. If the program under consideration contains a computation in the form of the term `A(i)+B(2*i)`, the read access `A(i)` may simply be inserted into `A(i)+B(2*i)` itself (thus enabling the compiler to use registers for `A(i)` and). However, if several occurrence instances may read from a single source occurrence instance α, the value represented by α has to be stored in order to be copied into each of the use sites, and it has to be available until all uses are done. This means that a space-time mapping has to be computed that puts the space-time coordinates of the occurrence instance at hand in relation to the execution time (and space) of the other occurrence instances. For write accesses, the same restrictions hold. The following definition establishes sets containing such occurrence instances as special sets that will be of further interest to us in Chapter 4.

Definition 3.17 (Interesting Set) *Consider a COIG* $(\mathfrak{OI}/\equiv_{(\Delta^{(s,f)} \cup \Delta^f, \mathrm{id})}, \Delta_{\equiv_{(\Delta^{(s,f)} \cup \Delta^f, \mathrm{id})}})$. *An* **interesting set** *of this COIG is a set* $\mathfrak{O} \subseteq \mathfrak{OI}/\equiv_{(\Delta^{(s,f)} \cup \Delta^f, \mathrm{id})}$ *of occurrence instances that contains at least one element* α *that either*

- *represents a write access*

$$\mathrm{head}(\mathrm{Occ}^{-1}(\mathrm{OccId}(\alpha))) = \texttt{write\_}.$$

or

- *is the source of several dependences in* $\Delta_{\equiv_{(\Delta^{(s,f)} \cup \Delta^f, \mathrm{id})}}$:

$$\left(\exists \beta, \beta' : \beta \neq \beta' \in \mathfrak{OI}/\equiv_{(\Delta^{(s,f)} \cup \Delta^f, \mathrm{id})} : \begin{array}{l} (\alpha, \beta) \in \Delta_{\equiv_{(\Delta^{(s,f)} \cup \Delta^f, \mathrm{id})}} \wedge \\ (\alpha, \beta') \in \Delta_{\equiv_{(\Delta^{(s,f)} \cup \Delta^f, \mathrm{id})}} \end{array} \right)$$

For example, in Figure 3.4 on page 74, the (singleton) set of occurrence instances for the term `2*n-A` is an interesting set; all sets of occurrence instances that are not supersets thereof are *not* interesting sets.

Theorem 3.13 in Section 3.1.2 states that $\equiv_{(\Delta^{(s,f)}, \Delta^i)} = \equiv_{(\Delta^{(s,f)} \cup \Delta^f, \mathrm{id})}$. So we would like to use $\Delta^e := \Delta^{(s,f)} \cup \Delta^f$, the set of structural and non-structural flow dependences, as the set of equivalence propagating dependences and the very easily computable relation id as basic equivalence relation. Note that, with this choice, Δ^{e+} and $(\Delta^{e+} \cup \Delta^s_{\mathrm{impl}})$ are strict partial orders. Moreover, together with the implicit structural flow dependences Δ^s_{impl}, input dependences only hold between read accesses (and between linear expressions), i.e., between occurrence instances that are minimal wrt. $(\Delta^e \cup \Delta^s_{\mathrm{impl}})^+$. Therefore, if we do not define any additional equivalences (for example in order to implement some user defined equivalence exceeding the power of the `Fortran EQUIVALENCE` statement already discussed above), we can compute the entire equivalence relation $\equiv_{(\Delta^e, \Delta^i)}$ according to Corollary 3.6 in a single sweep in direction of Δ^e. Here, we will assume that, indeed, we do not add equivalences artificially but are only interested in the equivalence $\equiv_{(\Delta^e, \Delta^i)}$. In this case, Corollary 3.6 and Theorem 3.13 (which states that we can compute input dependences from flow dependences) also tell us that we only need to compute the propagation of equivalences, i.e., a sequence $\Lambda_1 = \mathrm{id}, \ldots, \Lambda_i = \mathit{LiftRel}_{\Delta^e}(\Lambda_{i-1}), \ldots$ Thus, sweeping through the occurrence instances grouped according to Δ^e and always rewriting the current level and dependence information regarding this level appropriately, only requires the application of the *LiftRel* operator – we do not need to compute the full union as presented in Corollary 3.6. So the first step is to partition the set $\mathfrak{OI}$ into a sequence of subsets through a topological sort wrt. Δ^e. This is done by the following Algorithm 3.2.1.

Algorithm 3.2.1 [*GroupOccurrenceInstances*]:
Input:
$\mathfrak{OI}$: set of occurrence instances of the program fragment under consideration.

$\Delta^e \subseteq \mathfrak{OI} \times \mathfrak{OI}$:
equivalence propagating dependences (relation from source to target occurrence instances).

Output:
Level : topologically sorted array of sets of occurrence instances, with each element holding one level of incomparable elements wrt. Δ^e.

Procedure:

```
workset := 𝔒𝔍;
depWorkset := Δ^e;
Level := ∅;
i := 1;
while workset ≠ ∅
  /* Throw all occurrence instances with occurrence numbers that occur in */
  /* dependence targets into the next workset */
  /* occurrences that do not belong to dependence targets build */
  /* the current level */
  targets := depWorkset(workset)[occdim_src];
  worksetNew := ∅;
  depWorksetNew := ∅;
  /* Append element number i to Level and initialize to ∅ */
  Level := append(Level, ∅);
  Level[i] := {α ∈ workset | OccId(α) ∉ targets};
  worksetNew := {α ∈ workset | OccId(α) ∈ targets};
  depWorksetNew := {(α, β) ∈ depWorkset | α[occdim_src] ∈ worksetNew};
  i := i + 1;
  workset := worksetNew;
  depWorkset := depWorksetNew;
endwhile
return Level;
```

The levels of occurrence instances produced by Algorithm *GroupOccurrenceInstances* represent the different levels in the (combined) operator trees of the expressions found in the program fragment, as shown, e.g., in Figure 3.3 on page 73. Note that the level of an occurrence is defined by the longest path from a write access to that occurrence.

For this formulation of the algorithm, we suppose that the dependences between the different operand positions of occurrence instances with the same index vector and the same occurrence are *not* given explicitly. We can do so, because the definition of *LiftRel* (Definition 3.2) does not depend on these implicit dependences. Therefore, we view all occurrence instances with the same occurrence number as incomparable wrt. Δ^e and push them into the same element of the array *Level*. Note that, since Δ^{e+} is a strict partial order, as we have established in Section 3.1.2 on page 73, the resulting levels are connected by Δ^e, and an occurrence instance on level i may have a successor wrt. Δ^e on one or more levels $i' > i$, but *not* on any levels $i'' < i$.

Algorithm 3.2.2 represents the actual program transformation:

Algorithm 3.2.2 [*LCCPHighLevel*]:
Input:
OccurrenceLevels :

array of sets of occurrence instances, sorted by Algorithm 3.2.1.

$\Delta \subseteq \mathfrak{OI} \times \mathfrak{OI}$:

flow, anti, or output dependences holding between occurrence instances of $\mathfrak{OI}$ that also have to be respected by the transformed program.

$\Delta^e \subseteq \mathfrak{OI} \times \mathfrak{OI}$:

dependence relation on $\mathfrak{OI}$ along which the equivalences are to be propagated.

$\equiv_{\left(\Delta^{(s,f)} \cup \Delta^f, \mathrm{id}\right)} \subseteq \mathfrak{OI} \times \mathfrak{OI}$:

equivalence relation defining read accesses that are to be considered equivalent, $\equiv_{\left(\Delta^{(s,f)} \cup \Delta^f, \mathrm{id}\right)} = \left(\Delta^{i^-}\right)^*$.

Output:
CondensedOccurrenceLevels :

array of sets of equivalence classes representing the different values calculated by the program fragment – with n = size(*OccurrenceLevels*), $\bigcup_{i \in \{1,\ldots,n\}}$ *CondensedOccurrenceLevels* [*i*] = $\bigcup_{i \in \{1,\ldots,n\}}$ *OccurrenceLevels* [*i*]/$\mathit{LiftRel}^{\infty}_{\Delta^e}(\equiv_{\left(\Delta^{(s,f)} \cup \Delta^f, \mathrm{id}\right)})$.

Δ' : dependences between the occurrence instances of *CondensedOccurrenceLevels* – $\Delta' = \Delta / \mathit{LiftRel}^{\infty}_{\Delta^e}(\equiv_{\left(\Delta^{(s,f)} \cup \Delta^f, \mathrm{id}\right)})$.

Procedure:

```
Δ' := ∅;
EqRel[1] := ≡_(Δ^(s,f) ∪ Δ^f, id);
for i = 1 to size(OccurrenceLevels)
  CondensedOccurrenceLevels[i] := OccurrenceLevels[i]/EqRel[i];
  for j = 1 to i − 1
    forall 𝔄 ∈ CondensedOccurrenceLevels[i]
      forall 𝔅 ∈ CondensedOccurrenceLevels[j]
        Δ' := Δ' ∪ {(𝔄, 𝔅) | (∃α, β : α ∈ 𝔄, β ∈ 𝔅 : (α, β) ∈ Δ)};
        Δ' := Δ' ∪ {(𝔅, 𝔄) | (∃α, β : α ∈ 𝔄, β ∈ 𝔅 : (β, α) ∈ Δ)};
      end forall
    end forall
  endfor
  EqRel[i + 1] := LiftRel_Δ^e(EqRel[i]);
endfor
return (CondensedOccurrenceLevels, Δ');
```

Note that, formally, an occurrence instance graph is quite distinct from a condensed occurrence instance graph in that the vertices of the former are occurrence instances, while the vertices of the latter are sets of occurrence instances. In practice, the question arises how to represent a condensed occurrence instance graph in a computer. One point is again that a COIG can get infinitely large – for the same reason that a OIG can get infinitely large. Another point is that each single equivalence class building a vertex of a COIG may also be infinitely large. The simple solution is to create an isomorphism between the equivalence classes of occurrence instances to be represented and some other (finite) objects. In other words, we choose a representative of a complete equivalence class and, since it is possible that each occurrence instance of a OIG is the sole element of its equivalence class (i.e., $\equiv = \{(\alpha, \alpha) \mid \alpha \in \mathfrak{OI}\}$), the objects representing equivalence classes have to be occurrence instances. Thus, a COIG can be represented in the same way as an OIG: as a reduced OIG of occurrence instances, each of which is a representative for a whole set of equivalent occurrence instances.

3.3 Program Representation by Representatives of Equivalence Classes

Let us briefly recall what we have learned up to now. In the polyhedron model, the actual data structures that algorithms work on are never infinitely large but finite representations of possibly infinite sets. The reduced form of a possibly infinite graph is a finite graph that uses – possibly infinite – sets in order to represent possibly infinite homogenous collections of operations. An OIG $OIG = (\mathfrak{OI}, \Delta)$ that describes the original program is represented in its reduced form by a partitioning $\mathfrak{OI} = \bigcup_{i \in \{1,\ldots,n\}} \mathfrak{OI}_i$ with each partition representing a set of the form $\mathfrak{OI}_i = \{\alpha \mid M_{\mathfrak{OI}_i} \cdot \alpha \geq 0\}$ and each vertex being tagged with one such set. An edge in the reduced OIG is a labelled edge $(\mathfrak{A}, \mathfrak{B}, H), a, b \in \{\mathfrak{OI}_1, \ldots, \mathfrak{OI}_n\}$, where the exact dependence of an occurrence instance in $\mathfrak{B}$ from one in $\mathfrak{A}$ is given as a linear mapping $H : \mathfrak{D} \rightarrow \mathfrak{A}$ $(\mathfrak{D} \subseteq \mathfrak{B})$; $\mathfrak{D}$ also has the form described above.

Now we want to represent a reduced COIG in just the same way. In contrast to the sets represented in a COIG, the sets we want to represent in a reduced COIG are sets of the form $\mathfrak{OI}_{\equiv,i} = \{\Xi(\alpha) \mid M_{\mathfrak{A}} \cdot \alpha \geq 0\}$ with Ξ being a function such that

$$\Xi(\alpha) = \Xi(\beta) \Leftrightarrow \alpha \equiv \beta \tag{3.1}$$

That is, Ξ takes an occurrence instance α and returns a single unique representative for the equivalence class of α. Correspondingly, vertices in the reduced COIG have essentially the same form as the ones in the COIG – however, now they have to hold between occurrence instance sets as described above. Using representatives instead of equivalence classes, the code to be generated is then represented in exactly the same way as for a simple reduced OIG. From now on, we will assume that a COIG uses representatives of equivalence classes (correspondingly, a reduced COIG is actually represented by a reduced OIG). The only question that remains is: which representative do we choose? We will consider this question in detail in Sections 3.5.1 and 3.5.2. However, for the time being, let us suppose that we can find *some* representative for an equivalence class, for example the lexicographic minimum of the equivalence class, as suggested in the following example.

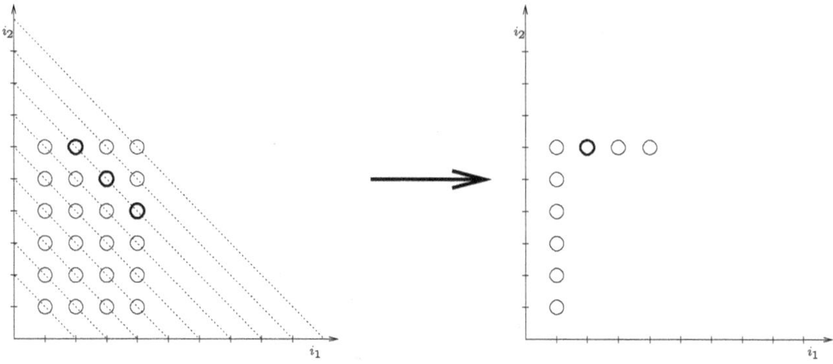

Figure 3.5: Choosing a unique representative for the read access `A(i1+i2)` in Example 3.18 by lexicographic minimum.

Example 3.18 *Consider the code fragment*

```
DO i1=1,4
  DO i2=1,6
    B(i2,i1)=A(i1+i2)**3
  END DO
END DO
```

The value $A(i_1+i_2)$ (and thus $A(i_1+i_2)^3$) does not change for increasing i_1 if i_2 is decreased by the same amount. Figure 3.5 shows the different access instances to the array A for the different $\begin{pmatrix} i_1 \\ i_2 \end{pmatrix}$-coordinates. The circles on a dotted line are in the equivalence relation $\equiv$. In other words, they represent the same value. As an example, the value $A(8)$ is distinguished by fat circles.

The right hand side of Figure 3.5 shows only the lexicographic minima of the corresponding equivalence classes.

An occurrence instance α in our new COIG now represents the complete set $[\alpha]_\equiv$ of equivalent occurrence instances. Therefore we have to rewrite the edges of the original OIG (and thus of the reduced OIG) so that only representatives are used. We have seen the result of such a procedure in Example 3.15.

In the reduced form, the edges of an OIG are labelled with one or more h-transformations (defined on disjunct domains), which map a given target occurrence instance to the unique occurrence instance it directly depends on. Correspondingly, redefining the dependence relations so that the representative occurrence instances are used instead of the original ones, is primarily done by defining new h-transformations from the original ones.

3.4 The Transformation Algorithm

Algorithm 3.2.2 sketches the calculation of a COIG. Since we are working with a reduced OIG that we want to condense into a reduced COIG, the actual algorithm looks a bit different. However, the actual handling of sets and mappings between these sets consists – for the most part – of standard operations that we will not discuss any further here. Implementations of these standard operations for integer vector spaces can be found in several specific libraries such as the Polylib [Wil93, Loe99], or the `Omega` library [Pug93, KMP+96a].

In this situation, an OIG is represented by finitely many vertices, each representing a set of occurrence instances (in our case usually with the restriction of a constant common occurrence number within that set) and finitely many edges between these vertices, each represented by h-transformations. In the algorithms below, only the representation with h-transformations is used – i.e., a connection between two sets $\mathfrak{O}_1$ and $\mathfrak{O}_2$ exists if there is an h-transformation whose domain intersects with $\mathfrak{O}_2$ and the image intersects with $\mathfrak{O}_1$. One may speedup these algorithms by actually using graph representations that employ explicit edges between the vertices, e.g. in the form of an adjacency matrix; however, these representations would act to obfuscate the algorithms below.

Actual implementations may of course vary, for example, in the use of representing any vertices at all, in the use of libraries representing polyhedra (which may use polyhedra or unions of polyhedra as basic structures), as well as in the use of one large set as occurrence instance set (as suggested in Algorithm 3.2.1) or an array or a hash table of sets whose elements only contain sets of occurrence instances of a certain occurrence number (note that every finite input program only contains a finite number of different occurrences). All the algorithms discussed below were implemented and incorporated into the LooPo system [GL97].

Along the same lines, we suppose in the following that dependences Δ are given as h-transformations as described in Section 2.3.2. Each h-transformation is of the form

$$\begin{array}{ll} H_1: & \mathfrak{D}_1 \to \mathfrak{E}: \quad \alpha \mapsto M_{H_1} \cdot \alpha \\ & \vdots \\ H_n: & \mathfrak{D}_n \to \mathfrak{E}: \quad \alpha \mapsto M_{H_n} \cdot \alpha \end{array}$$

with $\mathfrak{D}, \mathfrak{E} \subseteq \mathbb{Z}^{n_{src}+n_{blob}+2}$ being $\mathbb{Z}$-polyhedra. Although it is strictly speaking not necessary, it is helpful to view each set of occurrence instances in the following as being defined over a single pair of occurrence and operand number, i.e., $\#(\text{OccId}(\mathfrak{D})) = \#(\text{OpId}(\mathfrak{D})) = 1$. Correspondingly, the h-transformations are then restricted to these sets. In this representation, each load, store, and each computation operator is represented by an individual set. For example, an addition is represented by three sets: one set each for the left operand of the + operator, for its right operand, and, finally, for the computation itself. Different instances of this computation, however, are still represented by the same number of sets – albeit with an extended index space that defines several iterations. This representation is also implicitly assumed in the formulation of Algorithm 3.2.1 above.

The basic procedure is then as follows:

1. We suppose two occurrence instances to represent the same computation if the occurrence instances representing their operational arguments depend on the same computation, as described in Section 3.3. This equivalence is defined and implemented by the *LiftRel* operator (see Definition 3.2). Therefore, we first have to group occurrence instances – as already specified by Algorithm 3.2.1 – so that occurrence instances comparable wrt. the equivalence propagating dependences Δ^e – that *LiftRel* uses to determine equivalence – will be assigned to different groups (thus, equivalence in one group may be propagated to equivalence in the next group).

2. The previous step introduced a topologically sorted sequence $(\mathfrak{O}_1, \ldots, \mathfrak{O}_n)$ wrt. Δ^e so that $\bigcup_{i \in \{1,\ldots,n\}} \mathfrak{O}_i = \mathfrak{O}\mathfrak{I}$. Each set $\mathfrak{O}_i$ is represented as a union $\mathfrak{O}_i = \bigcup_{j \in \{1,\ldots,m\}} \mathfrak{O}_{i,j}$ of sets of occurrence instances (since $\mathfrak{O}_i$ is not necessarily a polyhedron). We apply $\mathit{LiftRel}_{\Delta^e}$ to each set $\mathfrak{O}_{i,j}$ of that level of occurrence instances and combine the result to produce a representative function that maps each occurrence instance of $\mathfrak{O}_i$ to an instance of a newly created occurrence.

The transformation algorithm of Step 2 is then given by Algorithm 3.4.1. The structure is the same as of Algorithm 3.2.2. A main loop iterates through all the levels of Δ^e bottom up, which are essentially the levels of the operator tree (plus the flow dependences at the bottom used to obtain basic equivalences). In this loop, the procedure is as follows:

1. **Represent occurrence instance level by representatives.** Here, we use a (piecewise linear) representative mapping Ξ, which maps an occurrence instance to its representative. Applying this function to the polyhedra comprising this occurrence instance level yields the desired result.

2. **Rewrite equivalence propagating dependences.** In order for a representative occurrence instance to represent the same calculation as all the occurrence instances in the equivalence class, the dependences have to be rewritten.

3. **Rewrite other dependence information.** Actually, this only occurs at the highest and lowest levels (i.e., the first and last iteration of the algorithm), since read and write accesses are minimal, resp., maximal elements wrt. $\Delta^f \cup \Delta^{(s,f)}$.

4. **Propagate equivalences.** This is done by applying the *LiftRel* operator – initially to the identity function. The actual computation is done according to Corollary 3.7.

Steps 2 and 3 assert that the computation represented by the COIG is still the same as for the OIG. We have seen this rewriting of dependences already in Example 3.15 and visualized in Figures 3.3 and 3.4. Essentially, a dependence relation given by an h-transformation H whose source occurrence instances are to be represented according to the representative mapping Ξ have to be rewritten to

$$H' = \Xi \circ H$$

On the other hand, if the dependence targets are to be represented according to a mapping $\equiv_{\left(\Delta^{(s,f)} \cup \Delta^{f}, \mathrm{id}\right)}$, the new dependence has the form

$$H' = H \circ \Xi^{g}$$

Note that this is only valid because our definition of equivalence asserts that $\alpha \equiv_{\left(\Delta^{(s,f)} \cup \Delta^{f}, \mathrm{id}\right)} \beta \Rightarrow H(\alpha) = H(\beta)$ and thus $\#\left(H(\Xi^{-1}(\alpha))\right) = 1$, i.e. two occurrence instances are only equivalent if they depend on the same source occurrence instance. This was exactly our intention when designing this equivalence and, now, this design allows us to pick *any* possible pre-image. The only exception to this rule consists of anti dependences, for which dependence targets are write accesses. However, for these, $\Xi = \mathrm{id}$ holds.[2]

In our method, the occurrence instances to be represented by equivalence classes are grouped into levels by Algorithm 3.2.1 (*GroupOccurrenceInstances*), while the dependence information is not (since this is a relation between instances of occurrences that might lie on completely different levels). Therefore, rewriting the dependence information according to Ξ is not as simple as applying Ξ to *all* available h-transformations. Since Ξ need not be defined on all occurrence instances that appear in h-transformations, this could lead to an empty result in the composition. What we intend for such a composition is of course to keep the dependence information unchanged as long as we do not have any representatives for the respective occurrence instances. Therefore, Algorithm 3.4.1 first expands Ξ to an identity mapping for the complete index space for which Ξ does not already define a different mapping.

The representative mapping Ξ (*repMaps* in the algorithm) that is to be applied is determined by Algorithm 3.5.2 (*CreateRepresentativeMappings*), which we shall discuss in Section 3.5.3. Note that *CreateRepresentativeMappings* in this algorithm only has to produce some mapping that can be used to produce new index space and dependence information and does not necessarily already have to be a mapping to the actual representatives. The latter is produced by an additional algorithm called *PrepareRepMap* which is used at the beginning of Step 4 in Algorithm 3.4.1. In Section 3.5.2, this strategy will turn out to be useful for the kind of representative selection employed. However, for the time being, we may think of *PrepareRepMap* as an identity function that just passes through the array of mappings passed to it.

[2]For an extension of this method to the case that write accesses *can* be equivalent to each other, one may use the union of all possible dependence sources – leading to *several* h-transformations.

Algorithm 3.4.1 [*LCCPLowLevel*]:
Input:
OccurrenceLevels :
array of sets of occurrence instances, sorted by Algorithm 3.2.1 (corresponds to *OccurrenceLevels* in Algorithm 3.4.1).

RestDeps :
array of dependences defined by h-transformations (corresponds to Δ in Algorithm 3.4.1).

EqPropDeps :
array equivalence propagating dependences defined by h-transformations (corresponds to Δ^e in Algorithm 3.4.1).

BasicRepMaps :
array of representative mappings for the occurrence instances of *OccurrenceLevels* [1]; a representative mapping is defined by an h-transformation (corresponds to $\equiv_{(\Delta^{(s,f)} \cup \Delta^f,\, \mathrm{id})}$ in Algorithm 3.4.1).
Output:
CondensedOccurrenceLevels :
array of sets of occurrence instances with each occurrence instance representing an equivalence class of the input occurrence instances (corresponds to *CondensedOccurrenceLevels* in Algorithm 3.4.1).

RestDepsNew :
array of dependences between the elements of *CondensedOccurrenceLevels* (corresponds to Δ' in Algorithm 3.4.1).

Procedure:
```
CondensedOccurrenceLevels := ∅;
RestDepsNew := RestDeps;
EqPropDepsNew := EqPropDeps;
RepMapsNew := BasicRepMaps;
/* main loop: */
for i = 1 to size(OccurrenceLevels)
  /* STEP 1: represent occurrence instance level by representatives */
  RepMaps := CreateRepresentativeMappings(OccurrenceLevels[i], RepMapsNew);
  CondensedOccurrenceLevels[i] := ∅;
  forall ≡_(Δ^(s,f) ∪ Δ^f, id) ∈ RepMaps
    forall 𝔓 ∈ OccurrenceLevels[i]
      𝔔 := 𝔓 ∩ dom(≡_(Δ^(s,f) ∪ Δ^f, id));
      if 𝔔 ≠ ∅ then
        CondensedOccurrenceLevels[i] :=
          append(CondensedOccurrenceLevels[i], ≡_(Δ^(s,f) ∪ Δ^f, id)(𝔔))
      endif
    end forall
  end forall
  /* STEP 2: rewrite equivalence propagating dependences */
  /* rewrite sources of original dependences */
  EqPropDepsNew := RepMaps ∘ EqPropDepsNew;
  /* rewrite targets of original dependences */
  /* (however, first we have to create an inverse mapping of repMaps) */
  InvRepMaps := ∅;
  for j = 1 to size(RepMaps)
    Let invMap be an arbitrary inverse mapping to RepMaps[j]:
    invMap = im(RepMaps[j]) → dom(RepMaps[j]) : α ↦ (RepMaps[j])^g(α);
```

```
    InvRepMaps := append(InvRepMaps, InvMap);
  endfor
  EqPropDepsNew := RepMaps ∘ EqPropDepsNew;
  /* STEP 3: rewrite other dependence information */
  /* rewrite sources of original dependences */
  RestDepsNew := RepMaps ∘ RestDepsNew;
  /* rewrite targets of original dependences */
  RestDepsNew := RestDepsNew ∘ InvRepMaps;
  /* STEP 4: propagate equivalences */
  /* we may have to incorporate a preprocessing phase to obtain a */
  /* representative mapping suitable for propagation */
  RepMapsToPropagate := PrepareRepMap(RepMaps);
  RepMapsNew := LiftRel_{EqPropDeps^-1}(RepMapsToPropagate);
endfor
return (CondensedOccurrenceLevels, RestDepsNew);
```

3.5 How to Choose Representatives

In the previous section, we have introduced an algorithm for computing a semantically equivalent COIG from an OIG. However, the algorithm, and thus the COIG, depend heavily on *CreateRepresentativeMappings* in order to decide, which representative to choose for a set of occurrence instances. In the following sections, we will develop an algorithm for this task. Of course, there are several possibilities to obtain representatives. Section 3.5.1 will introduce a straightforward method that includes quite difficult computations and may result in rather complex results. We will then go on to examine another method in Section 3.5.2 that is based on a change of basis.

3.5.1 A Straightforward Method: The Lexicographic Minimum

Probably the most intuitive choice of a representative of an equivalence class $[\alpha]_\equiv$ is some $\alpha' \in [\alpha]_\equiv$, i.e., choose some element of the equivalence class that represents the complete set. Within the equivalence class, a natural choice is the lexicographic minimum $\min_\prec([\alpha]_\equiv)$, since it is the first instance in the original program that computes the given value: if we compute the value at this time, it will be defined before every use. Therefore, enumerating the representatives in lexicographic order produces a correct program that is semantically equivalent to the input program. We only have to ensure that the value is stored in a place where it will not be deleted before the last use.

Example 3.19 *An example of a representation by the lexicographic minimum is depicted in Figure 3.5 on page 78, which depicts the selection of a representative for the read access* `A(i1+i2)` *of Example 3.18. The representatives are found along a parallel to the i_2-axis for the first six occurrence instances; the following four occurrence instances are enumerated at the maximal i_2-level along a parallel to the i_1-axis. Since the equivalence of the read accesses to array A propagates to the exponentiation operator, the maximal occurrence instances wrt. Δ^e that are combined into new equivalence classes are those that represent the exponentiation. Just as in Example 3.15, there are now several occurrence instances that depend on the same computation. Therefore, we have to introduce new arrays TMP_1 and TMP_2 that store the result values:*

```
! enumerate the parallel to the i2-axis
      DO i2=1,6
!        Statement S1:
         TMP1(i2)=A(1+i2)**3
!        Statement S2:
```

```
      B(i2,1)=TMP1(i2)
    END DO
! enumerate the rest of the rectangle
    DO i1=2,4
! within the rectangle, there is a lower triangle that
! reads from the parallel to the i2-axis
      DO i2=1,7-i1
!       Statement S3:
        B(i2,i1)=TMP1(i1+i2-1)
      END DO
! within the rectangle, there is an upper triangle that
! reads from the parallel to the i1-axis (once those values are written)
      DO i2=8-i1,5
!       Statement S4:
        B(i2,i1)=TMP2(i1+i2-6)
      END DO
! new values are produced along a parallel to the i1-axis
!     Statement S5:
      TMP2(i1)=A(i1+6)**3
!     Statement S6:
      B(6,i1)=TMP2(i1)
    END DO
```

In this example, the result is a program that computes the values to be assigned to B in statement S_1 *and* S_5 *and assigns them to their destinations in the other statements. These assignments are done in several statements, firstly in order to honour the lexicographic ordering wrt. the computation statements,* S_1 *and* S_5*, and secondly because they have to read from different memory locations. These read accesses from different sources are due to the fact that the computation statements represent occurrence instances enumerated along two different dimensions; therefore, they have to assign their result values to different arrays so as to not interfere with each other (for example, in this straightforward version of the program, both statements assign to element 1 of their respective output arguments –* S_1 *to* $TMP_1(1)$*, and* S_5 *to* $TMP_2(1)$*). This split of one polyhedron into two, on the other hand, is due to the fact that the lexicographic minimum of equivalent occurrence instances describes a "break": it is defined by a piecewise linear mapping and cannot be described by a closed linear form.*

Note that the output program in Example 3.19 represents only the occurrence instances of the COIG of the input program in Example 3.15, enumerated in lexicographic order. The choice of the lexicographic minimum actually yields a schedule for the statements S_1 and S_5 that have to be inserted in order to maintain the flow of data. Thus, it immediately leads to an output program that is equivalent to the input program.

However, the representation by the lexicographic minimum also has some drawbacks:

1. The calculation of the lexicographic minimum of an integer polyhedron is a complex calculation (it is NP-complete).

2. The representation of the resulting set is quite complicated, even if we only search for the lexicographic minimum of the equivalent points within a single polyhedron.

3. If points of several different polyhedra are equivalent, the representation becomes even more complicated, and these complicated representations propagate along the direction in which equivalences on larger terms are calculated.

Clearly, the first point represents a major drawback, since the number of polyhedra in a reduced OIG, and thus a reduced COIG, is greatly increased in comparison to a simple statement based dependence graph, as we have already seen in Section 2.3.2. Therefore, the run time of the algorithm will suffer a lot from choosing such a complex calculation for selecting representatives. Choosing the lexicographic minimum implicitly computes a schedule for the supplied occurrence instances simply in order to find representatives. This is especially important if we are aiming at obtaining a parallel loop program since, in this case, a schedule is usually computed anyway (as discussed at the beginning of this chapter with the presentation of Figure 3.1), thus, we do not really gain anything from determining representatives whose lexicographic order also defines a legal execution order.

The other two points above can in part be gleaned from Example 3.19. The partitioning of the computation occurrence instances into two areas that are mapped by different linear mappings to their corresponding lexicographic minima leads to two different polyhedra (and thus to the statements S_1 and S_5) that have to be enumerated. Additionally, these two source polyhedra also lead to a partitioning of the dependent polyhedron describing the assignment to B. Furthermore, since the rank of the linear subscript function $\begin{pmatrix} i_1 \\ i_2 \end{pmatrix} \mapsto i_1 + i_2$ for array A is 1, we only should need a single dimension to enumerate the resulting points. An alternative which addresses this issue is presented in the next section.

3.5.2 A Twisted Method: Change of Basis

One alternative to using the lexicographic minimum in Example 3.18 is depicted in Figure 3.6:

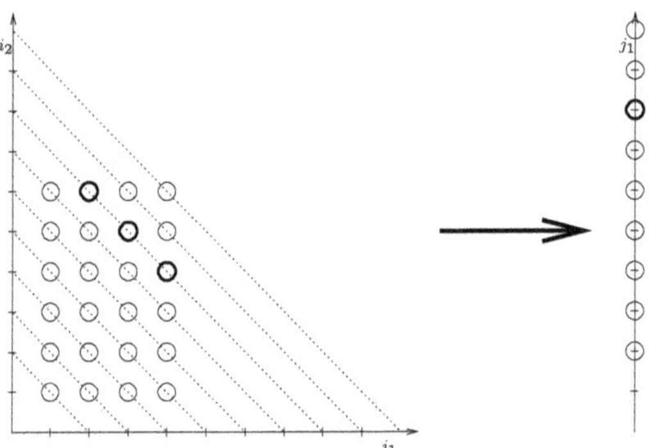

Figure 3.6: Choosing a unique representative for the read access `A(i1+i2)` in Example 3.18 by a simple linear function.

The dotted lines representing equivalent occurrence instances meet the i_2-axis in an interval of integer points. In this example, we could represent the equivalence classes by occurrence instances with an index space of $\{j_1 \mid 2 \leq j_1 \leq 10\}$. Thus, the "break" that we observed in the case of selecting the lexicographically minimal occurrence instance depicted in Figure 3.5 is removed and only one (relatively simple) polyhedron is generated. The drawback is, of course, that this approach leaves us with a completely different index set and thus without an *a priori* legal execution order.

Let us take one step back now. Why should we care at all about the original index space of α, when searching for a representative? For example, we could act as if the occurrence $\mathrm{OccId}(\alpha)$ were enumerated for all finite index space vectors – effectively forgetting about each and every loop bound, as long as we only enumerate those occurrence instances at the end that are equivalent to some occurrence instance that is actually enumerated in the original program. Consider the following example.

Example 3.20 *The left side of Figure 3.7 shows the index space of* `A(i2-i1)**3` *in the code fragment*

```
DO i1=1,10
```

```
    DO i2=MAX(1,5-i1),10
      B(i2,i1)=A(i2-i1)**3
    END DO
  END DO
```

Index vectors that define the same value for $(A(i_2 - i_1))^3$ *– and thus represent equivalent occurrence instances – are connected through dotted lines. The lexicographic minima of these equivalent index vectors are marked with fat circles. They can only be represented by several polyhedra:*

$$\left\{\left(\begin{array}{c}1\\i_2\end{array}\right)\middle|\, 4 \leq i_2 \leq 10\right\} \cup$$
$$\left\{\left(\begin{array}{c}i_1\\1\end{array}\right)\middle|\, 4 \leq i_1 \leq 10\right\} \cup$$
$$\left\{\left(\begin{array}{c}i_1\\i_2\end{array}\right)\middle|\, \max(2, 5 - i_2) \leq i_1 \leq 6 - i_2 \wedge i_2 \geq 2\right\}$$

This is because, in order to stay within the original index space, we have to enumerate points on the line $i_2 = 5 - i_1$ *(or the smallest integer above – the non-integer points on this line are marked as thin, dashed circles). This is due to the lower bound* $5 - i_1$ *for* i_2*. If we were allowed to also consider points outside the actual index space, as done on the right hand side of Figure 3.7, things would be easier. Here, we ignore the lower bound* $5 - i_1$ *for* i_2*, so that the set of index vectors representing the different equivalent classes is given just by two intercepts of the* i_1*- and the* i_2*-axis, respectively. I.e., it suffices to enumerate the sets*

$$\left\{\left(\begin{array}{c}1\\i_2\end{array}\right)\middle|\, 1 \leq i_2 \leq 10\right\} \quad \cup$$
$$\left\{\left(\begin{array}{c}i_1\\1\end{array}\right)\middle|\, 2 \leq i_1 \leq 10\right\}$$

Of course, neglecting the correct lower bound amounts to considering a slightly different program – in our case, this would read:

```
DO i1=1,10
  DO i2=1,10
    B(i2,i1)=A(i2-i1)**3
  END DO
END DO
```

However, this is irrelevant for the plain purpose of creating new occurrence instances, since we do not use this program to obtain the pre-image of our representative mapping, but only to construct an image *of that mapping – i.e., we do not use the occurrence instances of this second program to find out, which points were originally enumerated – for this, we still use the first program above.*

Selecting the lexicographic minimum appeared advantageous because it asserts a legal execution order. But – as already hinted in the previous section – if we employ a scheduling algorithm after LCCP anyway, we do not gain anything. So, if we decide to ignore bounds just for finding a representative, what consequences do we have to face?

Essentially, there are three points that have to be observed:

Regard Original Loop Bounds:

In the COIG, we have to a ascertain that the bounds of the polyhedra defined by the vertices encompass exactly those integer vectors that correspond to occurrence instance sets that are indeed enumerated in the original program. I.e., we have to make sure that the transformed index space does not include integer vectors that correspond to occurrence instances that are not executed by the original program. On the other hand, we also have to ensure that, for all occurrence instances in the original program, there is a representative in the transformed index space.

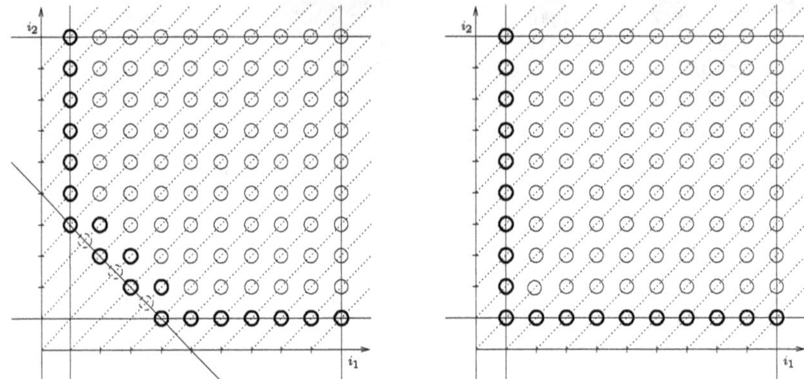

Figure 3.7: Left side: Index space of the expression `A(i2,i1)**3` in the program fragment of Example 3.20 with equivalent instances connected by dotted lines. Right side: lexicographic minima for relaxed lower i_2-bound $i_2 \geq 1$.

Stay Within the Integers:

Ignoring loop bounds – and thus linear inequalities – is akin to using an $(n_{src}+n_{blob}+2)$-dimensional vector space over the rational numbers instead of a $\mathbb{Z}$-module as the space containing our representatives, since linear equalities always hold independently of a non-zero denominator (as discussed in Chapter 2.2). However, we have to make sure that, in the end, we will only enumerate integer values (so as not to enumerate over an infinite set).[3]

Scheduler Issues:

If there is no order of the representatives that is related to the lexicographic order of the original occurrence instances (which define the execution order in the original program), dependence information may be inconclusive for a subsequent scheduler (see Figure 3.1 on page 60). In this case, dependence sources may appear lexicographically larger than their targets, which should *never* be the case for an input program – and is in fact guaranteed *not* to occur with the usual dependence analysis algorithms. A scheduler that only guarantees not to change the direction of any dependence will *not* be good enough, since such a scheduler cannot guarantee that all dependence sources are executed before their respective targets.

Actually the first two points here are but two sides of the same coin, considering the solvability of imposed constraints, since they both come directly from ignoring the loop bounds (and thus changing from integer spaces to rational ones). However, the first point considers the restrictions that define the function's domain within a free module ($\mathbb{Q}^{n_{src}+n_{blob}+2}$ or $\mathbb{Z}^{n_{src}+n_{blob}+2}$, respectively), and thus considers only the domain of that function. This is in contrast to the second point, which states the necessity for the functional matrix of the mapping to ensure that any vector of the free module $\mathbb{Z}^{n_{src}+n_{blob}+2}$ is mapped to an element of $\mathbb{Z}^{n_{src}+n_{blob}+2}$ (and never to some $\beta \in \mathbb{Q}^{n_{src}+n_{blob}+2} \setminus \mathbb{Z}^{n_{src}+n_{blob}+2}$), thus enabling the switch back to a module based on $\mathbb{Z}$ instead of a vector space based on $\mathbb{Q}$ by just reintroducing loop bounds (i.e., domain restrictions). We have to keep these compatibility issues in mind when pursuing an alternative way to obtain representatives. Note that it also suffices to satisfy these compatibility issues in order to obtain a 1-to-1 mapping from original occurrence instances to representatives. This means that for each value computed in the original OIG we obtain a representative in the target COIG. On the other

[3]Note that, in contrast to the rational numbers, which form a topological space that is dense-in-itself, the integers are nowhere dense within the rationals, so that any finitely bounded set of integers is a finite set, whereas any finitely bounded set of rationals is still infinite.

hand, the first two points also guarantee that our transformation does not introduce any additional computations that are not part of the original code fragment. In other words, heeding all of the above rules *ensures* the correctness of the transformation.

We will now first consider the case of a single polyhedron (rather than a union of polyhedra) as a representation of a set of occurrence instances. This is the case of instances of a single occurrence. We shall then go on to tackle the case of determining representatives for a union of occurrence instance sets with possibly different occurrences. In both these cases, we will have to consider the first two points above very closely. Section 3.6 is then devoted entirely to the last point.

Equivalences on Instances of the same Occurrence

The idea is to create a piecewise linear function $\Xi : \mathfrak{OI} \to \mathfrak{OI}/\equiv_{(\Delta^{(s,f)} \cup \Delta^f, \text{id})}$ (with $\mathfrak{OI}/\equiv_{(\Delta^{(s,f)} \cup \Delta^f, \text{id})} \subseteq \mathbb{Z}^{n_{src}+n_{blob}+2}$ being the set of representatives) that is defined by a union of linear functions $\Xi = \bigcup_{i \in \{1,\dots,n_\Xi\}} \Xi_i$. A linear function Ξ_i maps two vectors α and β to the same output vector, $\gamma = \Xi_i(\alpha) = \Xi_i(\beta)$ iff it maps the difference between α and β to 0, i.e., iff $\alpha - \beta \in \ker(\Xi_i)$. Correspondingly, α and β are equivalent wrt. $\equiv_{(\Delta^{(s,f)} \cup \Delta^f, \text{id})}$ iff their difference is equivalent to 0 ($\alpha \equiv_{(\Delta^{(s,f)} \cup \Delta^f, \text{id})} \beta \Leftrightarrow 0 \equiv_{(\Delta^{(s,f)} \cup \Delta^f, \text{id})} \beta - \alpha$). All in all, this means that the representative mapping Ξ just has to be constructed from linear mappings whose kernels are exactly the sets that are $\equiv_{(\Delta^{(s,f)} \cup \Delta^f, \text{id})}$-equivalent to 0, i.e.,

$$\bigcup_{i \in \{1,\dots,n_\Xi\}} \ker(\Xi_i) = \left\{\alpha \,\middle|\, \alpha \equiv_{(\Delta^{(s,f)} \cup \Delta^f, \text{id})} 0\right\} \tag{3.2}$$

We now aim at producing a single one of these linear functions Ξ_i above, namely one that deals with the simplest of possible equivalences: the equivalence of occurrence instances of the same occurrence. Let us call our prototype $\Xi_=$. Restrictions regarding the part of the index space in which $\Xi_=$ is to be valid are not of our concern at the moment, neither are we interested in the image of $\Xi_=$. Therefore, we assume the prototype

$$\Xi_= : \mathbb{Q}^{n_{src}+n_{blob}+2} \to \mathbb{Q}^{n_{src}+n_{blob}+2} : \alpha \mapsto M_{\Xi_=} \cdot \alpha \tag{3.3}$$

for now. Of course, $\Xi_=$ is not surjective, if it represents a non-trivial equivalence relation. Since we are not concerned with index space restrictions, we can leave them out of an underlying equivalence relation $\Lambda \subseteq \equiv_{(\Delta^{(s,f)} \cup \Delta^f, \text{id})}$. In the following, we will denote the equivalence relation without index space restrictions by $\Lambda_= \supseteq \Lambda$. Suppose

$$\equiv_{(\Delta^{(s,f)} \cup \Delta^f, \text{id})} = \left\{(\alpha, \beta) \in \mathbb{Z}^{2\cdot(n_{src}+n_{blob}+2)} \,\middle|\, M_{\Lambda,\geq} \cdot \begin{pmatrix} \alpha \\ \beta \end{pmatrix} \geq 0 \wedge M_{\Lambda_=} \cdot \begin{pmatrix} \alpha \\ \beta \end{pmatrix} = 0\right\} \tag{3.4}$$

We are now just interested in $M_{\Lambda_=}$. So, combining Equations (3.2) and (3.4), we seek a function $\Xi_=$, whose kernel is

$$\ker(\Xi_=) = \left\{\alpha \,\middle|\, M_{\Lambda_=} \cdot \begin{pmatrix} \alpha \\ 0 \end{pmatrix} = 0\right\} \tag{3.5}$$

The function matrix $M_{\Xi_=}$ can then be derived in part from

$$M := \begin{pmatrix} M_{\Lambda_=}[\cdot, 1] & \dots & M_{\Lambda_=}[\cdot, n_{src} + n_{blob} + 2] \end{pmatrix}$$

Let B be the matrix whose columns consist of base vectors of the kernel of M, $\ker(M)$. And let B' be the matrix whose columns consist of base vectors of the complement of $\ker(M)$ in $\mathbb{Q}^{n_{src}+n_{blob}+2}$, $\ker(M)^C$ (i.e., $\text{Span}(B) = \ker(M)$, $\text{Span}(B') = \ker(M)^C$). The one condition that $M_{\Xi_=}$ has to satisfy up to now is $M_{\Xi_=} \cdot \alpha = 0 \Leftrightarrow M \cdot \begin{pmatrix} \alpha \\ 0 \end{pmatrix} = 0$. This is the case iff $M_{\Xi_=} \cdot B = 0$ and

$M_{\Xi_=} \cdot B' \neq 0$, which holds iff $M_{\Xi_=}$ is a solution to

$$B^T \cdot M_{\Xi_=}{}^T = 0 \tag{3.6}$$

and for all $i \in \{1, \ldots, n_{src} + n_{blob} + 2 - \text{rk}(M)\}$:

$$(B'^T \cdot M_{\Xi_=}{}^T)[\cdot, i] = B'^T \cdot M_{\Xi_=}{}^T \cdot \iota_i = B'^T \cdot (\iota_i^T \cdot M_{\Xi_=})^T \neq 0 \tag{3.7}$$

Example 3.21 *The code of Example 3.18*

```
DO i1=1,4
  DO i2=1,6
    B(i2,i1)=A(i1+i2)**3
  END DO
END DO
```

contains the expression `A(i1+i2)**3` *whose equivalence classes are sketched in Figure 3.5. With the notation as above, the equivalence* $\equiv$ *of the corresponding occurrence instances is defined by the following system of (in)equalities:*

$$i_1 - 1 \geq 0 \tag{3.8}$$
$$-i_1 + 4 \geq 0 \tag{3.9}$$
$$i_2 - 1 \geq 0 \tag{3.10}$$
$$-i_2 + 6 \geq 0 \tag{3.11}$$
$$i'_1 - 1 \geq 0 \tag{3.12}$$
$$-i'_1 + 4 \geq 0 \tag{3.13}$$
$$i'_2 - 1 \geq 0 \tag{3.14}$$
$$-i'_2 + 6 \geq 0 \tag{3.15}$$

and

$$occ - 3 = 0 \tag{3.16}$$
$$op = 0 \tag{3.17}$$
$$occ' - 3 = 0 \tag{3.18}$$
$$op' = 0 \tag{3.19}$$
$$i_1 + i_2 - i'_1 - i'_2 = 0 \tag{3.20}$$
$$m_c - m'_c = 0 \tag{3.21}$$

which in turn is expressed by the matrices $M_{\Lambda,\geq}$*, representing inequations (3.8) to (3.15), and* $M_{\Lambda_=}$*, representing Equations (3.16) to (3.21):*

$$M_{\Lambda,\geq} = \begin{pmatrix} 1 & 0 & 0 & 0 & -1 & 0 & 0 & 0 & 0 & 0 \\ -1 & 0 & 0 & 0 & 4 & 0 & 0 & 0 & 0 & 0 \\ 0 & 1 & 0 & 0 & -1 & 0 & 0 & 0 & 0 & 0 \\ 0 & -1 & 0 & 0 & 6 & 0 & 0 & 0 & 0 & 0 \\ 0 & 0 & 0 & 0 & 0 & 1 & 0 & 0 & 0 & -1 \\ 0 & 0 & 0 & 0 & 0 & -1 & 0 & 0 & 0 & 6 \\ 0 & 0 & 0 & 0 & 0 & 0 & 1 & 0 & 0 & -1 \\ 0 & 0 & 0 & 0 & 0 & 0 & -1 & 0 & 0 & 6 \end{pmatrix}$$

$$M_{\Lambda_=} = \begin{pmatrix} 0 & 0 & 1 & 0 & -3 & 0 & 0 & 0 & 0 & 0 \\ 0 & 0 & 0 & 1 & 0 & 0 & 0 & 0 & 0 & 0 \\ 0 & 0 & 0 & 0 & 0 & 0 & 0 & 1 & 0 & -3 \\ 0 & 0 & 0 & 0 & 0 & 0 & 0 & 0 & 1 & 0 \\ 1 & 1 & 0 & 0 & 0 & -1 & -1 & 0 & 0 & 0 \\ 0 & 0 & 0 & 0 & 1 & 0 & 0 & 0 & 0 & -1 \end{pmatrix}$$

Note that the last row of $M_{\Lambda_=}$ *is due to the fact that* m_c *– the value 1 – does not change between the potentially equivalent occurrence instances, and neither does any other parameter. Instead of including this row, we could just have collapsed the parameter dimensions of the two index spaces into a single interval of dimensions to obtain a* 5×7 *matrix for* $M_{\Lambda_=}$*, in this case* $(2 \cdot n_{src} + n_{blob} + 2 = 7)$*.*[4]

The function matrix we want to produce only depends on the left part of $M_{\Lambda_=}$*, leading to*

$$M = \begin{pmatrix} 0 & 0 & 1 & 0 & -3 \\ 0 & 0 & 0 & 1 & 0 \\ 0 & 0 & 0 & 0 & 0 \\ 0 & 0 & 0 & 0 & 0 \\ 1 & 1 & 0 & 0 & 0 \\ 0 & 0 & 0 & 0 & 1 \end{pmatrix}$$

A possible matrix B *representing the kernel of* M *is then any matrix*

$$B \in \mathbb{Q} \cdot \begin{pmatrix} -1 \\ 1 \\ 0 \\ 0 \\ 0 \end{pmatrix}$$

Since it is more convenient to stay within the integers, we choose B *from* $\mathbb{Z}^{5 \times 1}$ *– and, for the sake of simplicity, we just use* $\begin{pmatrix} -1 & 1 & 0 & 0 & 0 \end{pmatrix}^T$*. For* B'*, we simply choose an appropriate part of the unit matrix whose columns span* $\operatorname{Span}(B)^C$*. In addition, we decide arbitrarily that the former unit vectors spanning* $\operatorname{Span}(B)^C$ *as a subspace of* $\mathbb{Q}^{n_{src}+n_{blob}+2}$ *shall also be mapped to unit vectors in the target space by our new mapping* $\Xi_=$*. We further deem arbitrarily the* i*-th index vector to be mapped to the* $(i-1)$*-st unit vector. We do so by using appropriate unit vectors for the rows of* N *and* B'^T *in* $B'^T \cdot {M_{\Xi_=}}^T = N$*. Thus, the conditions for* $M_{\Xi_=}$ *become:*

$$\begin{pmatrix} -1 & 1 & 0 & 0 & 0 \end{pmatrix} \cdot {M_{\Xi_=}}^T = 0 \quad \text{and} \quad \begin{pmatrix} 0 & 1 & 0 & 0 & 0 \\ 0 & 0 & 1 & 0 & 0 \\ 0 & 0 & 0 & 1 & 0 \\ 0 & 0 & 0 & 0 & 1 \end{pmatrix} \cdot {M_{\Xi_=}}^T = \underbrace{\begin{pmatrix} 1 & 0 & 0 & 0 \\ 0 & 1 & 0 & 0 \\ 0 & 0 & 1 & 0 \\ 0 & 0 & 0 & 1 \end{pmatrix}}_{=:N}$$

This is an inhomogenous linear equation system that is easily solved:

$$\begin{pmatrix} -1 & 1 & 0 & 0 & 0 \\ 0 & 1 & 0 & 0 & 0 \\ 0 & 0 & 1 & 0 & 0 \\ 0 & 0 & 0 & 1 & 0 \\ 0 & 0 & 0 & 0 & 1 \end{pmatrix} \cdot {M_{\Xi_=}}^T = \begin{pmatrix} 0 & 0 & 0 & 0 \\ 1 & 0 & 0 & 0 \\ 0 & 1 & 0 & 0 \\ 0 & 0 & 1 & 0 \\ 0 & 0 & 0 & 1 \end{pmatrix} \quad \Leftrightarrow \quad M_{\Xi_=} = \begin{pmatrix} 1 & 1 & 0 & 0 & 0 \\ 0 & 0 & 1 & 0 & 0 \\ 0 & 0 & 0 & 1 & 0 \\ 0 & 0 & 0 & 0 & 1 \end{pmatrix}$$

[4]Note that, even with the reduced dimensionality, we still use *all* source indices here, not only the ones bound by loops enclosing some given statement.

$M_{\equiv_=}$ above represents a possible function matrix for the mapping to representatives for the exponentiation of Example 3.18. However, we are completely free to choose the matrix N in Example 3.21, and this is not the only possible choice. In the following, we will take a closer look at what traits might be desirable for matrix $M_{\equiv_=}$ and how to achieve this.

Original Loop Bounds We are now able to create an endomorphism on $\mathbb{Q}^{n_{src}+n_{blob}+2}$ that maps exactly those occurrence instances to the same point in $\mathbb{Q}^{n_{src}+n_{blob}+2}$ that represent the same value wrt. $\equiv_{(\Delta^{(s,f)} \cup \Delta^f, \mathrm{id})}$. Going back to the first point identified above on page 87, we have to make sure that our mapping does not introduce occurrence instances that represent a value that was never computed in the original program. The following slight modification of the example above shows how something like this might happen.

Example 3.22 *As in Example 3.21, we consider the computation* `A(i1+i2)**3`*, in which only the value of $A(i_1 + i_2)$ changes. However, $A(i_1 + i_2)$ is only a one-dimensional expression, which can be considered completely independent of one of the loops (either i_1 or i_2). We choose to view this expression as independent of i_2. This means the kernel of the representative mapping is the complete i_2-dimension, which may therefore be totally ignored in the output code. However, if we replace one of the i_2-bounds of Example 3.21 by a value that is not available at compile time, ignoring this bound may lead to an unnecessary execution of exponentiation operations. Let us assume that the following code fragment is to be processed:*

```
DO i1=1,4
  DO i2=1,n
    B(i2,i1)=A(i1+i2)**3
  END DO
END DO
```

Since the structure of the bounds does not affect the choice of a representative mapping, we apply the same mapping as in Example 3.21, which defines the new loop bound j_1 for the execution of the exponentiation as $j_1 := i_1 + i_2$. The result is (depending on the scheduler etc.) code fragment similar to the following one.

```
DO j1=2,n+4
  TMP(j1)=A(j1)**3
END DO
DO j1=1,4
  DO j2=1,n
    B(j2,j1)=TMP(j1+j2)
  END DO
END DO
```

For $n = 0$, the original program fragment does not define any iteration, while the created code fragment still computes the set $\{A(j)^3 \mid j \in \{2, \ldots, 4\}\}$ – which is not even guaranteed to be well defined, since it includes a potentially illegal access to array A. Of course, we would like to avoid such a situation. On the other hand, if the original code fragment also contained another computation of $A(i_1)^3$, we would have expected *a transformation of*

```
DO i1=1,4
  DO i2=1,n
    B(i2,i1)=A(i1+i2)**3
  END DO
  C(i1)=A(n+i1)**3+A(i1)**3
END DO
```

into:

```
DO j1=1,n+4
  TMP(j1)=A(j1)**3
END DO
DO j1=1,4
  DO j2=1,n
    B(j2,j1)=TMP(j1+j2)
  END DO
  C(j1)=TMP(n+j1)+TMP(j1)
END DO
```

This means that, somehow, the original loop bounds still have to be observed. The original loops that introduce these bounds just do not need to be executed for the complete *index space they span, but for at most a single iteration of the computation statements in their body.*

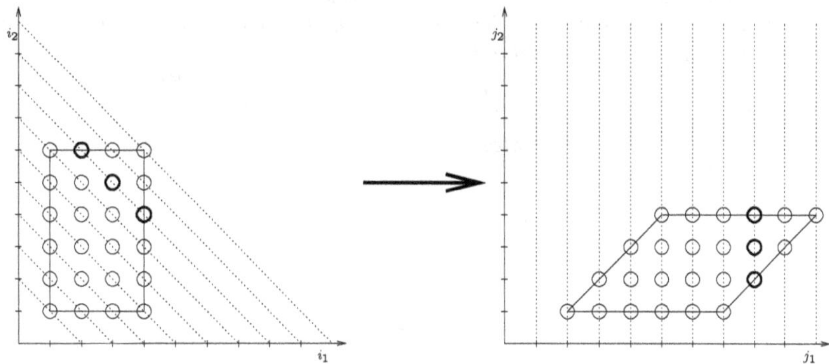

Figure 3.8: Original and transformed index space of A(i1+i2)**3 in Example 3.22 for $n = 6$.

Figure 3.8 depicts the transformation of the index space of A(i1+i2)**3 in this example, for $n = 6$. However, here the dimension along which occurrence instances are equivalent is not removed but instead used for the j_2-dimension of the new coordinate system. As a result, it features the dotted lines indicating equivalent occurrence instances along its direction. This also gives a hint of how to go about keeping the original index space: if we augment our representative mapping $\Xi_=$ to a complete base transformation, we obtain a loop nest that enumerates the same (original) occurrence instances using standard techniques [CF93, Len93, Xue93, Xue94, Wet95, Xue96, QRW00]. This is essentially the same idea as applied to singular space-time mappings in LooPo [Wet95, GLW98]. However, these dimensions represent a special case insofar as they do not actually have to be enumerated – they just have to be checked as to whether there exists an integer number between their respective lower and upper bound. This becomes even clearer if we use yet another – slightly unconventional – pair of bounds for i_2 in Example 3.21.

Example 3.23 *The rational inequation*

$$\frac{i_1}{3} \leq i_2 \leq \frac{i_1+1}{3} \tag{3.22}$$

may or may not have an integer solution for i_2 – in fact, enumerating i_1 with a stride of 1 *always yields two integer solutions in succession, then no integer solution, then again two successive solutions, and so on. To be more specific, an integer i_2 solving Inequation (3.22) exists iff*

$$\begin{array}{rcl} \lceil \frac{i_1}{3} \rceil & \leq & \lfloor \frac{i_1+1}{3} \rfloor \\ & \Leftrightarrow & \\ \lfloor \frac{i_1-1}{3} \rfloor + 1 & \leq & \lfloor \frac{i_1+1}{3} \rfloor \end{array} \tag{3.23}$$

Since, for non-negative integers, the Fortran *integer division is equal to the floor of the division, a loop nest that executes its body for solutions of Inequation (3.22) in* $\{1, \ldots, 6\}$ *is given by*

```
DO i1=1,6
  DO i2=(i1-1)/3+1,(i1+1)/3
    B(i2,i1)=A(i1+i2)**3+C(3)
  END DO
END DO
```

The index space of `A(i1+i2)**3` *in this code fragment is depicted on the left hand side of Figure 3.9. The right hand side shows the same transformation as before: the* j_2*-axis now represents the enumeration of equivalent values, while the* j_1*-axis is exactly the same as the* i_1*-axis before. Not checking for the existence* $\left(\exists i_2 : i_2 \in \mathbb{Z} : \frac{i_1}{3} \leq i_2 \leq \frac{i_1+1}{3}\right)$ *now corresponds to enumerating the complete orthogonal projection of that polyhedron onto the* j_1*-axis – indicated by the fat section of the axis. In particular, an occurrence instance with a* j_1*-coordinate of* 4 *is executed in this case – represented by the point with the dashed border in Figure 3.9. However, the original program does not define any occurrence instances equivalent to this point: while the original program enumerates* 4 *occurrence instances, the transformed one executes* 5*. This is because the additional occurrence instance,* $\begin{pmatrix} 4 \\ \frac{11}{2} \end{pmatrix}$*, although not an integer vector, features only integers in the coordinates checked by the loops in the transformed program (which do not check* j_2*).*

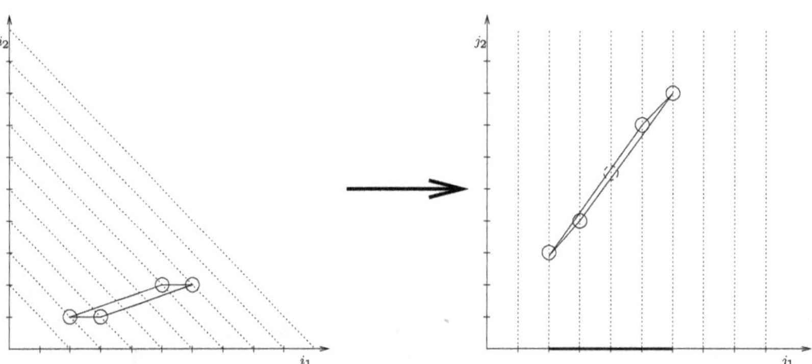

Figure 3.9: Original and transformed index space of `A(i1+i2)**3` in Example 3.23.

This set of examples shows that there still has to be a check of bounds of dimensions that actually need *not* be enumerated in a transformed program. Otherwise occurrence instances not executed by the original program may be executed by the transformed one. In order to do so, we may need to store example values of solutions within these bounds in pseudo parameters. This makes clear that the enumeration of pseudo parameters actually forms a third level of execution, different from both indices and parameters. We now have

- parameters that have a given value from the start of the code fragment,
- indices that change their value during execution of the code fragment, and
- pseudo parameters that do not actually define iterations to be enumerated by the target program, but represent restrictions on the target index space that have to hold to execute a given program fragment.

Pseudo Parameters Pseudo parameters are existentially quantified integer variables. They can also be used to model families of dependence relations [Kei97]. Usually, pseudo parameters represent a restriction of the index space that can be implemented by a simple check – an `if`-statement that tests, whether the coordinates enumerated by an enclosing loop nest represent a multiple of some given integer or belong to some half space that restrains the index space to be enumerated. However, in general, pseudo parameters also have to be enumerated, as Example 3.23 shows: if we view the access `C(3)` in the program fragment, *both* dimensions, i_1 and i_2, are irrelevant for the outcome of the lookup operation. However, as already shown, not every value of i_1 also leads to a solution for i_2. Therefore, several different possible values for i_1 have to be enumerated in order to find a solution for i_2. But as soon as such a solution has been found, there is no reason to enumerate any other value of a pseudo parameter – as long as there is not also a change in another dimension. So, pseudo parameters differ from parameters insofar as a pseudo parameter is not bound to a value on execution of the considered program fragment, i.e., a solution for its value has to be found at run time, possibly by enumerating through possible solutions with a loop. Pseudo parameters also differ from indices insofar as a pseudo parameter does *not* need to be enumerated for the complete range of values it may take, but only until *some* value is found that defines a non-empty polyhedron.

We have established that we will need a complete change of basis in order to represent our transformation. The fact that some occurrence instances are equivalent is then represented by the fact that these occurrence instances only differ in their coordinates of certain dimensions, namely those dimensions that we have set aside as LCCP pseudo parameter dimensions. Projecting two equivalent occurrence instances α and β with $\alpha \equiv_{(\Delta^{(s,f)} \cup \Delta^f, \mathrm{id})} \beta$ onto dimensions that do not correspond to LCCP pseudo parameters always results in the same points for both occurrence instances, α and β. We have not introduced these pseudo parameters to the program, yet. However, in Section 2.3, we already discussed pseudo parameters used to represent $\mathbb{Z}$-polyhedra with holes by higher-dimensional polyhedra without holes. We solved this by setting aside some pseudo parameter dimensions that were used to check whether a certain index is an integer multiple of some stride. Assuming n_{depth} to be the maximum depth of loop nests in the program to be processed, we only needed to assume the existence of n_{depth} such pseudo parameters. The same reasoning holds for these new pseudo parameters we meet here: since each of these pseudo parameters is only used as a replacement of an index of an embracing loop, we only need up to n_{depth} many. And, again, this means that these parameters can be added to each input program artificially. So, for each index in the deepest loop nest of the original program, we now have to introduce one pseudo parameter to check whether there is an integer number between the lower and upper bound of this index and one pseudo parameter to check, whether this solution is a multiple of some stride. Therefore, from now on, we assume *all* these $2 \cdot n_{\mathrm{depth}}$ pseudo parameters to be already among the n_{blob} parameters of the input program. Since pseudo parameters represent a combination of indices and parameters, they are represented by index space dimensions adjacent to both index and parameter dimensions. We suppose that the pseudo parameters introduced are the first $2 \cdot n_{\mathrm{depth}}$ parameters of the original program fragment and are thus represented by dimensions $n_{src} + 1$ through $n_{src} + n_{\mathrm{depth}}$ in the polyhedra in $\mathbb{Z}^{n_{src}+n_{blob}+2}$ representing the input program fragment.

In order to distinguish used pseudo parameters from unused ones, we will assume the initial values of these pseudo parameters to be m_∞ – just as with unbound indices. Thus, using the new pseudo parameters, we are able to reconstruct the original loop bounds. Since pseudo parameters, always represent integer values (just as indices and parameters), we also have to ascertain the representative mapping to map back into its domain, i.e., to be defined by a piecewise endomorphism on $\mathbb{Z}^{n_{src}+n_{blob}+2}$. Let us now take a closer look at how to accomplish this.

Stay Within the Integers One reason for choosing a specific base may be the first important property of our representative mapping that we reviewed on page 86: we need to map integer points to integer points. The easiest way to do so is to use a unimodular transformation. Unimodular transformations are linear mappings that guarantee the pre-image of any integer vector within its image to be an integer vector, too. These mappings can be characterized by their defining matrix M: $|\det(M)| = 1$ iff M is unimodular (and with it the function $\alpha \mapsto M \cdot \alpha$). Now, the crucial question is: is it always possible to get a unimodular transformation? A transformation is unimodular iff its inverse is unimodular, e.g., if we start with some subspace (the kernel $\ker(M)$), we have to find vectors $\nu_1, \ldots, \nu_m \in \mathbb{Z}^n$ that span this subspace and that can be expanded by integer vectors $\nu_{m+1}, \ldots, \nu_n$, so that the overall matrix $\begin{pmatrix} \nu_1 & \ldots & \nu_n \end{pmatrix}$ is unimodular. In the following theorem, we consider some base vectors $\mu_1, \ldots, \mu_m$ spanning the given subspace. Theorem 3.24 then tells us that it is not generally possible to find such vectors $\nu_1, \ldots, \nu_n$.

Theorem 3.24 *Let $\mu_1, \ldots, \mu_m \in \mathbb{Z}^n$ ($m \leq n$) be linearly independent integer vectors, $\mathfrak{M} = \mathrm{Span}(\mu_1, \ldots, \mu_m)$.*

1. *It is not generally possible to find some $\nu_1, \ldots, \nu_n \in \mathbb{Z}^n$ so that both Equation (3.24) and Property (3.25) hold:*

$$\mathfrak{M} = \mathrm{Span}(\nu_1, \ldots, \nu_m) \tag{3.24}$$

and

$$\begin{pmatrix} \nu_1 & \ldots & \nu_n \end{pmatrix} \text{ is unimodular} \tag{3.25}$$

2. *If $n = 2$, there are always vectors ν_1 and ν_2 so that both Equation (3.24) and Property (3.25) hold.*

3. *If $\mu_1, \ldots, \mu_m$ do not share a common non-zero dimension, i.e., $(\forall i, j : i \neq j \in \{1, \ldots, m\} : (\forall k : k \in \{1, \ldots, n\} : \mu_i[k] \neq 0 \Rightarrow \mu_j[k] = 0))$, then there are vectors $\nu_1, \ldots, \nu_n \in \mathbb{Z}^n$ so that both Equation (3.24) and Property (3.25), hold.*

Proof:

1. It suffices to consider the following counterexample: let $\mu_1 = \begin{pmatrix} 2 \\ 0 \\ 1 \end{pmatrix}$ and $\mu_2 = \begin{pmatrix} 0 \\ 2 \\ 1 \end{pmatrix}$. Let $\nu_1 = a \cdot \mu_1 + b \cdot \mu_2$ and $\nu_2 = a' \cdot \mu_1 + b' \cdot \mu_2$. Let $\nu_3 \in \mathbb{Z}^n$ be some arbitrary integer vector that can be adapted so as to make the matrix $N = \begin{pmatrix} \nu_1 & \nu_2 & \nu_3 \end{pmatrix} = \begin{pmatrix} 2 \cdot a & 2 \cdot a' & \nu_3[1] \\ 2 \cdot b & 2 \cdot b' & \nu_3[2] \\ a + b & a' + b' & \nu_3[3] \end{pmatrix}$ unimodular. Let us now determine the determinant of N:

$$\begin{array}{rcl} \det(N) & = & (2 \cdot a) \cdot (2 \cdot b') \cdot \nu_3[3] \\ & & +(a+b) \cdot (2 \cdot a') \cdot \nu_3[2] \\ & & +(2 \cdot b) \cdot (a' + b') \cdot \nu_3[1] \\ & & -(a+b) \cdot (2 \cdot b') \cdot \nu_3[1] \\ & & -(2 \cdot a) \cdot (a' + b') \cdot \nu_3[2] \\ & & -(2 \cdot b) \cdot (2 \cdot a') \cdot \nu_3[3] \\ & = & 2 \cdot (2 \cdot a \cdot b' \cdot \nu_3[3] \\ & & +(a+b) \cdot a' \cdot \nu_3[2] \\ & & +b \cdot (a' + b') \cdot \nu_3[1] \\ & & -(a+b) \cdot b' \cdot \nu_3[1] \\ & & -a \cdot (a' + b') \cdot \nu_3[2] \\ & & -2 \cdot b \cdot a' \cdot \nu_3[3]) \end{array}$$

This means, for any integer vector ν_3, the determinant is always even. Thus, the matrix N cannot be unimodular.

2. In the case of $n = 2$, we can easily enumerate the different possibilities for m:

$m = 0$: We may choose whatever vectors we like. The trivial solution is $\nu_1 = \iota_1$ and $\nu_2 = \iota_2$.

$m = 1$: Let $d = \gcd(\mu_1[1], \mu_1[2])$. Then we can safely set $\nu_1 = \frac{\mu_1}{d}$, since $\frac{\mu_1}{d} \in \mathbb{Z}^2$ and $\mathrm{Span}(\mu_1) = \mathrm{Span}(\frac{\mu_1}{d})$. Leaving ν_2 undetermined for the moment, we have:

$$\det\begin{pmatrix} \nu_1 & \nu_2 \end{pmatrix} = \det\begin{pmatrix} \frac{\mu_1[1]}{d} & \nu_2[1] \\ \frac{\mu_1[2]}{d} & \nu_2[2] \end{pmatrix} = \frac{\mu_1[1]}{d} \cdot \nu_2[2] - \frac{\mu_1[2]}{d} \cdot \nu_1[1]$$

Corresponding values for $\nu_2[1], \nu_2[2]$ solving

$$\frac{\mu_1[1]}{d} \cdot \nu_2[2] - \frac{\mu_1[2]}{d} \cdot \nu_1[1] = 1$$

can then be found by a generalized Euclidean Algorithm. Since $\frac{\mu_1[1]}{d}, \frac{\mu_1[2]}{d}$ are relatively prime, a solution always exists.

$m = 2$: Again, the solution is $\nu_1 = \iota_1$ and $\nu_2 = \iota_2$, because we may choose *both* base vectors as long as they span $\mathfrak{M} = \mathbb{Q}^2$.

3. We now assume that

$$\big(\forall i, j : i \neq j \in \{1, \ldots, m\} : \big(\forall k : k \in \{1, \ldots, n\} : \mu_i[k] \neq 0 \Rightarrow \mu_j[k] = 0\big)\big)$$

Additionally, the vectors μ_i are cancelled, i.e., the coefficients of each vector are relatively prime. We will now try to find integer vectors $\nu_1, \ldots, \nu_m$ that span the same rational vector space as $\mu_1, \ldots, \mu_m$ and integer vectors $\nu_{m+1}, \ldots, \nu_n$ so that the matrix N is unimodular, i.e., so that the absolute value of its determinant is 1, $\det(N) = \pm 1$. Thus, matrix N is roughly of the following from:

$$n\left\{\begin{pmatrix}
\nu_1 & & \nu_{m'} & & \nu_m & \nu_{m+1} & & \nu_n \\
\neq 0 & \cdots & 0 & \cdots & 0 & a_1 & \cdots & z_1 \\
0 & & \vdots & & \vdots & \vdots & & \vdots \\
\vdots & & \vdots & & 0 & \vdots & & \vdots \\
\vdots & & \vdots & & \neq 0 & \vdots & & \vdots \\
\vdots & & \vdots & & 0 & \vdots & & \vdots \\
\vdots & & 0 & & \vdots & \vdots & & \vdots \\
\vdots & & \neq 0 & & \vdots & \vdots & & \vdots \\
\vdots & & 0 & & 0 & \vdots & & \vdots \\
0 & \cdots & 0 & \cdots & \neq 0 & a_n & \cdots & z_n
\end{pmatrix}\right.$$

$$\underbrace{\underbrace{\underbrace{\quad\quad\quad}_{m'}\ \underbrace{\quad\quad}_{m-m'}}_{m}\ \underbrace{\quad\quad\quad}_{n-m}}_{n}$$

The role of m' above will become clear in a moment. First, let us recall how to obtain the determinant of this matrix – which has to evaluate to 1 or -1. Following Laplace's formula, we calculate the determinant by expansion along a column j:

$$\det(N) = \left(\sum i : i \in \{1, \ldots, n\} : \ (-1)^{i+j} \cdot N[i,j] \cdot \det(N_{i,j})\right) \tag{3.26}$$

Here, $N_{i,j}$ is the matrix obtained from N by omitting the i-th row and the j-th column. Any vector of $\mu_1, \ldots, \mu_m$ that contains only one non-zero entry can be replaced by the corresponding unit vector, without changing the generated space. Therefore, if a vector $\mu_i \in \{\mu_1, \ldots, \mu_m\}$ contains only a single non-zero entry $\mu_i[j] \neq 0$, we will use a vector $\nu_i = \iota_j$ in its place. This means that the absolute value of the determinant is the same as with $N_{i,j}$, i.e., it stays constant if we ignore column i and the corresponding row j, for which μ_i contains a non-zero entry. Let us now assume that, for some $m' < m$, the vectors $\mu_1, \ldots, \mu_{m'}$, were exactly those vectors that contained only one non-zero entry each and have been replaced by corresponding unit vectors $\nu_1, \ldots, \nu_{m'}$. Since we supposed that no two vectors of $\mu_1, \ldots, \mu_m$ share a non-zero position, which in particular holds for $\mu_1, \ldots, \mu_{m'}$, and since the matrix is square (i.e., there are at most n dimensions in which any vector of $\mu_1, \ldots, \mu_{m'}$ may show a non-zero entry), there are only $n - m'$ possible dimensions left, in which the other $m - m'$ vectors $(\mu_{m'+1}, \ldots, \mu_m)$ may feature non-zero entries. Since these $m-m'$ vectors have at least two non-zero entries each, they have to exhibit $2 \cdot (m-m')$ distinct dimensions with non-zero entries overall. This leaves us with the fact that the number of these at least $2 \cdot (m - m')$ non-zero dimensions cannot be greater than the number of the $n - m'$ dimensions that are not already occupied by non-zero entries in vectors $\mu_1, \ldots, \mu_{m'}$. In other words, we have:

$$\begin{aligned} 2 \cdot (m - m') &\leq n - m' && (3.27) \\ &\Leftrightarrow \\ m - m' &\leq n - m' - m + m' \\ &\Leftrightarrow \\ m - m' &\leq n - m && (3.28) \end{aligned}$$

Inequation (3.28) tells us directly that, for each of the $m - m'$ vectors left that may exhibit more than one non-zero entry, there is also at least one of the $n - m$ vectors $\nu_{m-n+1}, \ldots, \nu_m$ that we may choose ourselves.

In general, there may be more than $2 \cdot (m - m')$ non-zero dimensions covered by the vectors $\mu_{m-m'}, \ldots, \mu_m$, say $2 \cdot (m - m') + x$. But still, the number of all non-zero dimensions of these vectors taken together has to be smaller than the $n - m'$ dimensions left over from the vectors $\mu_1, \ldots, \mu_{m'}$:

$$\begin{aligned} 2 \cdot (m - m') + x &\leq n - m' && (3.29) \\ &\Leftrightarrow \\ (m - m') + x &\leq n - m' - (m - m') = n - m' - m + m' \\ &\Leftrightarrow \\ (m - m') + x &\leq n - m && (3.30) \end{aligned}$$

In other words, for each of the x additional non-zero dimensions of one or more of the vectors $\mu_{m-m'}, \ldots, \mu_m$, there is one of the $n - m$ vectors in $\nu_{m+1}, \ldots, \nu_n$ that we may choose for ourselves, plus there is also one of these freely choosable vectors for each μ_i from $\mu_{m-m'}, \ldots, \mu_m$ that may feature non-zero coordinates in at least two dimensions.

Let us assume we have simply copied the μ_i to the corresponding ν_i:

$$(\forall i : i \in \{m - m', \ldots, m\} : \nu_i := \mu_i)$$

We will now compute the determinant of N including the part of vectors featuring more than one non-zero coordinate. From the restriction defined in Inequation (3.30), we can deduce that we may apply row and column permutations to our matrix N (which only influences the sign, not the absolute value of the determinant) so that the resulting matrix has the form

$$\begin{pmatrix} \begin{pmatrix} 1 & & 0 \\ & \ddots & \\ 0 & & 1 \end{pmatrix} & & & 0 \\ & A_1 & & \\ & & \ddots & \\ 0 & & & A_{m-m'} \end{pmatrix}$$

where each A_i is a quadratic matrix containing the non-zero entries of vector μ_i in the first column and possibly non-zero elements of vectors from $\nu_{m+1}, \ldots, \nu_n$ in the remaining columns – Inequation (3.30) assures us that there are enough freely choosable vectors for extending each μ_i to such a quadratic matrix. The determinant of the overall matrix now evaluates to the product of the A_i. Therefore, we only have to prove that, given any of these A_i, we can determine values for the matrix entries in the columns behind the first one so that $\det(A_i) = \pm 1$.

Suppose A_i has the form

$$A_i = \begin{pmatrix} A[1,1] & \ldots & A[1,y] \\ \vdots & & \vdots \\ A[y,1] & \ldots & A[y,y] \end{pmatrix}$$

where y is the number of non-zero entries in the vector μ_i. We may then choose all values $A[k,j]$ freely, for which $j \geq 2$. Only the values $A[k,1]$ are given (and all are integer). With appropriate unimodular row transformations, we can even assert that, for all k, we have $A[k,1] > 0$. We can then transform A_i using unimodular transformations so that the resulting matrix has the form:

$$\begin{pmatrix} 1 & (\sum k : k \in \{1,\ldots,y\} : c_{1,k} \cdot A[k,2]) & \ldots & (\sum k : k \in \{1,\ldots,y\} : c_{1,k} \cdot A[k,y]) \\ 0 & (\sum k : k \in \{1,\ldots,y\} : c_{2,k} \cdot A[k,2]) & \ldots & (\sum k : k \in \{1,\ldots,y\} : c_{2,k} \cdot A[k,y]) \\ \vdots & \vdots & & \vdots \\ 0 & (\sum k : k \in \{1,\ldots,y\} : c_{y,k} \cdot A[k,2]) & \ldots & (\sum k : k \in \{1,\ldots,y\} : c_{y,k} \cdot A[k,y]) \end{pmatrix}$$

The procedure here is again a kind of generalized Euclidean algorithm: we always pick two rows k_1 and k_2 in turn and subtract the row with the greater entry in the first column from the other one until we reach the greatest common divisor of the corresponding entries $A[k_1,1]$ and $A[k_2,1]$ in the first column (this is an unimodular transformation). The result – possibly after row permutation – is a matrix whose only non-zero entry in the first column is the gcd of the original column entries. In our case, this is 1, because we assume the μ_i to be cancelled, and their non-zero entries are therefore relatively prime.

For each column j, we now create the following system of linear equations:

$$\begin{array}{rcl}
(\sum k : k \in \{1,\dots,y\} :\ c_{1,k} \cdot A\texttt{[}k,j\texttt{]}) & = & 0 \\
& \vdots & \\
(\sum k : k \in \{1,\dots,y\} :\ c_{j-1,k} \cdot A\texttt{[}k,j\texttt{]}) & = & 0 \\
(\sum k : k \in \{1,\dots,y\} :\ c_{j,k} \cdot A\texttt{[}k,j\texttt{]}) & = & 1 \\
(\sum k : k \in \{1,\dots,y\} :\ c_{j+1,k} \cdot A\texttt{[}k,j\texttt{]}) & = & 0 \\
& \vdots & \\
(\sum k : k \in \{1,\dots,y\} :\ c_{y,k} \cdot A\texttt{[}k,j\texttt{]}) & = & 0
\end{array} \tag{3.31}$$

with the given values $c_{\cdot,k}$ and variables $A\texttt{[}k,j\texttt{]}$. A solution to these equations yields a determinant of 1 for the submatrix A_i, because with such a solution we have shown that A_i can be transformed to a unit matrix using unimodular transformations. These equation systems (one for each value of j) are all solvable, because they all consist of y equations with y variables. Therefore, the determinant of N, as the product of the determinants of the submatrices A_i (aside from the sign due to row and column permutations) evaluates to 1. We have thus devised a way to create a unimodular matrix N.

✓

So, it is not in general possible to create a unimodular representative mapping. However, the third point of Theorem 3.24 asserts that such a mapping *can* be created if $\ker(M)$ can be spanned by integer vectors that do not share any non-zero dimension. The proof of this fact also sketches a way to create a corresponding matrix. Since applying non-unimodular matrices to a polyhedron results in holes in the transformed polyhedron, which may be more difficult to enumerate, it is a valid decision to restrict oneself to this simpler case. A secondary effect is that, if only vectors that do not share any non-zero dimension build the subspace of equivalent occurrence instances, the dimensions in the resulting polyhedron are less likely to depend on each other. This in turn means that the loops enumerating these dimensions in which only *one* integer solution has to be found may be simplified.[5] However, we will *not* restrict ourselves to this special case and instead take the risk of choosing a non-unimodular transformation. Therefore, we have to find some other way so that the image of our mapping is still contained in $\mathbb{Z}^{n_{src}+n_{blob}+2}$. We will now pursue this goal for arbitrary integer matrices.

Scaling the index space A way to circumvent this problem – at the price of creating holes in the space to be enumerated – consists of scaling the index space so that the points to be enumerated – viewed as integer points embedded in a rational space – also lie on integer coordinates in the transformed space. This approach is illustrated in the following example.

Example 3.25 *Figure 3.10 shows the occurrence instances of* `A(2*i1+3*i2)` *in the following code fragment:*

```
DO i1=1,4
  DO i2=1,6
    B(i2,i1)=A(2*i1+3*i2)+C(i2,i1)
  END DO
END DO
```

Equivalent occurrence instances are, again, connected by dotted lines. The occurrence instances equivalent to $A(20)$ are marked by fat circles. On the right hand side, all occurrence instances are replaced by occurrence instances representing their respective equivalence classes, varying only in

[5]In such cases, we may be able to replace a `for`-loop by an `if`-statement.

a single dimension. In the upper part, we chose the i_2-dimension as variable while, in the lower part, the i_1-dimension is used. I.e., the right hand side of Figure 3.10 corresponds to a projection along the direction in which reuse happens (along the dotted lines) onto either one of the original index space dimensions. Note that we have already established that we will not use a projection in this case, but instead a base transformation that asserts that reuse occurs along a well defined base vector – which we just ignore in this representation. It turns out that, no matter which dimension we choose to keep, we always get both integer points (large circles), and non-integer points (small circles) that have to be enumerated in order to represent all the different equivalence classes (if we keep the i_1-dimension, we have to enumerate multiples of $\frac{1}{2}$, while keeping the i_2-dimension leads to multiples of $\frac{1}{3}$).
Note that we have already shown in Theorem 3.24 that, in a two-dimensional space, it is always possible to find a unimodular mapping that asserts that the resulting points to enumerate are all integer – in this case, a projecting onto $\begin{pmatrix} 1 \\ -1 \end{pmatrix}$ yields the desired result. However, since it is not generally possible to find a unimodular mapping (which is also stated by Theorem 3.24), we choose to ignore this fact for the sake of the argument. Figure 3.11 shows another solution: if we scale the original polyhedron by a factor of 2, we can retain the i_1-dimension while the resulting projection, shown on the right of Figure 3.11 only enumerates integer values. However, this comes at the price of holes in the polyhedron, which have to be modelled by pseudo parameters, as described above.

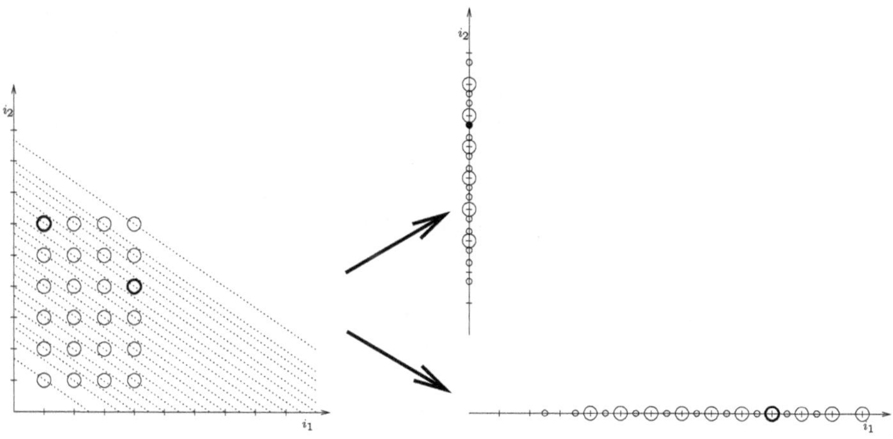

Figure 3.10: Choosing a unique representative for the read access `A(2*i1+3*i2)` by change of basis and projection onto either i_1 or i_2 in Example 3.25.

The need for the scaling in Example 3.25 emerges from the inversion of a matrix M_{base} (defining the new base of the index space) that is done in order to compute a base transformation from the base represented by the columns of M_{base}: although the matrix itself is integer, its inverse does not have to be. Let us suppose that we have such a non-singular matrix $M_{\text{base}} \in \mathbb{Z}^{(n_{src}+n_{blob}+2)\times(n_{src}+n_{blob}+2)}$. Inverting M_{base} is finding a solution X to the equation

$$M_{\text{base}} \cdot X = I_{n_{src}+n_{blob}+2,n_{src}+n_{blob}+2}$$

We can bring this matrix into echelon form by determining the least common multiple $m = \text{lcm}(M_{\text{base}}\texttt{[}\cdot, c\texttt{]})$ of the rows in each column c in turn. Picking one row r with a non-zero entry in column c, we add that row $\frac{m}{M_{\text{base}}\texttt{[}r,c\texttt{]}}$ times to each other row i, multiplied by $-\frac{m}{M_{\text{base}}\texttt{[}i,c\texttt{]}}$, leaving only row r with a non-zero entry in column c of the resulting matrix. After iterating through all columns, the matrix is in echelon form, up to a row permutation. Having the matrix – that we

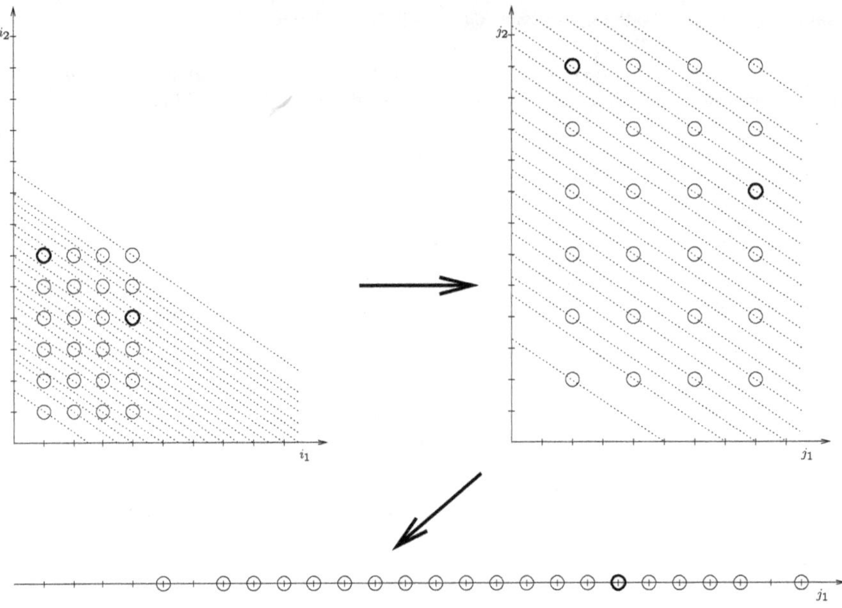

Figure 3.11: Original index space of `A(2*i1+3*i2)` in Example 3.25 (left), the index space scaled by 2 (right), and (below) the projection onto the first dimension, relabelled j_1 to indicate the change of basis.

now call $M_{\text{base}}{}'$ – in echelon form, we can solve for X by solving column by column (for column d) the equation system

$$M_{\text{base}}{}' \cdot X\,[\cdot, d] = \iota_d$$

Without loss of generality, we assume that no row permutation was necessary in order to arrive at an echelon form. Viewing our single row r in column c, again, the result for $X\,[c, d]$ is computed from the equation

$$M_{\text{base}}\,[r, c] \cdot X\,[c, d] = \iota_d\,[c] - (\textstyle\sum i : i \in \{r+1, \ldots, n_{src} + n_{blob} + 2\} : \; M_{\text{base}}\,[r, i] \cdot X\,[i, d])$$

In other words, the integer on the right hand side is divided by $M_{\text{base}}\,[r, c]$ – a common denominator for the complete row r of matrix X. Therefore, multiplying by the product of coefficients that can be chosen as leading coefficients in a row of $M_{\text{base}}{}'$ always leaves an integer matrix.

So, let us assume some given vectors $\kappa_1, \ldots, \kappa_n$ that have to span one part of our transformed index space (because, in the program, reuse occurs along their directions). I.e., M_{base} consists in part of the column vectors κ_i. For each of these vectors, we choose a dimension r_i to obtain a scaling factor. In this case, the scaling factor introduced by $\kappa_1, \ldots, \kappa_n$ evaluates to $\prod_{i=1}^{n} \kappa_i\,[r_i]$. The vectors for the other part we may choose ourselves. We cannot do anything about the resulting matrix being non-unimodular, so a first idea is to just use simple vectors – such as unit vectors. This will also reduce the scaling factor s we have to apply, since we will have to pick these unit vectors for the calculation of s in an appropriate row of the matrix M_{base}, and there is no smaller non-trivial integer factor than the 1 found in a unit vector (in terms of its absolute value). Thus, we have already reduced the possible maximum size of the holes generated by our technique. This is far from optimal (as we have already seen for the case of a two-dimensional index space) but it may reduce overhead in real applications. With just a little more work, we can reduce this factor even further, as the next example shows.

Example 3.26 *Example 3.25 introduced an array access* `A(2*i1+3*i2)`. *The instances of this array access represent values that are reused along the vector* $\begin{pmatrix} 3 \\ -2 \end{pmatrix}$. *If we want to exploit this reuse by merging all occurrence instances along this vector into one occurrence instance, we first need to scale the original occurrence instance set so that we can restrict ourselves to the integer points in the target space of the transformation to describe the same array accesses as in the original program. Of course, the scaling factor we have to apply here may depend on the vectors we choose in addition to* $\begin{pmatrix} 3 \\ -2 \end{pmatrix}$ *as a new basis for* $\mathbb{Q}^2$. *However, we have restricted ourselves to selecting unit vectors for this task, which leaves us with a choice of* ι_1 *and* ι_2. *In the previous example, we arrived at a scaling factor of* 2 *using* ι_1 *and thus a transformation matrix* $\begin{pmatrix} 1 & 3 \\ 0 & -2 \end{pmatrix}^{-1} \cdot 2$. *The other possibility using unit vectors is* $\begin{pmatrix} 0 & 3 \\ 1 & -2 \end{pmatrix}^{-1} \cdot 3$ *with its scaling factor of* 3. *Both of these possibilities are depicted in Figure 3.12: a scaling factor of* 3 *leads to integer values on the* i_2*-axis, while a factor of* 2 *leads to integer values on the* i_1*-axis. The minimal scaling factor* 1, *however, is only reached if we use integer multiples of the non-unit vector* $\begin{pmatrix} 1 \\ -1 \end{pmatrix}$, *leading to* $\begin{pmatrix} 1 & 3 \\ -1 & -2 \end{pmatrix}^{-1}$.

Example 3.26 shows that the decision to restrict ourselves to unit vectors may prevent us from selecting a base that would lead to an optimal scaling factor. As the next example shows, this may even lead to quite huge factors in general.

Example 3.27 *Let us just look at some general subspaces for a moment, without considering particular program fragments. Another example where our approach may lead to several different possible scaling factors is the – quite arbitrary – subspace spanned by the vectors* $\begin{pmatrix} 15 \\ 10 \\ 6 \\ 0 \\ 0 \end{pmatrix}$, $\begin{pmatrix} 2 \\ 0 \\ 5 \\ 3 \\ 0 \end{pmatrix}$, *and* $\begin{pmatrix} 7 \\ 0 \\ 0 \\ 15 \\ 3 \end{pmatrix}$. *Assuming that we do not have to do any row permutation, the algorithm we will use later (Algorithm 3.5.4) completes this basis to a transformation matrix*

$$\begin{pmatrix} 15 & 2 & 7 & 1 & 0 \\ 10 & 0 & 0 & 0 & 1 \\ 6 & 5 & 0 & 0 & 0 \\ 0 & 3 & 15 & 0 & 0 \\ 0 & 0 & 3 & 0 & 0 \end{pmatrix}^{-1}$$

leading to a scaling factor of $6 \cdot 3 \cdot 3 = 54$, *whereas a different possibility for choosing unit vectors to complete the matrix is*

$$\begin{pmatrix} 15 & 2 & 7 & 0 & 0 \\ 10 & 0 & 0 & 1 & 0 \\ 6 & 5 & 0 & 0 & 0 \\ 0 & 3 & 15 & 0 & 0 \\ 0 & 0 & 3 & 0 & 1 \end{pmatrix}^{-1}$$

which would lead in our algorithm to a scaling factor of $5 \cdot 15^2 = 1125$ *(although, as should be mentioned, in this case a factor of* 1071 *also suffices to finally obtain an integer matrix).*

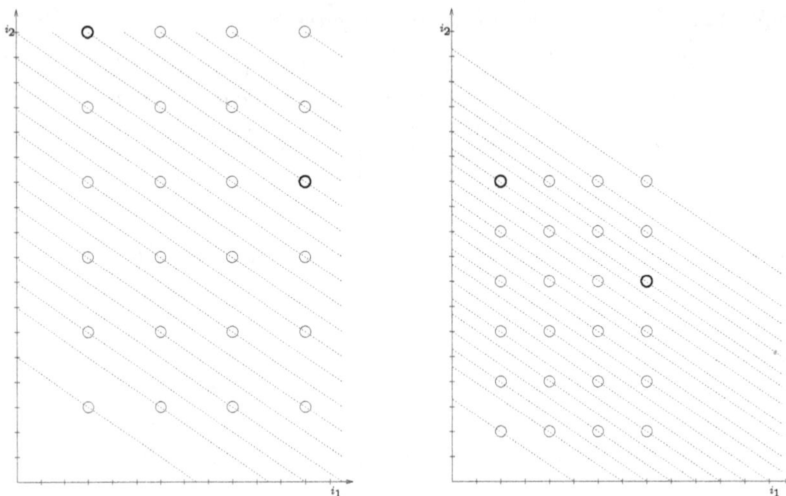

Figure 3.12: Scaled index spaces of `A(2*i1+3*i2)` from Example 3.25 with equivalent occurrence instances marked. Left: scaled by 3. Right: scaled by 2.

Example 3.26 and Example 3.27 illustrate that an intelligent choice of unit vectors may reduce the scaling factor. For example, we could reserve a dimension r_i for each vector κ_i so that

1. only κ_i and no other κ_j $(j \neq i)$ has a non-zero coordinate in dimension r_i and
2. among these dimensions that are only non-zero in κ_i, $|\kappa_i[r_i]|$ is minimal.

Then we use only unit vectors of not already reserved dimensions in order to extend $M = \begin{pmatrix} \kappa_1 & \dots & \kappa_n \end{pmatrix}$ to a full rank matrix. This can be easily achieved: we can find dimensions satisfying the first point by bringing the columns of M into echelon form (we may change M as long as we do not change $\mathrm{Span}(M)$), and among these it is easy to select a dimension with minimal coordinate value in a vector. Although this procedure does not guarantee minimal scaling factors, it is is quite effective, as Example 3.27 demonstrates.

We have now devised a way to obtain a function which maps several distinct but equivalent instances of the same occurrence to new occurrence instances that only differ from each other in certain reserved dimensions – conversely, occurrence instances that only differ in these reserved dimensions are equivalent. Projecting out the reserved dimensions from the resulting occurrence instance sets then produces occurrence instances that represent the equivalence classes of the original ones. However, this only holds for instances of the *same* occurrence. We will now tackle equivalences between different sets of occurrence instances.

Going Multi-Occurrence

In order to merge instances of different occurrences, it is necessary to create a representative mapping that maps them to the same representative. Suppose two occurrence instance sets $\mathfrak{O}_1$ and $\mathfrak{O}_2$. Suppose further that we have already established equivalences within $\mathfrak{O}_2$, which are represented by a function $\Xi'_2 : \mathfrak{O}_2 \to \mathbb{Z}^{n_{src}+n_{blob}+2}$, i.e., two equivalent occurrence instances are mapped to the same coordinates in $\mathbb{Z}^{n_{src}+n_{blob}+2}$ – aside from the coordinates in LCCP pseudo parameter dimensions. Correspondingly, there may be a mapping $\Xi'_1 : \mathfrak{O}_1 \to \mathbb{Z}^{n_{src}+n_{blob}+2}$. However, both mappings will map to different occurrence numbers, i.e., we may have the case that $\mathrm{im}(\Xi'_1) \cap \mathrm{im}(\Xi'_2) = \emptyset$. Instead, we aim at two mappings, $\Xi_1 : \mathfrak{O}_1 \to \mathbb{Z}^{n_{src}+n_{blob}+2}$ and $\Xi_2 : \mathfrak{O}_2 \to \mathbb{Z}^{n_{src}+n_{blob}+2}$, whose images at least overlap. The strategy is here simply to produce Ξ_1 from Ξ_2 – which, in turn, we define directly by Ξ'_2, i.e., $\Xi_2 := \Xi'_2$. Suppose an equivalence relation $\Lambda_{1,2} \subseteq \mathfrak{O}_1 \times \mathfrak{O}_2$ between $\mathfrak{O}_1$ and $\mathfrak{O}_2$. As before (for Ξ_1, Ξ_2), this relation can be represented by a function $\Xi_{1,2}$. Then we only have to create a mapping from the composition of these relations in order to create the representative mapping for $\mathfrak{O}_1$: $\Xi_1 := \Xi_2 \circ \Xi_{1,2}$. The common representative mapping for $\mathfrak{O}_1$ and $\mathfrak{O}_2$ is then $\Xi_1 \cup \Xi_2$. It is not even necessary that $\mathrm{im}(\Xi_{1,2}) \subseteq \mathfrak{O}_2$, as hinted in the above description: since Ξ'_2 is based on the equivalences in $\mathfrak{O}_2$, *ignoring* domain restrictions, it will map equivalent occurrence instances that exceed the boundaries of $\mathfrak{O}_2$ to the same points as occurrence instances that actually belong to $\mathfrak{O}_2$. The only important point here is that equivalent occurrence instances of both sets, $\mathfrak{O}_1$ and $\mathfrak{O}_2$, are mapped in the same way, i.e., there has to be a decision for a final set of occurrence instances that is good for representing both $\mathfrak{O}_1$ and $\mathfrak{O}_2$. Using an adjacency matrix for the relation between these sets with the exact relations occupying the elements of the matrix, these representative mappings can be determined by first creating a representative mapping that takes the equivalences within each set itself into account, i.e., the entries on the diagonal of the adjacency matrix, and then calculating compositions of this mapping with all relations in the same row of the matrix.

When all sets of occurrence instances have associated functions that map their elements to possible representatives, we only have to choose these mappings so that equivalent occurrence instances are mapped to the same point (up to projection onto index dimensions). It suffices to sort all possible representative mappings for a set of occurrence instances $\mathfrak{O}_i$ in the same way for all the $\mathfrak{O}_j$, $j \neq i$. "In the same way", here means "according to the last function applied", i.e., the function Ξ_j on the diagonal element of the adjacency matrix that was applied to obtain the current representative mapping (since using the same mapping here implies equivalent result values). Then we go through this sorted sequence in turn and use the corresponding mapping in its complete domain, (aside from the already considered parts) until the union of all mapping domains covers the occurrence instance set $\mathfrak{O}_i$ to be represented.

A possible sorting criterion is the rank of the projection of Ξ_j onto the index dimensions: the larger the rank, the more different occurrence instance dimensions lie in its image. Note that the ranks of the functions on the diagonal define a bound on how many dimensions may be omitted for a given computation, since these ranks are by definition minimal. If we prefer functions with higher rank, we represent an occurrence instance by a more general expression – for example, we may express `A(2)` by `A(i)`. One may also choose the opposite approach and identify each tiny bit of the index space that may be computed in a nest with lower dimensionality than the other occurrence instances as an individual special case. We will assume the first strategy, since this approach tends to lead to simpler code. If the ranks are equal, we sort for the occurrence number. This is completely arbitrary, but it guarantees a unique choice.

Example 3.28 *In the following example code, the accesses* `A(2)` *and* `A(i)` *– and thus the terms* `A(2)**2` *and* `A(i)**2` *– can be equivalent; however, this equivalence only holds for* $i = 2$*:*

```
      DO i=0,10
!       [11]     [10] [1]       [3]     [2]   [7]   [4]       [6]     [5]
        B(i)      =   A(2)      **      2     +     A(i)      **      2
      END DO
```

The loop nest above contains the number of each occurrence in the comment above the corresponding piece of code. Interesting for us are the occurrences number 3 *and* 6*, because they represent the possibly equivalent exponentiations.*

The equivalences of Example 3.28 can be represented in an adjacency matrix like the following:

$\mathfrak{D}_1 \equiv \mathfrak{D}_2$	$\mathfrak{D}_1 = \left\{ \begin{pmatrix} i \\ 3 \\ \infty \end{pmatrix} \middle\vert\, 0 \leq i \leq 10 \right\}$	$\mathfrak{D}_2 = \left\{ \begin{pmatrix} i \\ 6 \\ \infty \end{pmatrix} \middle\vert\, 0 \leq i \leq 10 \right\}$
$\mathfrak{D}_1$	$\Lambda_{1,1} = \left\{ \begin{pmatrix} i \\ 3 \\ \infty \\ i' \\ 3 \\ \infty \end{pmatrix} \middle\vert\, i, i' \in \mathbb{Z} \right\} \cap \left\{ \begin{pmatrix} i \\ 3 \\ \infty \\ i' \\ 3 \\ \infty \end{pmatrix} \middle\vert\, 0 \leq i, i' \leq 10 \right\}$	$\Lambda_{1,2} = \left\{ \begin{pmatrix} i \\ 3 \\ \infty \\ 2 \\ 6 \\ \infty \end{pmatrix} \middle\vert\, i \in \mathbb{Z} \right\} \cap \left\{ \begin{pmatrix} i \\ 3 \\ \infty \\ 2 \\ 6 \\ \infty \end{pmatrix} \middle\vert\, 0 \leq i \leq 10 \right\}$
$\mathfrak{D}_2$	$\Lambda_{2,1} = \left\{ \begin{pmatrix} 2 \\ 6 \\ \infty \\ i' \\ 3 \\ \infty \end{pmatrix} \middle\vert\, i' \in \mathbb{Z} \right\} \cap \left\{ \begin{pmatrix} 2 \\ 6 \\ \infty \\ i' \\ 3 \\ \infty \end{pmatrix} \middle\vert\, i = 2 \wedge 0 \leq i' \leq 10 \right\}$	$\Lambda_{2,2} = \left\{ \begin{pmatrix} i \\ 6 \\ \infty \\ i \\ 6 \\ \infty \end{pmatrix} \middle\vert\, i \in \mathbb{Z} \right\} \cap \left\{ \begin{pmatrix} i \\ 6 \\ \infty \\ i \\ 6 \\ \infty \end{pmatrix} \middle\vert\, 0 \leq i \leq 10 \right\}$

The vectors in the adjacency matrix here only present the interesting dimensions:

1. The loop index i.
2. The occurrence number.
3. The pseudo parameter n_{LCCP} for the dimension that replaces i.

The diagonal elements of the adjacency matrix hold relations with constant occurrence number. In the case of $\mathfrak{D}_1$, *all* elements of the complete index space are equivalent. For $\mathfrak{D}_2$, occurrence instances are only equivalent if their coordinate in the i-dimension is the same $(i = i')$ (i.e., an occurrence instance is equivalent only to itself). Furthermore, *all* elements of $\mathfrak{D}_1$ are equivalent to the single elements of $\mathfrak{D}_1$ that is enumerated by the i-iteration 2 (i.e., $i' = 2$) and, correspondingly, this element of $\mathfrak{D}_1$ is equivalent to all elements of $\mathfrak{D}_2$. The diagonal relations can then be replaced by functions as follows:

$\mathfrak{D}_1 \equiv \mathfrak{D}_2$	$\mathfrak{D}_1 = \left\{ \begin{pmatrix} i \\ 3 \\ \infty \end{pmatrix} \mid 0 \leq i \leq 10 \right\}$	$\mathfrak{D}_2 = \left\{ \begin{pmatrix} i \\ 6 \\ \infty \end{pmatrix} \mid 0 \leq i \leq 10 \right\}$
$\mathfrak{D}_1$	$\Xi_1 : \mathfrak{D}_1 \rightarrow \mathbb{Z}^3 : \begin{pmatrix} i \\ 3 \\ \infty \end{pmatrix} \mapsto \begin{pmatrix} \infty \\ 3 \\ i \end{pmatrix}$	$\Lambda_{1,2} = \left\{ \begin{pmatrix} i \\ 3 \\ \infty \\ 2 \\ 6 \\ \infty \end{pmatrix} \mid i \in \mathbb{Z} \right\} \cap \left\{ \begin{pmatrix} i \\ 3 \\ \infty \\ 2 \\ 6 \\ \infty \end{pmatrix} \mid 0 \leq i \leq 10 \right\}$
$\mathfrak{D}_2$	$\Lambda_{2,1} = \left\{ \begin{pmatrix} 2 \\ 6 \\ \infty \\ i' \\ 3 \\ \infty \end{pmatrix} \mid i' \in \mathbb{Z} \right\} \cap \left\{ \begin{pmatrix} 2 \\ 6 \\ \infty \\ i' \\ 3 \\ \infty \end{pmatrix} \mid i = 2 \wedge 0 \leq i' \leq 10 \right\}$	$\Xi_2 : \mathfrak{D}_2 \rightarrow \mathbb{Z}^3 : \alpha \mapsto \alpha$

Note that there is no reason to leave the occurrence numbers in $\operatorname{im}(\Xi_i)$ at the same value as in the pre-image. The important point here is that, while Ξ_1 exchanges the loop dimension with the pseudo parameter dimension, Ξ_2 leaves these dimensions just as they were. This expresses the equivalence of all elements in $\mathfrak{D}_1$ (whereas there is no equivalence between distinct elements of $\mathfrak{D}_2$). We can now create actual representative mappings by calculating the composition of the mappings in the diagonal element with the relations in the corresponding row of the adjacency matrix. Representing these relations by a pair of functions (Ξ_L, Ξ_R), as in Section 2.4.2, we may represent this relation as $\Xi_R^g \circ \Xi_L$ with an *arbitrary* generalized inverse Ξ_R^g, as we have established above (since all points obtained by any such mapping are equivalent and thus result in the same point when projecting onto the first two dimensions – ignoring the n_{LCCP}-dimension); with the same reasoning, the domain of Ξ_2 is irrelevant in the composition. We decide for Ξ_R^{go}, which uses the coordinate 0 for a dimension that may be chosen freely. This results in the following adjacency matrix:

$\mathfrak{D}_1 \equiv \mathfrak{D}_2$	$\mathfrak{D}_1 = \left\{ \begin{pmatrix} i \\ 3 \\ \infty \end{pmatrix} \mid 0 \leq i \leq 10 \right\}$	$\mathfrak{D}_2 = \left\{ \begin{pmatrix} i \\ 6 \\ \infty \end{pmatrix} \mid 0 \leq i \leq 10 \right\}$
$\mathfrak{D}_1$	$\Xi_1 : \mathfrak{D}_1 \rightarrow \mathbb{Z}^3 : \begin{pmatrix} i \\ 3 \\ \infty \end{pmatrix} \mapsto \begin{pmatrix} \infty \\ 3 \\ i \end{pmatrix}$	$\Xi_{1,2} : \mathfrak{D}_1 \rightarrow \mathbb{Z}^3 : \begin{pmatrix} i \\ 3 \\ \infty \end{pmatrix} \mapsto \begin{pmatrix} 2 \\ 6 \\ \infty \end{pmatrix}$
$\mathfrak{D}_2$	$\Xi_{2,1} : \left\{ \begin{pmatrix} 2 \\ 6 \\ \infty \end{pmatrix} \right\} \rightarrow \mathbb{Z}^3 : \begin{pmatrix} 2 \\ 6 \\ \infty \end{pmatrix} \mapsto \begin{pmatrix} \infty \\ 3 \\ 0 \end{pmatrix}$	$\Xi_2 : \mathfrak{D}_2 \rightarrow \mathbb{Z}^3 : \alpha \mapsto \alpha$

Note that the choice of 0 as last coordinate in the target of $\Xi_{2,1}$ is arbitrary: this coordinate can be chosen at will due to the fact there is an equivalence relation between instance $i = 2$ of Occurrence 6 and all instances of occurrence 3. Following the definition of Ξ_R^{go}, we choose coordinate 0. As discussed above, the sorting step yields sequences of the following form:

$\mathfrak{O}_1$	$\Xi_{1,2} : \mathfrak{O}_1 \to \mathbb{Z}^3 : \begin{pmatrix} i \\ 3 \\ \infty \end{pmatrix} \mapsto \begin{pmatrix} 2 \\ 6 \\ \infty \end{pmatrix}$	$\Xi_1 : \mathfrak{O}_1 \to \mathbb{Z}^3 : \begin{pmatrix} i \\ 3 \\ \infty \end{pmatrix} \mapsto \begin{pmatrix} \infty \\ 3 \\ i \end{pmatrix}$
$\mathfrak{O}_2$	$\Xi_2 : \mathfrak{O}_2 \to \mathbb{Z}^3 : \alpha \mapsto \alpha$	$\Xi_{2,1} : \left\{ \begin{pmatrix} 2 \\ 6 \\ \infty \end{pmatrix} \right\} \to \mathbb{Z}^3 : \begin{pmatrix} i \\ 6 \\ \infty \end{pmatrix} \mapsto \begin{pmatrix} \infty \\ 3 \\ 0 \end{pmatrix}$

Note that $\Xi_{2,1}$ is only defined for $i = 2$. Since Ξ_2 is the mapping with the higher rank, it defines the preferred representative mapping. The representatives of *both* sets are thus:

$$\left\{ \begin{pmatrix} i \\ 6 \\ \infty \end{pmatrix} \middle| \, 0 \leq i \leq 10 \right\} \cup \left\{ \begin{pmatrix} 2 \\ 6 \\ \infty \end{pmatrix} \right\} = \left\{ \begin{pmatrix} i \\ 6 \\ \infty \end{pmatrix} \middle| \, 0 \leq i \leq 10 \right\}$$

There is just one further point to be observed. With the change of basis, we have just exchanged dimensions. However, the unique representative of an occurrence instance is given by a subsequent projection that simply removes the dimensions n_{LCCP} through $n_{\mathrm{LCCP}} + n_{\mathrm{depth}}$. Thus, all occurrence instances that only differ in the coordinates of dimensions n_{LCCP} to $n_{\mathrm{LCCP}} + n_{\mathrm{depth}}$ are finally mapped to the same point and thus identified with a single unique representative. This is implemented by the function *PrepareRepMap* that we have encountered during the discussion of Algorithm 3.4.1 in Section 3.4. It removes the last dimensions in which representatives may have differed to produce a unique representative. In Algorithm 3.4.1 this is necessary in order to be able to decide that two occurrence instances α and α', $\alpha \neq \alpha'$, reading from different, but equivalent, occurrence instances $\beta \neq \beta'$, $(\beta, \beta') \in \equiv_{(\Delta^{(s,f)}, \Delta^{i})}$, are actually equivalent: *PrepareRepMap* produces a representative mapping Ξ so that $\Xi(\beta) = \Xi(\beta')$, and the equivalence of α with α' can be followed.

We have now sketched a way to find representatives using a change of basis. The next section is devoted to a more detailed discussion of the actual algorithms implementing this method.

3.5.3 The Algorithms for the Change of Basis

The basic procedure for selecting representatives is now clear. For the following detailed discussion, we will proceed in the reverse order of Section 3.5.2 and start with an algorihm that can be plugged into Algorithm 3.4.1 on page 82 and work from there towards a more precise description of the selection process.

For the sake of a clean representation, we use the (trivial) algorithm *CreateIDAliases* (Algorithm 3.5.1) to create an enumeration of sets of occurrence instances. The point here is just to have *some* enumeration of occurrence instance sets. Which algorithm to use is of no importance to the method itself. Algorithm 3.5.1 (*CreateIDAliases*) implements a mapping that represents an enumeration of all the sets of occurrence instances (defined by unions of polyhedra) that make up the original loop program to be processed by sorting according to occurrence number.

Algorithm 3.5.1 [*CreateIDAliases*]:
Input:
$\mathfrak{O} \subseteq \mathbb{Z}$: finite set of not necessarily contiguous integer numbers.
Output:
$F : \{1, \ldots, \#(\mathfrak{O})\} \to \mathfrak{O}$:
strictly monotonous enumeration of $\mathfrak{O}$.

Procedure:

```
workset := 𝔒;
i := 1;
while workset ≠ ∅
  o := min(worksett);
  F[i] := o;
  worksett := worksett \ {o};
  i := i + 1;
endwhile
return F;
```

This enumeration is used for an adjacency matrix in Algorithm 3.5.2 (*CreateRepresentativeMappings*). The algorithm uses *ReduceImageDimension* (Algorithm 3.5.4) to create a linear function $\Xi : \{\alpha \in \mathbb{Z}^{n_{src}+n_{blob}+2} \mid \mathrm{OccId}(\alpha) = o \wedge \mathrm{OpId}(\alpha) = 0\} \to \mathbb{Z}^{n_{src}+n_{blob}+2}$ that maps two instances of occurrence o to new occurrence instances differing only in their coordinates in the reserved dimensions $n_{\mathrm{LCCP}}, \ldots, n_{\mathrm{LCCP}} + n_{\mathrm{depth}} - 1$ iff they are deemed equivalent by the input equivalence relation $\equiv_{(\Delta^{(s,f)} \cup \Delta^{f}, \mathrm{id})}$. Applying Ξ to the index space of o creates, in principle, a new polyhedron with minimal dimensionality. I.e., if the (projection of the) new polyhedron covers more than a single point in some given dimension, then this polyhedron defines several different values (wrt. the Herbrand universe).

Algorithm 3.5.2 composes polyhedra in $\mathbb{Z}^n$.[6] For the sake of clarity, we do not go into the details of how to compute these relations. Instead, we refer to the literature [KMP+96b, Sch86]. Note that, due to the restriction to the integer numbers as the underlying ring of $\mathbb{Z}^{(n_{src}+n_{blob}+2)\times(n_{src}+n_{blob}+2)}$, the description for the resulting set may include the requirement that this newly defined set is not only a polyhedron, but a $\mathbb{Z}$-polyhedron; i.e., the relation may only be expressible as an intersection with an integer lattice – which can be represented by additional constraints in existentially quantified variables (called exist variables in `Omega` [Pug93] and pseudo parameters in LooPo [Kei97]) that can be used to express the fact that a certain linear combination of indices and parameters should be a multiple of some given integer. The need for these parameters arises independently of those needed for the LCCP transformation itself (see page 94 in Section 3.5.2). Note, however, that the number of these additional parameters is always bounded by the number of indices of the deepest loop nest in the original program. Therefore, we can add these parameters already at the beginning of the program analysis and view them as a part of the n_{blob} parameters of the program. Since, furthermore, we never need more index dimensions in any target program than in the input program in order to express the same calculations, we can always stay within $\mathbb{Z}^{n_{src}+n_{blob}+2}$. In addition, we suppose that the functions dom and im are defined not only for functions, but for general relations:

[6]Note that the polyhedra used here are special cases in that they have distinct source and target dimensions. Source and target dimensions are only related by equations, while inequations only hold among source dimensions or among target dimensions.

$$\begin{array}{ll}\text{dom}: & \mathcal{P}(\mathbb{Z}^{r\times s}) \to \mathcal{P}(\mathbb{Z}^{r\times s}): \\ & \mathfrak{R} = \left\{ \left(\begin{array}{c}\alpha\\ \beta\end{array}\right) \in \mathbb{Z}^{r+s} \middle| M_= \cdot \left(\begin{array}{c}\alpha\\ \beta\end{array}\right) = 0 \wedge M_\geq \cdot \left(\begin{array}{c}\alpha\\ \beta\end{array}\right) \geq 0 \right\} \to \\ & \left\{ \alpha \in \mathbb{Z}^r \,\middle|\, (\exists \beta : \beta \in \mathbb{Z}^s : (\alpha,\beta) \in \mathfrak{R}) \right\} \\ \text{im}: & \mathcal{P}(\mathbb{Z}^{r\times s}) \to \mathcal{P}(\mathbb{Z}^{r\times s}): \\ & \mathfrak{R} = \left\{ \left(\begin{array}{c}\alpha\\ \beta\end{array}\right) \in \mathbb{Z}^{r+s} \middle| M_= \cdot \left(\begin{array}{c}\alpha\\ \beta\end{array}\right) = 0 \wedge M_\geq \cdot \left(\begin{array}{c}\alpha\\ \beta\end{array}\right) \geq 0 \right\} \to \\ & \left\{ \beta \in \mathbb{Z}^s \,\middle|\, (\exists \alpha : \alpha \in \mathbb{Z}^r : (\alpha,\beta) \in \mathfrak{R}) \right\} \end{array}$$

We can now formulate the Algorithm 3.5.2 (*CreateRepresentativeMappings*), which determines the representation of a set of occurrence instances.

Algorithm 3.5.2 [*CreateRepresentativeMappings*]:
Input:
$\mathfrak{O}$: set of occurrence instances – one level as produced by *GroupOccurrenceInstances* (Algorithm 3.2.1).

Λ : equivalence relation on $\mathfrak{O}$ defined as a union of polyhedra $\Lambda = \bigcup_{i\in\{1,\dots,n\}} (\Lambda_{i,=} \cap \Lambda_{i,\geq})$ with $\Lambda_{i,=} = \left\{(\alpha,\beta) \in \mathfrak{O}\times\mathfrak{O} \middle| M_{\Lambda_{i,=}} \cdot \left(\begin{array}{c}\alpha\\ \beta\end{array}\right) = 0\right\}$ and $\Lambda_{i,\geq} = \left\{(\alpha,\beta) \in \mathfrak{O}\times\mathfrak{O} \middle| M_{\Lambda_{i,\geq}} \cdot \left(\begin{array}{c}\alpha\\ \beta\end{array}\right) \geq 0\right\}$. Each Λ_i only holds between instances of a single input and a single output occurrence.

Output:
Ξ : piecewise linear mapping $\Xi = \bigcup_{i\in\{1,\dots,n_\Xi\}} \Xi_i$, $\Xi_i : \mathfrak{O}_i \subseteq \mathbb{Z}^{n_{src}+n_{blob}+2} \to \mathfrak{O}'_i \subseteq \mathbb{Z}^{n_{src}+n_{blob}+2}$, $\bigcup_{i\in\{1,\dots,n_\Xi\}} \mathfrak{O}'_i = \mathfrak{O}$ mapping each occurrence instance of $\mathfrak{O}$ to its representative wrt. equivalence relation Λ.

Procedure:
```
𝔒ccs := OccId(𝔒);
/* Create a hash function for an adjacency matrix */
Hash := CreateIDAliases(𝔒ccs);
/* Create an adjacency matrix representing the equivalence relation */
/* The adjacency matrix is a two-dimensional matrix with each element representing */
/* the relation between the instances of the corresponding occurrences */
/* as an array of polyhedra. */
/* STEP 1: initialization */
for i = 1 to #(𝔒ccs)
  for j = 1 to #(𝔒ccs)
    AdjMatrix[i,j] := ∅;
  endfor
endfor
for i = 1 to n
  AdjMatrix[Hash(OccId(dom(Λi))), Hash(OccId(im(Λi)))] :=
    append(AdjMatrix[Hash(OccId(dom(Λi))), Hash(OccId(im(Λi)))], Λi);
endfor
/* STEP 2: dimension minimization */
for i = 1 to #(𝔒ccs)
  /* reassign AdjMatrix to a version whose image dimensionality is minimized */
```

$MinVersion := \emptyset$;
/* note that since $\equiv_{(\Delta^{(s,f)} \cup \Delta^f, \mathrm{id})}$ is an equivalence relation, */
/* $AdjMatrix\,[i,i] \neq \emptyset$ is guaranteed */
forall $(\Lambda_{i,=} \cap \Lambda_{i,\geq}) \in AdjMatrix\,[i,i]$
$\Xi'_{=} := ReduceImageDimension(\Lambda_{i,=})$;
$MinVersion :=$ append$(MinVersion, \Xi'_{=}|_{\Lambda_{i,\geq}})$;
end forall
$AdjMatrix\,[i,i] := MinVersion$;
endfor
/* STEP 3: transformation of output occurrence instance sets to minimized versions */
for $i = 1$ **to** $\#(\mathfrak{Occs})$
$NewVersion := \emptyset$;
for $j = 1$ **to** $\#(\mathfrak{Occs})$
forall $\Lambda_i \in AdjMatrix\,[i,j]$
forall $\Xi' \in AdjMatrix\,[j,j]$
Let $\Lambda_i = \Lambda_{i,\geq} \cap \Lambda_{i,=_R}^{-1} \circ \Lambda_{i,=_L}$;
Let $\Xi' : \mathrm{dom}(\Xi') \rightarrow \mathbb{Z}^{n_{src}+n_{blob}+2} : \alpha \mapsto \Xi'_{=}(\alpha)$
with $\Xi'_{=} : \mathbb{Z}^{n_{src}+n_{blob}+2} \rightarrow \mathbb{Z}^{n_{src}+n_{blob}+2}$;
$NewVersion :=$ append$(NewVersion, \Xi'_{=} \circ \Lambda_{i,=_R}^{g} \circ \Lambda_{i,=_L})$;
end forall
endfor
$AdjMatrix\,[i,j] := NewVersion$;
endfor
/* STEP 4: choose the actual representative mapping for each partition of $\mathfrak{OI}$ */
$n_{\Xi} := 0$;
for $i = 1$ **to** $\#(\mathfrak{Occs})$
$PossibleMappings := \emptyset$;
for $j = 1$ **to** $\#(\mathfrak{Occs})$
$PossibleMappings :=$ append$(PossibleMappings, AdjMatrix\,[i,j])$;
endfor
Sort $PossibleMappings$ in increasing order according to
the comparison function $CompareRelations$ (Algorithm 3.5.3);
$stillUnmapped := \mathbb{Z}^{n_{src}+n_{blob}+2}$;
for $k = 1$ **to** size$(PossibleMappings)$
$\Xi_i := PossibleMappings\,[k]\,|_{stillUnmapped}$;
$stillUnmapped := stillUnmapped \setminus \mathrm{dom}(PossibleMappings\,[k])$;
$n_{\Xi} := n_{\Xi} + 1$;
endfor
endfor
$\Xi := \bigcup_{j \in \{1,\ldots,n_{\Xi}\}} \Xi_i$;
return Ξ;

Step 1 of Algorithm 3.5.2 initializes an adjacency matrix wrt. Λ, whose element (i,j) contains the part of the given equivalence relation that holds between the *occurrenceinstanceset* number i and the set number j – where set number i is thought of as the domain of the relation and set number j as its image. In other words, the entry (i,j) represents the possibility to select an occurrence instance of set i by an occurrence instance of set j.

Step 2 replaces the part of the equivalence relation Λ that defines equivalences on the same occurrence by a representative mapping – yielding a simple base transformation that maps equivalent occurrence instances to points that only differ in some well-defined pseudo parameter dimensions.[7]

[7]Note that the different values for $\Xi'_{=}$ in each iteration denote mappings whose image does not overlap, since

In Step 3, the mappings generated in the previous step are composed with the parts of the equivalence relation that connect different sets of occurrence instances: if there is a connection between set number i and set number j, the algorithm computes the composition $\Xi' \circ \Lambda_{i,=_R}^{g} \circ \Lambda_{i,=_L}$ of the corresponding relation Λ_i connecting set i with set j and the representative mapping for set j itself. Note that, as we have established in Section 3.5.2 (page 104), we may use *any* generalized inverse of $\Lambda_{i,=_R}$ in this composition, since every element of $\Lambda_{i,=_R}^{g} \circ \Lambda_{i,=_L}(\alpha)$ is equivalent not only to α, but to every other element of $\Lambda_{i,=_R}^{g} \circ \Lambda_{i,=_L}(\alpha)$. The only respect, in which the selection of the inverse makes a difference is the fact that, when Algorithm 3.4.1 collects the occurrence instances to be executed, this particular inverse may lead to unusual bounds in some dimensions, since the specific pre-image point that $\Lambda_{i,=_R}^{g}$ maps to does not necessarily lie in the original domain of $\Lambda_{i,=}$. It may therefore create new occurrence instances that were not enumerated before. However, this does not change equivalence relations between occurrence instances. Therefore, the image of Ξ' only changes in that there are additional polyhedra that are possibly offset from the previous image in LCCP pseudo parameter dimensions. No additional occurrence value is added to the image of Ξ'. In the collection of new occurrence instances in Step 1 of Algorithm 3.4.1, this leads to a disjunction of polyhedra that differ in coordinates in the pseudo parameters dimensions for the LCCP parameters. The result is a function that maps the original occurrence instances of set number i to occurrence instances that use the reserved LCCP pseudo parameters wherever possible – due to the final application of a mapping produced by Algorithm 3.5.4 (*ReduceImageDimension*. Here, it suffices to iterate only once through the adjacency matrix, because the relation stored in it is already transitive, so that *AdjMatrix* $[i, \cdot]$ represents all possible mappings we could use to find representatives for occurrence instance set *Hash*(i). We can therefore go straight to the next step to select the appropriate representative function.

At the beginning of Step 4, there are several possible representative functions stored in the adjacency matrix for each set of occurrence instances. In order to represent equivalent occurrence instances that come from different sets by the same representative, the same mapping has to be selected for these different sets of occurrence instances. The algorithm does this by sorting the possible representative mappings found in the adjacency matrix according to an unambiguous criterion, which is supplied by Algorithm 3.5.3 (*CompareRelations*). The suitable mappings are stored in the sequence *possibleMappings*, sorted increasingly, and the complete set of occurrence instances is mapped according to the first matching function that in that sorted sequence.

This sequence of possible mappings can be sorted in different ways. The only important thing is that equivalent occurrence instances will be ultimately mapped to the same set. Algorithm 3.5.3 presented here is just one possible solution.

Algorithm 3.5.3 [*CompareRelations*]:
Input:
$((\Xi_1 \circ \Lambda_{1,=}) \cap \Lambda_{1,\geq})$:
polyhedron to be compared to $(\Lambda_{2,=}, \Lambda_{2,\geq})$.

$((\Xi_2 \circ \Lambda_{2,=}) \cap \Lambda_{2,\geq})$:
polyhedron to be compared to $(\Xi_1 \circ \Lambda_{1,=}, \Lambda_{1,\geq})$.
Output:
$\{-1, 0, 1\}$:
-1, if $(\Lambda_{1,=}, \Lambda_{1,\geq})$ is supposed to be *less* than $(\Xi_2 \circ \Lambda_{2,=}, \Lambda_{2,\geq})$, $+1$, if its to be considered *greater*, 0 otherwise.

Procedure:
```
if rk(Ξ1) > rk(Ξ2) then
  return −1;
else
```

they all come from minimizing the dimensionality of instances of the same occurrence. Therefore, Algorithm 3.5.4 returns either the same mapping (with different domain) for the given relation $\equiv_{(\Delta^{(s,f)} \cup \Delta^{f}, \mathrm{id})_{i,=}}$ or a mapping with one or the other unit vector mapped to itself instead of an LCCP pseudo parameter, which creates a completely different target index space with different dimensionality.

```
  if rk(Ξ1) < rk(Ξ2) then
    return 1;
  else
    /* the image has the same dimensionality – we use the occurrence numbers for arbitration */
    return sgn(OccId(im(Ξ1)) – (OccId(im(Ξ2))));
  endif
endif
```

Note that Algorithm 3.5.3 (*CompareRelations*) depends on the way in which we compute the relations that are to be compared. These relations come from Algorithm 3.5.2. However, since all relations in the adjacency matrix in the algorithm are essentially computed in the same way, this is not really a problem. Algorithm 3.5.3 decides that a relation $\mathfrak{R}_1$ is to be sorted before a relation $\mathfrak{R}_2$ if the final linear function used for building $\mathfrak{R}_1$ has a larger overall image (in terms of dimensionality) than the function used for building $\mathfrak{R}_2$. Since both functions have to map equivalent occurrence instances in the same way, this only happens if the set $\mathfrak{O}_1$ of occurrence instances for which $\mathfrak{R}_1$ is computed is larger (enumerates more dimensions) than the one leading to $\mathfrak{R}_1$ ($\mathfrak{O}_2$). In other words, $\mathfrak{O}_2$ can be viewed as a special case of $\mathfrak{O}_1$. In this case, we want the occurrence instances to be represented by the set representing $\mathfrak{O}_1$, because this opens up the opportunity to have *all* occurrence instances represented by a *single* set – in the reverse case, one might be left with a partition of $\mathfrak{O}_1$ that still has to be represented by some other relation. In terms of code generation, this means that not every piece of code is really minimized to the smallest possible dimension: if we can deduce that there are more occurrence instances doing a similar computation, we create only one set doing this computation everywhere instead of one set responsible for the special case of the calculations that are used twice and another set for the calculations that are only used once. Although this is not a fundamental problem, we may find ourselves creating unnecessarily large and inefficient programs, if we do not consider this point.

However, one can also make a point for preferring large code with lots of special cases – which can be implemented by sorting for smaller image dimensions: this represents the pure idea behind LCCP– the dimensionality of every computation is minimized as far as possible, resulting in several small sets that can be computed in a way that utilizes the resources in an optimized fashion (for example by using a larger, one-dimensional processor array than a smaller, two-dimensional one). Therefore, one might also choose to replace this algorithm by one in which the comparison is done in exactly the opposite manner.

The remaining comparison step in Algorithm 3.5.3 – comparing the occurrence number of the corresponding sets – is only a way finally to achieve an unambiguous mapping in the case that the first criterion did not yield a decision.

The algorithms above rely on an algorithm *ReduceImageDimension* to obtain a representative mapping in which the dimensions are rearranged in such a way that enumeration in a direction in which equivalent values are computed is somehow removed. In Algorithm 3.5.4, we reduce the dimensionality (actually, only the number of index dimensions that potentially enumerate several different index values) of a polyhedron by producing a function that maps occurrence instances α and β to the same point (after projection onto index dimensions), if they are equivalent ($(\alpha,\beta) \in \equiv_{(\Delta^{(s,f)} \cup \Delta^f,\,\mathrm{id})}$). This mapping is computed from a polyhedron defining a relation $\Lambda_= \subseteq \Lambda$ that represent a part of the equivalence relation $\Lambda \subseteq \equiv_{(\Delta^{(s,f)} \cup \Delta^f,\,\mathrm{id})} \subseteq \mathfrak{O}\mathfrak{I} \times \mathfrak{O}\mathfrak{I}$. With matrices M', M and $M_{\Lambda,\geq}$ defining the relation $\Lambda_= = \left\{ (\alpha,\beta) \middle| \begin{pmatrix} -M' & | & M \end{pmatrix} \cdot \begin{pmatrix} \alpha \\ \beta \end{pmatrix} = 0 \wedge M_{\Lambda,\geq} \cdot \begin{pmatrix} \alpha \\ \beta \end{pmatrix} \geq 0 \right\}$, Algorithm 3.5.5 produces a base of the kernel of M, $\ker(M)$. Ignoring the restrictions defined by $M_{\geq}$, $\ker(M)$ characterizes those occurrence instances β that are related to the same occurrence instance α. Therefore, it suffices to create a mapping Ξ so that its kernel – after projection onto dimensions that actually are to be enumerated (i.e., onto index dimensions) – is the same as $\ker(M)$, i.e., $\ker(\pi_{1,\ldots,n_{src}} \circ \Xi) = \ker(M)$.

Algorithm 3.5.4 [*ReduceImageDimension*]:
Input:

$\Lambda_= :$ linear relation $\Lambda_= \left\{ (\alpha, \beta) \middle| M_{\Lambda_=} \cdot \begin{pmatrix} \alpha \\ \beta \end{pmatrix} = 0 \right\}$ defining an equivalence relation ($M_{\Lambda_=} \in \mathbb{Z}^{q \times (n_{src}+n_{blob}+2)}$).

Output:

$\Xi_= : \mathbb{Z}^{n_{src}+n_{blob}+2} \rightarrow : \mathbb{Z}^{n_{src}+n_{blob}+2} :$
linear function mapping occurrence instances to their representatives wrt. $\Lambda_=$.

Procedure:

```
/* STEP 1: preparation */
/* we may ignore the target space restrictions, since we build a new target space ourselves. */
/* Let M ∈ Z^{q×(n_src+n_blob+2)} be the source space portion of matrix M_{Λ=}: */
for i = 1 to q
  for j = 1 to n_src + n_blob + 2
    M[i,j] := M_{Λ=}[i,j];
  endfor
endfor
/* STEP 2: compute basis of the kernel of M; */
(C, KernelBase) := CreateKernelBase(M, 0, −1);
/* we suppose the vectors returned by CreateKernelBase to be */
/* arranged in a specific manner (see page 114); */
/* in particular, the actual base vector for a vector κ ∈ KernelBase */
/* is given by the row permutation C · κ */
/* STEP 3: build matrices defining a complete change of basis; */
M_base := I_{n_src+n_blob+2, n_src+n_blob+2};
Let n_LCCP ∈ {1, ..., n_blob} be the number of the first pseudo parameter added
to hold the new dimensions needed for LCCP (see discussion on page 94);
currLCCPParam := n_LCCP;
scaleFactor := 1;
for k = 1 to size(KernelBase)
  /* we call the current kernel base vector κ */
  κ := KernelBase[k];
  /* swap the kernel base vector with a */
  /* unit vector for a one of the pseudo parameter dimensions */
  Let lastDim be the last dimension with a non-zero coefficient in κ,
    (∀j : j ∈ {lastDim + 1, ..., n_src + n_blob + 2} : κ[j] = 0) ;
  M_base[·, currLCCPParam] := κ;
  M_base[·, lastDim] := ι_{currLCCPParam};
  currLCCPParam := currLCCPParam + 1;
  /* in addition, if κ has two non-zero components, we may have to scale; */
  scaleFactor := scaleFactor · κ[lastDim];
endfor
/* STEP 4: build the actual mapping */
Let Ξ= : Z^{n_src+n_blob+2} → Z^{n_src+n_blob+2} : α ↦ scaleFactor · (C · M_base)^{−1} · α;
return Ξ=;
```

Step 1 of Algorithm 3.5.4 (*ReduceImageDimension*) represents an initialization using the input matrix $M_{\Lambda_=}$, ignoring scoping information of the linear relation. The kernel of matrix M represents the part of the equivalence relation to be represented by the result function.

Step 2 then uses Algorithm 3.5.5 (*CreateKernelBase*) to produce a matrix $KernelBase \in \mathbb{Z}^{(n_{src}+n_{blob}+2)\times \mathrm{rk}(M)}$, whose columns consist of base vectors of $\ker(M)$. This is the central optimization step, since it determines the shape of the transformed index space. We can think of the set $\ker(M)$ as being cut out of the index space of the instances of the particular occurrence at hand, reducing the dimensionality of this index space and thus the number of occurrence instances to be enumerated. The arguments in the call represent just an initialization and are not relevant at this point. The matrix $C \in \mathbb{Z}^{(n_{src}+n_{blob}+2)\times(n_{src}+n_{blob}+2)}$ represents a row permutation that has to be applied to the column vectors of *KernelBase* to obtain the actual base vectors. This is because Algorithm 3.5.5 guarantees the matrix *KernelBase* to be of a specific form, which is used in the next step. Intuitively, each row number r of a matrix in this form consists of three areas:

1. Row r starts with zeroes (which go as far as possible).

2. The next section consists of non-zero entries – below each such non-zero entry, there are only zeroes in the matrix (so we can solve for the non-zero entries of row r when creating base vectors for the kernel).

3. The last section starts at the column that represents the first non-zero entry in row number $r+1$ and may hold non-zero entries as well as zeroes. This last section determines a possible non-zero value v resulting from the current binding obtained from solving rows $r' > r$. The coefficients for the entries in the second section of row r have to be chosen so that they add up to $-v$ for each solution generated.

Formally, this matrix form can be described as follows:

- With $\mathit{KernelBase} = \begin{pmatrix} \kappa_1 & \dots & \kappa_n \end{pmatrix}$, for each c with associated vector κ_c, we have:

$$\big(\exists r_c : r_c \in \{1, \dots, n_{src} + n_{blob} + 2\} : (\forall c' : c' \in \{1, \dots, c-1\} : \kappa_{c'}[r_c] = 0)\big) \tag{3.32}$$

- With c, κ_c, r_c as above, r_{c-1} correspondingly, we have

$$\begin{aligned} &\big(\forall r : r \in \{r_c + 1, \dots, n_{src} + n_{blob} + 2\} : \kappa_c[r] = 0\big) \wedge \\ &|\kappa_c[r_c]| = \min(\{|\kappa_c[r']| \mid r' \in \{1, \dots, r_{c+1} - 1\} \wedge \kappa_c[r'] \neq 0\}) \end{aligned} \tag{3.33}$$

We have already encountered this form as basis of a simple heuristics to select a small scaling factor in the discussion of Example 3.27.

Step 3 makes use of this form by picking the dimension with the smallest absolute coordinate value as the last non-zero element $\kappa[\mathit{lastDim}]$ of this vector. Note that this is the point at which we could also decide to consider only base vectors that do not share any non-zero dimension (which is easily done by filtering the vectors appropriately). As we have already observed on page 99 in Section 3.5.2, we could then use unimodular transformations – at the expense of not detecting all possible reuses. However, this case is not considered in this thesis. Note further that the scaling factor gets unnecessarily large if the base vectors, i.e., the single columns of M_{base}, are not cancelled (i.e., if the row entries of the respective column vectors are not relatively prime). Additionally, in Step 3 of Algorithm 3.5.4, we assume that the LCCP pseudo parameters introduced for index space dimensions along which recomputations of the same values are encountered are coded as consecutive dimensions in $\mathbb{Z}^{n_{src}+n_{blob}+2}$ and start at some dimension number n_{LCCP}.

Step 4 finally computes the actual mapping. As we have seen before in Section 3.5.2 (on pages 99 and following), we may need to apply a scaling of the mapping produced in the previous step in order to ascertain that the transformed index space is still an integer lattice and can thus be enumerated using loops. The mapping produced in Step 3 is scaled by *scaleFactor* as computed above, and M_{base} is inverted. Due to the use of the column transformation C, however, calculating the inverse of M_{base} now means solving Equation (3.34) in

$$\begin{aligned} M_{\text{base}} \cdot X &= I_{n_{src}+n_{blob}+2, n_{src}+n_{blob}+2} \\ &\Leftrightarrow \\ (M_{\text{base}} \cdot C) \cdot (C^{-1} \cdot X) &= I_{n_{src}+n_{blob}+2, n_{src}+n_{blob}+2} \\ &\Leftrightarrow \\ {M_{\text{base}}}' \cdot Y &= C \text{ (with } Y = C^{-1} \cdot X) \end{aligned} \tag{3.34}$$

Algorithm 3.5.4 makes use of Algorithm 3.5.5 (*CreateKernelBase*) to obtain a matrix consisting of base vectors so that it is easy to obtain the right unit vectors for completing the base transformation in a way that heuristically reduces the scaling needed. This algorithm is presented next.

Algorithm 3.5.5 [*CreateKernelBase*]:
Input:
$M \in \mathbb{Z}^{q\times(n_{src}+n_{blob}+2)}$:
integer matrix.

$\epsilon \in \mathbb{Z}^{n_{src}+n_{blob}+2}$:
vector defining the current binding that guarantees that the product of all rows of M below row r with ϵ equals zero: $\big(\forall r' : r' \in \{r+1, \ldots, q\} : M[r', \cdot] \cdot \epsilon = 0\big)$.

$r \in \mathbb{Z}$: -1 on the first call; otherwise the row in matrix M for which we have to find further constraints in order to create new kernel vectors of M.

Output:
KernelBase :
array of base vectors of $\ker(M \cdot C)$, in the form required by Equation (3.32) and Equation (3.33)

C : matrix representing a row permutation that has to be applied to the result vectors in *KernelBase* in order to obtain the actual base vectors of $\ker(M)$.

Procedure:

```
if r < 0 then
/* STEP 1: initialization – assert that each row of M consists of */
/*  - a contiguous part filled with zeroes and */
/*  - a coniguous part filled with non-zero values */
   Let M' be the echelon form of matrix M;
   C := ColumnTransformation(M');
   (KernelBase, D) := CreateKernelBase(C · M', 0, rk(M));
   return (KernelBase, C);
else
   KernelBase := ∅;
   /* let end be the column before the first non-zero column of M[r+1,·], or, if r = rk(M), */
   /* the last column (n_src + n_blob + 2) of M */
   end := { min({c | M[r+1,c] ≠ 0}) − 1   if r < rk(M)
          { (n_src + n_blob + 2)            if r = rk(M) ;
   /* let start be the first non-zero column of M[r,·], or, if r = 0, the first column (1) of M */
   start := { min({c | M[r,c] ≠ 0})   if r > 0
            { 1                        if r = 0 ;
   /* STEP 2: decide which case we have for creating new non-trivial solutions */
   /* if we need a non-zero entry in the current row, we take the one in in column start */
   /* if r = 0 (which is no legal row), we either have to:   */
```

```
  /*  - return the current binding (if it is non-zero) or */
  /*  - create new non-trivial solutions by giving each coefficient of the columns start to end − 1 */
  /*    (which are the zero-columns of M) arbitrary values (we choose 1) */
  if r = 0 then
    if ε ≠ 0 then
      /* return the cancelled binding */
      return (ε / gcd(ε[1],...,ε[n_src+n_blob+2]), I_{n_src+n_blob+2, n_src+n_blob+2});
    else
      for c = start to end
        /* a non-trivial solution is given by the c-th base vector */
        KernelBase := append(KernelBase, ι_{(n_src+n_blob+2),c});
      endfor
    endif
  else
    /* STEP 3: ascertain this row to add up to 0 */
    /* in any case, a possible solution is to maintain the old binding; */
    /* get the value of this row – without the coefficients that we may still choose */
    restOfRow := M[r,·] · ε;
    /* if restOfRow evaluates to ≠ 0, we have to find */
    /* a non-zero entry whose coefficient we may still choose */
    if restOfRow ≠ 0 then
      lcmOfCoefs := lcm(restOfRow, M[r, start]);
      ε' := (lcmOfCoefs / restOfRows) · ε;
      ε'[start] := − lcmOfCoefs / M[r,start];
      (K, D) := CreateKernelBase(M, ε', r − 1);
      KernelBase := append(KernelBase, K);
    endif
    /* STEP 4: start the backward substitution with a new non-zero value; */
    /* if we have not yet chosen non-zero values for the binding */
    /* in the rows below, we need solutions that assert that this row adds up to 0 */
    /* – we always need two columns to obtain 0, and we have to use a coefficient ≠ 0 */
    /* once for each column between start and end */
    if ε = 0 then
      for colOffset = 0 to end − start − 1
        lcmOfCoefs := lcm(M[r, end − colOffset], M[r, start + colOffset]);
        /* Create a new binding */
        ε' := 0;
        ε'[start + colOffset] := lcmOfCoefs / M[r,start+colOffset];
        ε'[start] := − lcmOfCoefs / M[r,start];
        (K, D) := CreateKernelBase(M, ε', r − 1);
        KernelBase := append(KernelBase, K);
      endfor
    endif
  endif
  return (KernelBase, I_{n_src+n_blob+2, n_src+n_blob+2});
endif
```

Algorithm 3.5.5 (*CreateKernelBase*) is a recursive algorithm that should be called on a matrix in echelon form. Therefore, the first step consists of transforming the given matrix into echelon form. Step 1 does this initialization – asserting that the given matrix is in echelon form and computing a column transformation so that the resulting matrix will be in a form suitable for choosing the correct completion with a low scaling factor. This is done by calling Algorithm 3.5.6 presented below. The second argument of Algorithm 3.5.5 is a binding for the actual values in the vector currently under consideration as new base vector; it is therefore initialized to 0.

The echelon form of the matrix is then traversed row by row from bottom to top. The corresponding code is marked as Step 2 in the algorithm. For each recursive invocation, the current solution vector ϵ is passed through, with *KernelBase* consisting of the sequence of all solution vectors finally returned as column vectors.

Steps 3 and 4 are more tightly coupled. Let us assume that $M \in \mathbb{Z}^{q\times(n_{src}+n_{blob}+2)}$ is already in echelon form. Then rows 1 through $\mathrm{rk}(M)$ are non-zero, and the number of base vectors the algorithm has to return is $\dim(\ker(M)) = c - \mathrm{rk}(M)$. Therefore, we may compute a kernel by setting each coordinate of the $c - \mathrm{rk}(M)$ dimensions that do not start any step in the echelon form to a non-zero value in turn and always use the first non-zero entry of a row (the one that does start a step) to an appropriate value so that the row still adds up to 0. This is done in the loop in Step 4. In the following recursive invocations of *CreateKernelBase*, this new binding of ϵ may lead to one of the upper rows to evaluate to a non-zero value (for the summands considered so far). In this case, we set the coefficient for the first non-zero entry of this row in Step 3 so that the complete row adds up to 0 again.

In order to obtain this useful matrix form discussed above, Algorithm 3.5.5 permutes the columns of its input matrix with the column transformation obtained from Algorithm 3.5.6 (*ColumnTransformation*). As discussed above, this form defines three well defined sections per row so that one section defines the dimensions that can be used to create new base vectors and one section defines a possible non-zero value that has to be cancelled out to obtain a solution. In addition, the column transformation reduces the scaling factor heuristically by sorting the non-zero entries of a row according to their absolute values (so that the first non-zero value of a row has the smallest absolute value of the entries of that row section). We have discussed the problem of minimizing the scaling factor in Section 3.5.2 (page 99 and following).

Algorithm 3.5.6 [*ColumnTransformation*]:
Input:
$M \in \mathbb{Q}^{q\times(n_{src}+n_{blob}+2)}$:

matrix in echelon form, with each row cancelled as far as possible.

Output:
$C \in \mathbb{Z}^{(n_{src}+n_{blob}+2)\times(n_{src}+n_{blob}+2)}$:

matrix defining a column transformation C for M so that for all $r \in \{1,\ldots,q+1\}$, there is a $c_r \in \{1,\ldots,(n_{src}+n_{blob}+2)+1\}$ with

$$\begin{aligned}&\big(\forall c : c \in \{1,\ldots,c_r-1\} : (M\cdot C)[r,c] = 0\big) \land\\&\big(\forall c : c \in \{c_r,\ldots,c_{r+1}-1\} : (M\cdot C)[r,c] \neq 0\big)\end{aligned}$$

and $c_1 < \cdots < c_{q+1}$. In addition, with notation as above, $M\cdot C$ satisfies

$$|(M\cdot C)[r,c_r]| = \min(|(M\cdot C)[r,1]|,\ldots,|(M\cdot C)[r,c_{r+1}-1]|)$$

Procedure:

```
Let C ∈ ℤ^((n_src+n_blob+2)×(n_src+n_blob+2));
/* for each column, find the maximal row that has a non-zero coefficient */
/* in this column (or 1, if the whole column is 0) */
for c = 1 to (n_src + n_blob + 2)
  LastNonZeroRow[c] := max({r | M[r,c] ≠ 0} ∪ {1});
endfor
/* each column c with the same row entry LastNonZeroRow[c] now competes to be */
/* the first non-zero column of row r; */
/* we may sort them arbitrarily, but we choose to sort them according to |M[r,c]| */
newColumn := 1;
for r = 1 to q
```

```
    /* get the columns that compete for row r */
    𝔈 := ∅;
    for c = 1 to (n_src + n_blob + 2)
      if LastNonZeroRow[c] = r then
        𝔈 := 𝔈 ∪ {(|M[r,c]|, c)};
      endif
    endfor
    while 𝔈 ≠ ∅
      /* sort them according to their coefficient */
      /* (second sorting criterion is their original column number */
      /*  – i.e. we sort the pairs in 𝔈 lexicographically) */
      minColPair := min_≺(𝔈);
      /* we just need to set column newColumn of C to the minColPair[2]-th unit vector */
      C[·, newColumn] := ι_((n_src+n_blob+2), minColPair[2]);
      𝔈 := 𝔈 \ {minColPair};
      newColumn := newColumn + 1;
    endwhile
  endfor
  return C;
```

Algorithm 3.5.6 takes a matrix in echelon form. Its main purpose is to permute the columns of this matrix so that, for each row r, there is a contiguous area of non-zero elements, namely the elements for which r is the last row featuring non-zero coefficients, so that their value may be chosen freely when the bottom-up traversal of the matrix in Algorithm 3.5.5 reaches row r. On the other hand, the scaling factor is optimized if the smallest coefficient in such a contiguous non-zero area within a row of the resulting matrix is found in the leftmost column of that area. This is, because a larger coefficient will result in a solution with a smaller factor in the sum $restOfRow + \epsilon'[start] \cdot M[r, start]$ that has to add up to 0 in Algorithm 3.5.5 – and in this algorithm, every dimension up to the leftmost one in the non-zero area is in turn set to a non-zero value. Therefore, the entries of a row of M are sorted in increasing order of their absolute value in the `while`-loop.

Note that the algorithms presented here leave room for optimizations in many places. They are intensionally kept simple for the sake of the argument. Nonetheless, they implement reliable procedures for merging equivalent occurrence instances based on the theory presented in the previous sections. However, there is still one problem we have not yet addressed in detail – namely finding a correct execution order. We will consider this problem in the next section.

3.6 Scheduler Issues

The previous sections introduced methods for creating a COIG, a condensed representation of the values computed by a program fragment, in which occurrence instances representing the same value are merged to a single occurrence instance, or – as is the case for the change of basis – to a set of occurrence instances that differ only in well-defined coordinates, so that an appropriate projection finally represents them as a single point. The next step – as sketched in Figure 3.1 at the beginning of this chapter – is to create a space-time mapping. In particular, we need to define a legal execution order to obtain a semantically equivalent transformation. Therefore, it is necessary to compute a schedule that ensures that the resulting program defines a legal execution order so that all dependences are obeyed and the transformed program is semantically equivalent to the original one.

Section 3.5 introduced two different possibilities to find representatives for this new program representation. The first one, using the lexicographic minimum of equivalent points, as detailed in Section 3.5.1, already leads to a schedule by itself, so that finding a space-time mapping for the program is no problem. Even if the computed schedule may not be the best choice, it is clear that a schedule exists and that a space-time mapping for the program can be computed.

For the representative selection strategy presented in Section 3.5.2, this is not immediately clear, as we have hinted at in the last point that we have identified to be crucial for representative selection on page 87. On the contrary, it is evident that some scheduling strategies are bound to fail to create a legal schedule for the output program. One such scheduler is Lamport's hyperplane method [Lam74], which relies on all dependence vectors (i.e., the difference $\beta - \alpha$ between target β and source α of a dependence $(\alpha, \beta) \in \Delta$) to be positive, since it only ensures to leave the lexicographic order of dependence vectors constant. However, with the method presented here, the direction of a dependence may very well change, as the following example shows.

Example 3.29 *Consider the following code fragment:*

```
   DO i=1,100
!    Statement S1:
!    [5]      [4]  [1]      [3]    [2]
     B(i)      =   A(i+1)   **     2
!    Statement S2:
!    [10]     [9]  [6]      [8]    [7]
     C(i)      =   A(i)     **     2
   END DO
```

The original occurrence instance sets for the exponentiations (with only the interesting dimensions shown – index, then operand, and occurrence numbers) are $\left\{ \begin{pmatrix} i \\ 3 \\ 0 \end{pmatrix} \middle| 1 \leq i \leq 100 \right\}$ *(the the instances of* `A(i+1)` *in* S_1*) and* $\left\{ \begin{pmatrix} i \\ 8 \\ 0 \end{pmatrix} \middle| 1 \leq i \leq 100 \right\}$ *(the instances of* `A(i)` *in* S_2*). Assuming that the original occurrence numbers are retained, the rules laid out in Sections 3.5.2 and 3.5.3 transform these sets to:*

$$\left\{ \begin{pmatrix} i \\ 3 \\ 0 \end{pmatrix} \middle| 1 \leq i \leq 100 \right\} \quad \text{and} \quad \left\{ \begin{pmatrix} 1 \\ 8 \\ 0 \end{pmatrix} \right\}$$

This means that, for $i < 100$*, iteration* i *of Occurrence 9 now reads from iteration* $i+1$ *of Occurrence 3 (instead of iteration* i *of Occurrence 8). Thus, for this case, the dependence vector is*

$$\begin{pmatrix} i \\ 9 \\ -1 \end{pmatrix} - \begin{pmatrix} i+1 \\ 3 \\ 0 \end{pmatrix} = \begin{pmatrix} -1 \\ 6 \\ -1 \end{pmatrix}$$

and thus negative, in contrast to the distance vector for Occurrence 4, which evaluates to the usual vector

$$\begin{pmatrix} i \\ 4 \\ -1 \end{pmatrix} - \begin{pmatrix} i \\ 3 \\ 0 \end{pmatrix} = \begin{pmatrix} 0 \\ 1 \\ -1 \end{pmatrix}$$

This means that a scheduling method like Lamport's hyperplane method is not applicable to a reduced COIG generated by our method. However, our method still ascertains that the COIG does not contain any cycles (in other words, the reduced COIG does not contain any cycles with a weight of zero – when we adorn the edges in the reduced COIG with the h-transformations as weight). This is because the definition of the *LiftRel* operator (Definition 3.2) excludes write accesses from being equivalent. This means that any dependence cycle – which has to go through a read and a write access – can only exist in the COIG, if a corresponding cycle exists in the

OIG. The following expresses the same idea in closer detail. The following lemma states that, if there exists an h-transformation in the transformed program (for which we ignore dimensions of LCCP pseudo parameters) mapping from a representative of an occurrence instance α to the representative of a write access instance β, then the original program featured an h-transformation mapping α to β.

Lemma 3.30 *Let $(\mathfrak{O}\mathfrak{I}, \Delta)$ be an OIG. Let Ξ be a representative mapping obtained from the equivalence relation $\equiv_{(\Delta^{(s,f)} \cup \Delta^f, \mathrm{id})}$ by Algorithm 3.5.2 as described in Section 3.5.3. Further, let H be a piecewise linear mapping such that $H = \Delta^{-1}$, $\pi : \gamma \mapsto \pi_{1,\dots,n_{LCCP}-1,\dots,n_{LCCP}+n_{depth},\dots,n_{src}+n_{blob}+2}(\gamma)$ a projection that eliminates the LCCP pseudo parameters. We extend $\cdot^g$ canonically to linear relations as a piecewise linear version of the generalized inverse. With $\alpha, \beta \in \mathfrak{O}\mathfrak{I}$, where β is a write access, we have*

$$(\pi \circ \Xi \circ H \circ (\pi \circ \Xi)^g)((\pi \circ \Xi)(\alpha)) = (\pi \circ \Xi)(\beta) \Rightarrow H(\alpha) = \beta$$

Proof:

Without loss of generality, we assume only one dependence on which a target of a write access may depend. Definition 3.2 guarantees that

$$(\pi \circ \Xi)(\alpha) = (\pi \circ \Xi)(\alpha') \Rightarrow H(\alpha) = H(\alpha')$$

We now have

$$(\pi \circ \Xi \circ H \circ (\pi \circ \Xi)^g)((\pi \circ \Xi)(\alpha)) \quad = \quad (\pi \circ \Xi)(\beta)$$

β is a write access, and write accesses are never equivalent to one another – they are thus represented by themselves. Therefore, we have write accesses on the left hand side and the right hand side of the above equation. Thus, we may omit the application of $\pi \circ \Xi$ in both. Therefore, the equation transforms into

$$(H \circ (\pi \circ \Xi)^g)((\pi \circ \Xi)(\alpha)) \quad = \quad \beta$$

According to the definition of $(\pi \circ \Xi)^g$, this is equivalent to:

$$H(\alpha) \quad = \quad \beta$$

✓

Lemma 3.30 asserts that any occurrence instance α directly depending on a write access β in the original OIG is represented by an occurrence instance that, again, directly depends on the very same write access β. This does not only hold for an immediate successor of a write access, but also for occurrence instances that depend on that successor through equivalence propagating dependences, since, as the following Lemma states, a sequence of equivalence propagating dependences essentially stays just the same in the COIG as in the OIG.

Lemma 3.31 *With the notation as in Lemma 3.30, for each path in the COIG that consists of structural flow dependences, there is also a path in the original OIG. Formally:*

$$\pi \circ \Xi(\alpha) = \alpha' \Rightarrow (\pi \circ \Xi \circ H \circ (\pi \circ \Xi)^g)^n(\alpha') = \pi \circ \Xi(H^n(\alpha))$$

Proof:

$n = 0$ is the trivial base case. So, let $n > 0$. We recall that if α and β are targets of structural flow dependences, the definition of *LiftRel*, and thus the definition of equivalence on occurrence instances, guarantees that

$$\pi \circ \Xi(\alpha) = \pi \circ \Xi(\beta) \Rightarrow H(\alpha) = H(\beta) \tag{3.35}$$

With this in mind, we simply rewrite the above formula:

$$
\begin{array}{rcl}
(\pi\circ\Xi\circ H\circ(\pi\circ\Xi)^g)^n(\alpha') & = & (\pi\circ\Xi\circ H\circ(\pi\circ\Xi)^g)((\pi\circ\Xi\circ H\circ(\pi\circ\Xi)^g)^{n-1}(\alpha')) \\
 & = & (\pi\circ\Xi\circ H\circ(\pi\circ\Xi)^g)(\pi\circ\Xi(H^{n-1}(\alpha))) \\
 & & \text{due to induction hypothesis}
\end{array}
$$

Thus, we have:

$$
\begin{array}{rcl}
 & (\pi\circ\Xi\circ H\circ(\pi\circ\Xi)^g)(\pi\circ\Xi(H^{n-1}(\alpha))) = & \\
\pi\circ\Xi(H(& \underbrace{\underbrace{\underbrace{\underbrace{(\pi\circ\Xi)^g(\pi\circ\Xi(H^{n-1}(\alpha)))}_{\substack{=\eta \text{ with } \pi\circ\Xi(\eta)=\pi\circ\Xi(H^{n-1}(\alpha) \\ \text{(due to the definition of } (\pi\circ\Xi)^g)}}}_{\substack{=\eta \text{ s.t. } H(\eta)=H(H^{n-1}(\alpha))=H^n(\alpha) \\ \text{(due to Equation 3.35)}}}}_{\underbrace{=H(\eta) \text{ with } H(\eta)=H^n(\alpha)}_{=H^n(\alpha)}}}_{=\pi\circ\Xi(H^n(\alpha))} &)) = \pi\circ\Xi(H^n(\alpha))
\end{array}
$$

✓

With these lemmata, each cycle in the COIG can be traced back to a cycle in the OIG.

Theorem 3.32 *Let* $(\mathfrak{OI}/\equiv_{(\Delta^{(s,f)}\cup\Delta^f,\,\mathrm{id})}, \Delta_{\equiv_{(\Delta^{(s,f)}\cup\Delta^f,\,\mathrm{id})}})$ *be the COIG corresponding to the OIG* $(\mathfrak{OI},\Delta)$ *condensed via equivalence relation* $\equiv_{(\Delta^{(s,f)}\cup\Delta^f,\,\mathrm{id})}$. *If* $(\mathfrak{OI}/\equiv_{(\Delta^{(s,f)}\cup\Delta^f,\,\mathrm{id})}, \Delta_{\equiv_{(\Delta^{(s,f)}\cup\Delta^f,\,\mathrm{id})}})$ *contains a cycle, there is also a corresponding cycle in* $(\mathfrak{OI},\Delta)$.

Proof:

With the notation as in Lemma 3.30, we first observe that the mapping Ξ, as produced by Algorithms 3.5.2 to 3.5.6, cannot introduce any cycle in a subgraph that contains edges from $\Delta^e = \Delta^{(s,f)}\cup\Delta^f$ only, since Δ^{e+} is a strict partial order and occurrence instances that are mapped to the same point via $\pi\circ\Xi$ are always incomparable wrt. Δ^{e+}. Input dependences are completely eliminated by the transformation and cannot be part of a path (and, thus, a cycle) in the condensed graph; all other forms of dependences necessarily contain a write access as a vertex.

Without loss of generality, we assume again that each occurrence instance in the original OIG depends on at most one other occurrence instance due to a equivalence propagating dependence (there may be additional dependences of other classes).[8] [9] Let us now assume a cycle

$$\alpha' = (\pi\circ\Xi\circ H\circ(\pi\circ\Xi)^g)^n(\alpha')$$

According to Lemma 3.31, for each path consisting of structural flow dependences

$$(\pi\circ\Xi\circ H\circ(\pi\circ\Xi)^g)^m(\beta') = \gamma'$$

[8] This is not a problem since, the only point where this is not the case – the implicit dependences between the different operand positions of an occurrence and its operand 0 – are again either incomparable to each other wrt. Δ^{e+}, which builds the only dependence connection between them, or can be examined by simply chosing the right predecessor relation (e.g., consider the i-th input argument as a source for the j-th output argument of an occurrence instance).

[9] There is only one case in which a target occurrence instance α depends on several different source occurrence instances, and this is the execution of an operator (i.e., $\mathrm{OpId}(\alpha)=0$). In this case, the different sources are either incomparable wrt. Δ^{e+} or can be examined by simply chosing the right predecessor relation (e.g., consider all input arguments before output arguments). Therefore, this assumption does not introduce any restrictions.

and for all $\gamma \in \pi \circ \Xi^{-1}(\gamma')$, there is a $\beta \in \pi \circ \Xi^{-1}(\beta')$ and a corresponding path $H^m(\beta) = \gamma$ in the original OIG. Now, let $\delta' = (\pi \circ \Xi \circ H_0 \circ (\pi \circ \Xi)^g)(\gamma')$ be an edge that does *not* belong to a structural flow dependence. This leaves the following cases:

1. γ' is a read access, δ' is a write access (the dependence is a flow dependence).
2. γ' is a write access, δ' is a write access (the dependence is an output dependence).
3. γ' is a write access, δ' is a read access (the dependence is an anti dependence).
4. γ' is a write access, δ' is an operator execution (the dependence is a structural output dependence).

However, for each of these cases and all $\gamma \in \pi \circ \Xi^{-1}(\gamma')$, there is, again, an edge $\delta = H_0(\gamma)$ in the OIG with $\delta \in \pi \circ \Xi^{-1}(\delta')$:

1. With γ' a read access and δ' a write access, this edge exists in the OIG, according to Lemma 3.30 (with δ' being the write access β in the lemma).
2. With both γ' and δ' write accesses, again, the edge exists in the OIG according to Lemma 3.30.
3. With γ' a write access and δ' a read access, we again have $\gamma' = \pi \circ \Xi(\gamma) = \gamma$. There has to be some read access δ so that $\delta = H_0(\gamma)$ (otherwise, Algorithm 3.4.1 does not create a dependence relation). In fact, since all the occurrence instances in $\pi \circ \Xi^{-1}(H_0(\gamma))$ read from the same occurrence instance (according to the definition of equivalence), there has to be a path from γ to *all* the elements of $\pi \circ \Xi^{-1}(H_0(\gamma))$ in the OIG.
4. With γ' a write access and δ' an operator execution, we have once more the case that γ' did not get replaced: $\gamma' = \pi \circ \Xi(\gamma) = \gamma$. However, δ' may now be different from $H_0(\gamma)$. Still, if there were a direct cycle involving δ', Lemma 3.30 asserts that we would also have $H(\delta) = \gamma$ for all $\delta \in \pi \circ \Xi^{-1}(\delta')$, and thus the cycle $H(H_0(\gamma)) = \gamma$ in the OIG. If there is an indirect cycle, the next edge in the COIG would have to be a (sequence of) structural flow dependences. And, for these, Lemma 3.31 again asserts that there is a corresponding path in the original OIG.

✓

Theorem 3.32 guarantees that a COIG is acyclic, if the original OIG is acyclic. Therefore any scheduling method that needs only consistent (acyclic) dependence graphs as input will work with our method. The scheduling methods by Feautrier [Fea92a, Fea92b] and by Darte and Vivien [DV94] belong to this class [DRV00].

Note that the only reason that the transformation into an OIG can never introduce cycles is the fact that the definition of equivalence on occurrence instances specifically does *not* include write accesses. If write accesses could be equivalent to each other, there could be problems, since it may then not be clear, when to execute the resulting (merged) write access, as the following example shows.

Example 3.33 *In the assignment sequence*

```
!       Statement S1:
        A=B
!       Statement S2:
        A=C
!       Statement S3:
        D=A
!       Statement S4:
        A=B
```

the read access A *in Statement* S_3 *has to refer to the value* C *(otherwise a wrong value for* D *may leave the program fragment). On the other hand, after both statement* S_1 *and* S_4*, an access* A *has to refer to* B *(there may well be further computations involving* A *– and possibly also* D *– between* S_1 *and* S_2*), one may be inclined to view the occurrence instances representing these write accesses as equivalent. Figure 3.13 illustrates this situation. The left hand side depicts the OIG for the program fragment. Flow dependences are represented by solid arrows, the output dependence by a dashed arrow, and the anti dependence by a dashed and dotted arrow. The image on the right shows a COIG as it could result from application of LCCP– the two read accesses to B are equivalent, so that everything but the write access can be replaced in one of the statements. Figure 3.14 then shows an* illegal *COIG due to identifying write accesses with each other: the two write accesses of statements* S_1 *and* S_4*, are represented by a single occurrence instance there. The effect is a dependence cycle (fat lines) consisting of an output, a flow, and an anti dependence. Therefore, this dependence graph cannot be scheduled at all.*

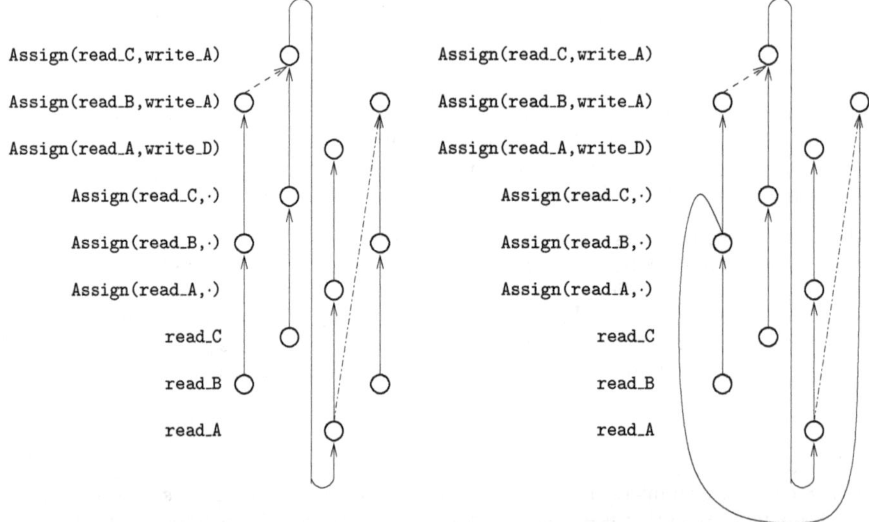

Figure 3.13: Left hand side: OIG of the code in Example 3.33. Right hand side: COIG for the same example..

When introducing additional, e.g., user defined, equivalences, this fact has to be observed. An implementation may still be possible, e.g., by node splitting, or by implementing only safe features, such as the `Fortran` `EQUIVALENCE` statement (which only needs equivalence on statements in dummy loops, and thus does not interfere with calculations). Nevertheless, this problem has to be taken into account when introducing such an extension. In this thesis, we will only consider the simple case and not handle such an extended notion of equivalence.

3.7 Code Generation

So far, the LCCP transformation has produced a COIG in the form of an optimized OIG that represents a loop program with reduced recomputations. It is now time to examine how this representation can be turned back into program code. It is not always straightforward how to

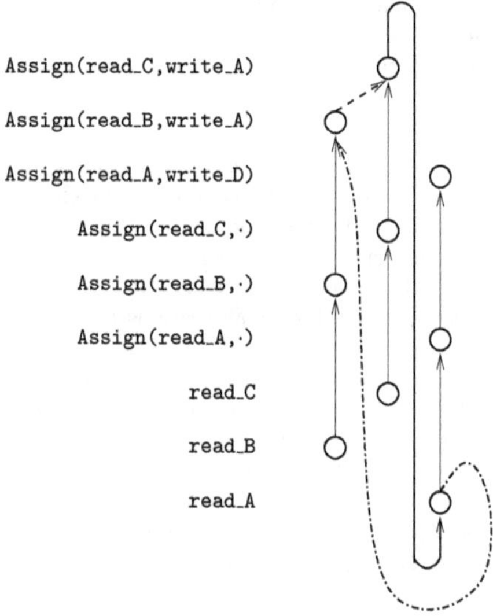

Figure 3.14: COIG with illegally merged write accesses.

produce code that can be compiled efficiently into an executable parallel program. Some central points of code generation have been identified in earlier work [FGL01b]. The two most important points are the regularity of the subscript functions and the regularity of the loop bounds, as pointed out in Section 2.1.2. We will now describe the principle of code generation in LCCP and some techniques to obtain efficient code.

We have established that a space-time mapping has to be computed for interesting sets in order to decide when and where to execute the newly created occurrence instances. And we have established that, when selecting representatives according to the change of basis approach presented in Section 3.5.2, it is absolutely *necessary* to employ a scheduling algorithm, and that scheduling methods like Feautrier's [Fea92a, Fea92b] and the scheduling algorithm by Darte and Vivien [DV94] can be applied to get the "time" part of this mapping. Chapter 4 will consider the placement (i.e., the "place" part) in more detail. For now, we will just assume to have some placement relation given. However, we will assume both the schedule and the placement to be defined by linear relations, i.e., neither necessarily has to be a linear function.

Most standard techniques in code generation in the polyhedron model are based on the assumption that the space-time mapping is a function rather than a linear relation [CF93, Len93, Xue93, Xue94, Wet95, Xue96, QRW00]. Bastoul extended the work of Quilleré, Radjopadhye and Wilde [QRW00] to a loop scanning algorithm that is capable of processing any linear relation as a space-time mapping [Bas03, Bas04]. Due to its flexibility and the low overhead in the produced code, this method, implemented in the tool CLooG, has been chosen for the code generation in the implementation of LCCP in LooPo. Nevertheless, any loop scanning technique can be used in conjunction with LCCP and the code generation technique described here.

In our case, the transformation is defined by a space-time mapping T and a generalized inverse of T, which we denote by $T^{g<}$. $T^{g<}$ is not only a generalized inverse but also a bijection between source and target index space. I.e., for each occurrence instance representative, there is a unique point in the target index space and, for each point in the target index space, there is a unique occurrence instance representative. This corresponds to Bastoul's extended enumeration method [Bas03] with the exception that Bastoul's method may select indices from the original index space in order to obtain a function from the given relation, viewing source and target space as one large polyhedron, while we make an explicit distinction between source and target space. Therefore, we have to introduce new index dimensions to the target space, here. We will take a look at the specific form of the function $T^{g<}$ later. For the moment, let us just assume that it is a generalized inverse of T.

General Loop Structure As pointed out in Section 2.1.2, `HPF` compilers strongly prefer rectangular loop nests. In the case of Version 7.1 of ADAPTOR, which we used for our experiments, the compiler is not able to create any optimized communication for non-rectangular nests of parallel (`INDEPENDENT`) loops. Restricting oneself to such simple loop nests is not a good approach for transformations in the polyhedron model. An inspection of ADAPTOR's run time library, DALIB, revealed the possibility of specifying array sections in DALIB communication routines that are larger than the referenced arrays. We were therefore able to create artificial rectangular `INDEPENDENT` loop nests and introduce comments that trigger a textual replacement of loop bounds created by ADAPTOR to the actual (non-rectangular) form in a postprocessing step. With the run time library's communication routines being able to handle out-of-bounds array accesses (which may now appear), the compiler is thus able to create quite efficient code, even for non-rectangular parallel loop nests. We also experimented with a more elaborate technique of tiling the index space with rectangular loop nests, until only loop nests enumerating the borders induce expensive communication. However, this latter technique turned out to have no advantage over the former.

The target index space defines $n_{tgt} = n_{time} + \#pdims + n_{\mathrm{ILP}} + n_{\mathrm{depth}}$ indices. These indices are divided into

1. n_{time} sequential dimensions as indicated by the scheduler,
2. #pdims parallel dimensions used by the placement method,
3. n_{ILP} additional potentially parallel dimensions,
4. n_{depth} pseudo parameters as introduced in Section 3.5.2 on page 94; these are partly enumerated in a loop and partly represented by an `if`-statement.

We propose to produce loop nests with a nesting that corresponds to this enumeration. For example, parallel dimensions have to be enumerated within sequential dimensions, since otherwise we cannot guarantee the correctness of the result program (without a sophisticated message passing scheme). The target code resulting from this nesting is sketched in Figure 3.15. In the following, the role of each index kind is be examined more closely.

n_{time} Schedule Loops The outer sequential loops represent the schedule. Since neither our target language `HPF` nor other prominent languages or language extensions like `OpenMP` offer explicit synchronization mechanisms short of explicitly specifying all communication using a communication library like `MPI` [Mes97] or `PVM` [Sun90], the output program has to be synchronous, i.e., synchronization occurs at least at the end of a sequential loop. Since the execution time of all the occurrence instances is defined wrt. the same time coordinates, all statements enumerate the same index variables.

***#pdims* Explicit Placement Loops** The loops within the sequential loops required by the schedule can be executed in any order insofar as correctness is concerned. However, they still enumerate (universally quantified) indices, i.e., they have to be enumerated completely. There may still be different levels on which it may be beneficial to aggregate the execution of these different occurrence instances to the same point in space:

1. On a distributed memory system, data is shared between different cluster nodes (resulting in very expensive communication).

```
      DO t_1=t_1_low,t_1_up
       ⋮
        DO t_n_time=t_n_time_low,t_n_time_up
!HPF$ INDEPENDENT, NEW(p_2,...,ilp_1,...,lccp_n_depth)
          DO p_1=p_1_low,p_1_up
           ⋮
!HPF$ INDEPENDENT, NEW(ilp_1,...,lccp_n_depth)
            DO p_#pdims=p_#pdims_low,p_#pdims_up
!HPF$ INDEPENDENT, NEW(ilp_2,...,lccp_n_depth)
              DO ilp_1=ilp_1_low,ilp_1_up
               ⋮
!HPF$ INDEPENDENT, NEW(lccp_1,...,lccp_n_depth)
                DO ilp_n_depth=ilp_n_depth_low,ilp_n_depth_up
                  DO lccp_1=lccp_1_low,lccp_1_up
                   ⋮
                    IF lccp_n_depth_low<=lccp_n_depth_up THEN
                     ⋮
! occ_op_dim is the occurrence number, operand number, and dimensionality of the
! occurrence instance set enumerated here
                      TMP_occ_op_dim(lccp_n_depth,ilp_n_depth,...,p_#pdims,...,t_1)=...
                     ⋮
                    END IF
                   ⋮
                  END DO
                END DO
               ⋮
              END DO
            END DO
           ⋮
          END DO
        END DO
       ⋮
      END DO
```

Figure 3.15: Basic structure of a generated loop nest.

2. Within a single node, there may be several CPUs, not necessarily uniformly connected to the memory banks.

3. Within a single CPU, modern processors may still feature different cores and processing units (e.g., integrated GPUs or support for multimedia extensions).

On all these levels, communication between dependent occurrence instances can slow down the computation, so that placing dependent occurrence instances on the same element of the respective structure may be beneficial – although the extreme case, using only one ressource, is most probably not advisable either, because it eliminates all parallelism on this level. Therefore, we suggest to produce the placement part of the space-time mapping through a sequence of runs of placement algorithms for each level of granularity – starting from the most expensive cost factor (which will be the inter-node communication in most cases). However, this approach goes beyond the scope of this thesis, so that we shall restrict ourselves to the inter-node communication usually in the focus of `HPF` parallelizations. A corresponding placement algorithm will be presented in Chapter 4. Since `HPF` is our primary target language, the placement algorithm is expected to use different target indices for the enumeration of occurrence instances that are placed on different processor arrays. Additionally, the bounds of the parallel loops directly indicate the position in this processor array at which the loop body is to be executed. This avoids the problems with inconsistent homes of loop bodies described in Section 2.1.2 on page 14 and produces efficient `HPF` code.

$n_{\mathbf{ILP}}$ Implicit Placement Loops Dimensions of the original source index space that are not associated with space or time dimensions have to be enumerated, nevertheless. Since they are independent of the schedule, these dimensions are marked as independent in Figure 3.15. However, since they are not considered by the placement, they can be specified as parallel only outside of the placement method, such as by multimedia instructions or by ILP that is automatically handled by the CPU itself. For our purpose, they are enumerated sequentially.[10]

$n_{\mathbf{depth}}$ LCCP Pseudo Parameters As mentioned in Section 3.5.2 (on page 94), LCCP pseudo parameters in general do lead to loops in the target code. Possible solution values for pseudo parameters may have to be enumerated until the first time that the statement corresponding to the considered occurrence instance set is executed for any given value of embracing loop indices. This is although pseudo parameters represent existentially quantified variables, because the values of these variables do not have to be defined by the environment of the program, and we may have to iterate through several possible solutions in order to find an example solution prooving the non-emptyness of the corresponding index space. Example 3.34 illustrates this case.

Example 3.34 *Reconsider Example 3.23. Let us now suppose that the value computed in the loop body of the code fragment is independent of the indices of the enclosing loops; in addition, we generalize the bounds of the outermost loop to m and n, respectively:*

```
DO i1=n,m
   DO i2=(i1-1)/3+1,(i1+1)/3
    B(i2,i1)=A(m)**3
  END DO
END DO
```

For the computation of $A(m)^3$, it suffices to know that there are indices i_1 and i_2 that comply with the loop bounds, execute the computation once and then copy the result value out to B. In Figure 3.9, where we need to determine whether an instance of `A(i1+i2)**3` *is to be executed by the program, this corresponds to searching a point with an integer i_2-coordinate on the dotted*

[10] Actually, in the implementation in LooPo, these loops are not only enumerated by sequential loops, but also *outside* of the parallel loops discussed in the previous paragraph. This is because, depending on the `HPF` compiler, declaring `INDEPENDENT` loops that do not enumerate an index in a subscript expression of an array access may confuse the compiler. Since none of the parallel loops carry any dependence, this is a legal transformation.

```
TEST_S1=.TRUE.
⋮
TEST_Sn=.TRUE.
⋮
⋮
i=o
DO WHILE(i<=u.AND.(TEST_S1.OR.....OR.TEST_Sn))
  ⋮
  IF(TEST_Sk) THEN
!   Statement Sj:
    ...
    TEST_Sk=.FALSE.
  END IF
  ⋮
END DO
```

Figure 3.16: General structure of a loop enumerating an LCCP pseudo parameter.

```
IF(CEILING(o)<=FLOOR(u)) THEN
  ⋮
END DO
```

Figure 3.17: A loop enumerating an LCCP pseudo parameter may be simplified to an `if`-statement under certain circumstances.

vertical lines originating from the area on the i_1-axis marked with a fat line. This area covers the values from m to n. However, as we have seen, the existence of such a point – within the region defined by the i_2-bounds, i.e., within the area between two straight lines through the long sides of the parallelogram of Figure 3.9 – depends on the exact value of i_1: it does not suffice to know that there is some i_1 within the defined bounds m and n.

Therefore LCCP pseudo parameters are, in general, enumerated just like indices. The general structure of such a loop enumerating $i \in \{o + j \cdot s \,|\, j \in \mathbb{N} \wedge o + j \cdot s \leq u\}$ is given in Figure 3.16. In order to be able to stop execution of these loops as soon as all statements have been executed once, they are enumerated as the innermost loops of the loop nest. For each statement S_k in the body of the loops, there is a test variable $TEST\_S_k$ that indicates whether S_k still needs to be executed. These variables are first set to $TRUE$, and reset to $FALSE$ in the same context in which S_k is executed; the following test in the (transformed) LCCP pseudo parameter loop then may shortcut this loop. Here, the initialization of the variables $TEST\_S_k$ is done only before the outermost loops binding a LCCP pseudo parameter (otherwise statements may get executed unnecessarily often, because the guarding variable has been reset). Note that this scheme leaves much room for further optimization. For example, it is not really necessary to guard each statement with an `if`-statement and an own variable – it suffices to use one variable for each basic block that is to be executed in the same context. In addition, if the body of a loop binding index i (with i as above) only contains loops that are independent of i, the i-loop can be simplified to an `if`, as shown in Figure 3.17. This case is actually very common (it appears, e.g., if the original code enumerates a rectangular index space). If all these loops can be replaced by `if`, there is no need for the test variables S_k and the corresponding control structures. These cases can be detected easily and taken advantage of in the code generation phase.

One Additional Placement Dimension We assume one additional dimension in the output of the placement algorithms. This dimension does not appear in the enumeration of loops above. The reason for this is that its natural representation is rather a variable declaration than a loop: in the model of nested placement methods, one placement method has to be responsible for the placement of the result value in the memory of the process executing a given occurrence instance. Aside from the loop indices, the place in memory depends on the array the computed value is to be stored in. Of course, there must not be any conflict between write accesses to such an array element within the lifetime of these accesses. Therefore, a safe choice is, for example, to allot the arrays according to the occurrence and operand numbers and dimensionalities of the occurrence instance sets to be stored. In Figure 3.15, the corresponding numbers are appended to the prefix `TMP_` to accomplish this allotment. A more elegant method may be, for example, a greedy colouring scheme as proposed in Cohen's dissertation [Coh99, Section 5.3.5].

Computation Statements Code generation for LCCP consists of generating the loops via one of the established polyhedron scanning techniques (with LCCP pseudo parameters disguised as the innermost loop indices), replacing and simplifying the loops enumerating pseudo parameters as discussed above, and finally filling placeholders for each statement with the code to be executed. Note that only dimensions with finite bounds have to be enumerated (with an infinite value representing dimensions of non-embracing loops, as in Section 2.3). In general, all values calculated by an interesting set have to be assigned to new array variables. Remember that we assume a single polyhedron representing a set of occurrence instances to have a unique occurrence number (see Section 2.3.1). Therefore, an interesting set, which also has only one occurrence number, defines the application of the same operator on different sets of data, depending on the iteration. So, the general form of the code for an interesting set $\mathfrak{O}$ with $o \in \mathrm{OccId}(\mathfrak{O})$ and $\mathrm{Occ}^{-1}(o) = G(\mathrm{i}_1), \ldots, \mathrm{i}_{\mathrm{ArityIn}(G)}, \mathrm{o}_1, \ldots, \mathrm{o}_{\mathrm{ArityOut}(G)})$ with n finite index coordinates (i.e., embraced by n loops with loop indices $i_1, \ldots, i_n$) is basically the same for all $\alpha \in \mathfrak{O}$. Depending on the operator G, the output looks roughly as follows:

read access $\mathtt{A(}F(i_1, \ldots, i_n)\mathtt{)}$:
: A read access is always represented in the straightforward way.

write access to $\mathtt{A(}F(i_1, \ldots, i_n)\mathtt{)}$:
: A write access, in general, reads from an interesting set, which creates new variables for storing intermediates (see below). Therefore, the implementation is to look up that newly created auxiliary variable:

  $\mathtt{A(}F(i_1, \ldots, i_n)\mathtt{)=TMP\_occ\_op\_dim'(}\mathrm{Idx}(H(\alpha))\mathtt{)}$

  However, if there is a unique write access that depends on each of the elements of a given interesting set, we can optimize this assignment away by replacing the write access to the auxiliary variable by the access to `A`.

general operator application:
: For each output operand (including operand 0 – the function's return value), a new array is created that stores the corresponding output value:[11] Note that this generates essentially a program in single assignment form for the intermediate values computed in the COIG.

  $\mathtt{TMP\_occ\_op\_dim(in,\ldots,i1)=G(}\mathit{createTerm}(\mathrm{OccId}(H(\mathit{OpndSel}(-\mathrm{ArityIn}(G), o))))\mathtt{,}$
  $\quad \mathtt{\ldots,}$
  $\quad \mathit{createTerm}(\mathrm{OccId}(H(\mathit{OpndSel}(-1, o))))\mathtt{,}$
  $\quad \mathtt{TMP\_occ\_1\_dim(in,\ldots,i1),}$

[11] Note that each output operand is an element of an interesting set, so that there will be a space-time mapping for it, with a placement defining the array to be written to. Note further that, up to this point, the space-time mapping can be copied from α.

```
...,
TMP_occ_ArityOut(G)_dim(in,...,i1))
```

For each input operand, the corresponding h-transformation is used to obtain the (unique) source. If this source is itself an element of an interesting set, a lookup to the corresponding auxiliary variable is created. Otherwise, we simply reiterate the above scheme.

createTerm is here a recursive call of the function creating output code for a given set of occurrence instances – whose properties we have just described. Note that the output code here is completely defined by the COIG. In particular, although instances of write accesses represent interesting sets and although the target program features assignments and write accesses for every interesting set, there are no additional write accesses defined in the reduced COIG for the interesting sets. They are merely *implemented* in this way in the target program. A write access α in a reduced COIG represents the programmer's wish to store the value from which α depends in a variable, probably for future use outside the considered program fragment.

The h-transformations above may not hold on the complete index space in which the code is executed. Thus, the corresponding code may have to be guarded with conditionals that test for the domain of the h-transformation. This can be achieved by something similar to the `?`-operator in `C`. However, operators like this are likely to irritate compilers (and in the case of `HPF`, there is not even such an operator). Therefore, we choose to create a hierarchy of `if`-statement around the code to be generated that creates a version for each possible combination of sources.

The h-transformations above are defined in the target index space of the program (consisting mainly of time and placement dimensions). This holds for *all* linear expressions in the program text. When creating such linear expressions defined by a linear function F in the produced code following a space-time mapping T, the new expression is defined by

$$F' = F \circ T^{g<}$$

This new function F' then defines linear expressions in time and processor coordinates. Therefore, the way $T^{g<}$ is formed for a space-time mapping T has an impact on the final expression generated: with the target space also containing implicit placement loops as discussed above, T is invertible. However, it may be overdetermined as in the following example.

Example 3.35 *Consider a space-time mapping defined by the (simplified) matrix*

$$\begin{array}{c} \\ t \\ p \\ n \\ 1 \end{array} \begin{array}{c} \begin{array}{ccc} i & n & 1 \end{array} \\ \begin{pmatrix} 1 & 0 & 0 \\ 1 & 0 & 0 \\ 0 & 1 & 0 \\ 0 & 0 & 1 \end{pmatrix} \end{array}$$

for a code fragment like the following:

```
DO i=1,n
  A(i)=B(i,i)*A(i-1)
END DO
```

I.e., the subscript function in the read access to B is defined by a matrix $\begin{pmatrix} 1 & 0 & 0 \\ 1 & 0 & 0 \end{pmatrix}$. *The corresponding code in the output program may well read* `B(p,p)`, `B(t,t)`, *or even* `B(t,p)`. *Which* *one to choose now depends on $T^{g<}$:*

$$\begin{pmatrix} 1 & 0 & 0 \\ 1 & 0 & 0 \end{pmatrix} \cdot \begin{pmatrix} 1 & 0 & 0 & 0 \\ 0 & 0 & 1 & 0 \\ 0 & 0 & 0 & 1 \end{pmatrix} = \begin{pmatrix} 1 & 0 & 0 & 0 \\ 1 & 0 & 0 & 0 \end{pmatrix}$$

corresponds to `B(t,t)`

$$\begin{pmatrix} 1 & 0 & 0 \\ 1 & 0 & 0 \end{pmatrix} \cdot \begin{pmatrix} 0 & 1 & 0 & 0 \\ 0 & 0 & 1 & 0 \\ 0 & 0 & 0 & 1 \end{pmatrix} = \begin{pmatrix} 0 & 1 & 0 & 0 \\ 0 & 1 & 0 & 0 \end{pmatrix}$$

corresponds to `B(p,p)`

Note that the options `B(t,p)` and `B(p,t)` do not occur here, although they are also valid representations of the same expression. This is because both appearences of i (which are both represented as $\begin{pmatrix} 1 & 0 & 0 \end{pmatrix}$) are subject to the same space-time mapping defined by T. Therefore, the target expression for both is determined by the same multiplication $\begin{pmatrix} 1 & 0 & 0 \end{pmatrix} \cdot T^{g<}$ and, therefore, both yield the same target expressions.
This is an opportunity for the code generation phase to tune subscript expressions to the need of the subsequent compiler run. We construct the matrix for $T^{g<}$ so that each dimension that can be chosen freely prefers parameters to sequential loop indices, which, themselves, are prefered to indices of implicit placement loops, which are in turn prefered to genuine parallel loop indices. This alleviates the work for an `HPF` compiler further, as it simplifies the subscript functions as discussed in Section 2.1.2, because it reduces the possibilities for several parallel indices in a target subscript expression.

In a similar approach, the order of the dimensions of the auxiliary arrays can be chosen so that the write accesses to auxiliary arrays are executed in a cache-friendly manner (i.e., for `Fortran`-like languages, the time dimensions should be enumerated in the last subscript dimensions, for `C`-like languages in the first subscript dimensions).

Further Optimizations A further straightforward optimization consists of the composition of subscript functions for the auxiliary variables: since the order of loops is always the same, the write accesses to auxiliary variables can always be subscribed with the embracing loop indices in a cache-friendly manner (in the order of the embracing loops for `C`-style languages and in the reverse order for `Fortran`-like languages). In addition, it is not necessary to introduce any dimension in the auxiliary array for pseudo parameters, since each pseudo parameter is existentially quantified and thus always defines a single coordinate.

As mentioned already, the introduction of new auxiliary arrays, as sketched in Figure 3.15, leads to an single assignment form of the portion of the COIG that computes the intermediate values. This also means that vast space is occupied by these auxiliary arrays. In order to reduce space consumption, storage memory optimizations as the one proposed by Lefebvre [LF98] (with Feautrier) and improved in joint work with Cohen [CL99, Coh99] should be applied to the auxiliary arrays created.[12]

Finally, the control structures enumerating LCCP pseudo parameters may be completely eliminated if the existence of an integer solution can be shown. For example, if an index space $\left\{ \begin{pmatrix} i \\ j \end{pmatrix} \middle| \, 1 \leq i, j \leq n \right\}$ is to be enumerated, where j is actually a pseudo parameter, it suffices to enumerate the space $\{i \,|\, 1 \leq n\}$, since a solution for i implies a solution for j. An optimization like this can be achieved by quantifier elimination using a feasibility test such as PIP [Fea03] or the `Omega` test [Pug92] (for further discussion of quantifier elimination applications, see the work by Größlinger, Griebl and Lengauer [GGL06]). This optimization may be part of the scanning method, or a pre- or postprocessing step thereof.

[12]Unfortunately, our implementation of LCCP in LooPo does not include this optimization, since work on this module could not be finished in time.

Let us conclude with a simple – synthetic – example to see the basic LCCP transformation at work.

Example 3.36 *Consider the following example:*

```
!HPF$ INDEPENDENT
      DO i1=1,n
!HPF$ INDEPENDENT
        DO i2=1,n
          A(i1,i2)=A(i1,i2)-B(i1)*B(i1)+B(i1)
        END DO
      END DO
```

The read accesses `B(i1)` *in the same* i_1*-iteration all represent the same value, because they read from the same write access. This write access takes place in a dummy loop, an artificially added loop representing initial input data introduced in Section 2.3.2.* `B(i1)` *always represents the same value, because the* i_2*-dimension is a kernel dimension of the function* $\begin{pmatrix} i_1 \\ i_2 \end{pmatrix} \mapsto i_1$. *One level higher along the structural flow dependences, it is clear that the different* i_2*-iterations of* $B(i_1) \cdot B(i_1)$*, correspondingly, represent the same value, since they both read from the same (unchanged) array elements. At this point, going one level higher in the operator tree, again, the precedence used by the compiler determines the final result:*

- *One may observe that the whole expression* `-B(i1)*B(i1)+B(i1)` *stays constant for different* i_2*-iterations and arrive at a loop program like the following one:*

```
!     time step 0
      t1=0
!HPF$ INDEPENDENT
      DO p1=1,n
          TMP1(p1)=B(p1)
      END DO
!     time step 1
      t1=1
!HPF$ INDEPENDENT
      DO p1=1,n
          TMP2(p1)=-TMP1(p1)*TMP1(p1)+TMP1(p1)
      END DO
!     time step 2
      t1=2
!HPF$ INDEPENDENT
      DO p1=1,n
!HPF$ INDEPENDENT
        DO p2=1,n
          A(i1,i2)=A(i1,i2)+TMP2(p1)
        END DO
      END DO
```

- *There is also the possibility that the compiler parses the right hand side of the assignment as* `(A(i1,i2)-(B(i1)*B(i1)))+B(i1)`*, in which case the term* `A(i1,i2)-(B(i1)*B(i1))` *is already two-dimensional so that different* i_2*-iterations never represent the same value. In this case, a program like the following is generated:*

```
!     time step 0
      t1=0
```

```
!HPF$ INDEPENDENT
      DO p1=1,n
          TMP1(p1)=B(p1)
      END DO
!     time step 1
      t1=1
!HPF$ INDEPENDENT
      DO p1=1,n
          TMP2(p2)=TMP1(p2)*TMP1(p2)
      END DO
!     time step 2
      t1=2
!HPF$ INDEPENDENT
      DO p1=1,n
!HPF$ INDEPENDENT
        DO p2=1,n
          A(i1,i2)=A(i1,i2)-TMP2(p1)+TMP1(p1)
        END DO
      END DO
```

Note that, in both versions, B is copied into an auxiliary variable, although this is not necessary. But, in both cases, there are several distinct targets of flow dependences from the same occurrence instance reading B – still, they appear in the same statement in the first case. This identifies the occurrence instances that represent read accesses to B as an interesting set and forces us to scrutinize the possible execution times and places for these occurrence instances more closely.

In this particular case, let us assume the following initial distribution:

```
!HPF$ TEMPLATE TEMP1(1:n,1:n)
!HPF$ DISTRIBUTE TEMP1(BLOCK,BLOCK)
!HPF$ ALIGN A(i1,i2)       WITH TEMP1(i1,i2)
!HPF$ ALIGN B(i1)          WITH TEMP1(i1, 1)
```

The placement method presented in the next chapter determines a distribution according to the following scheme for the second version of the LCCP output:

```
!HPF$ ALIGN ARRAY1(i1,i2) WITH TEMP1(i2, *)
!HPF$ ALIGN ARRAY2(i1,i2) WITH TEMP1(i2, *)
```

I.e., the computation of $B(i_1) \cdot B(i_1)$, which is independent of i_2, is replicated along the processor array dimension onto which the different i_2-coordinates are distributed. Thus, the value is still computed for several i_2-coordinates. Nevertheless, the second LCCP version indeed performs better than the original version by about 3.5%–7% for $n = 7000$ and 8 down to 2 nodes in our experiments. This is because the replication along this axis means that only one such computation has to be done for each block *of i_2-coordinates computed on a specified physical processor, which may still be less than the original computations due to the tiling done by the* `HPF` *compiler.*

3.8 Summary and Future Work

The basic goals of LCCP are the optimization of the workload in a deep loop nest and a better utilization of parallel processors by reducing the depth of transformed loop nests containing parts of the original computation. These are important opportunities for optimizations in cluster (and other parallel) architectures for high performance computing, multicore CPUs [Ram06] that need parallel computing directives to make full use of their computation capabilities, and can also be used in stream processing compilers [TKA02, GTA06, GR05], which in turn can contribute to performance improvement, especially of processors with integrated GPUs as a SIMD-extension unit [Hes06].

The interesting sets identified by LCCP represent the vertices defined by a reduced OIG that deserve closer examination. As a first step, they represent sets with a minimized dimensionality whose results may be used by other – possibly higher-dimensional – interesting sets. The initial optimization – computing the occurrence instances of an interesting set and assigning the result values to auxiliary variables for future use – leaves plenty of room for further optimizations, especially since the storing of data is an expensive operation itself – in both memory space and execution time – and should not be employed without reason.

One possible solution for performance degragation due to this introduction of additional references to main memory is to bring source and destination of dependences closer together again. One such way will be examined in the next chapter. We have already seen the result in Example 3.36. Although the LCCP version in this example actually performed better, there are also cases in which it is better to undo the work of LCCP completely and avoid the overhead of storing data. This may be achieved, e.g., by inspecting the transformed dependences and removing the status of interesting set from a set of occurrence instances if the dependence relation is "too simple" wrt. some metric (e.g., choosing a projection with which the h-transformation reduces to an identity function).

An additional variation would be to clone interesting sets into several collections of almost identical sets (with only the occurrence number changed to indicate the difference). This would in part undo the transformations achieved by LCCP, so there should be some mechanism to determine whether a set should be cloned or not.

One iteration of LCCP corresponds to the compilation of a conflict equation system in the usual dependence analysis in the polyhedron model. Although the results provided by LCCP are directly only useful for a *must conflict* approximation, an iterative dependence analysis based on the result of LCCP could increase the applicability of code motion techniques further so that at least some irregular computations may be handled efficiently by automatic parallelizers. Another way to increase the applicability of LCCP– and at the same time reduce the number of occurrence instance sets produced by LCCP– is to ignore domain restrictions even further during the selection of a common representative mapping in Section 3.5.2 (see pages 103*ff*). Occurrence instances that are originally mapped to different sets may thus be mapped to the same set of representatives.

Probably most important, however, is the fact that LCCP does not introduce equivalences on different instances of write accesses. This restriction prevents many possible optimizations. In particular, LCCP can be used to identify all the points automatically that lend themselves to the storage of intermediate values. As discussed in Section 3.6, this restriction is not to be overlooked, since the correctnes of the transformation depends on it. However, it can be lifted if we introduce new occurrences for occurrence instances that write only intermediate values (i.e., those that are not sources of dependences to dummy loops at the end of the program fragment). The LCCP algorithm can then be called repreatedly with the set of equivalence propagating dependences alternating between structural output dependences on the one hand and (structural and non-structural) flow dependences on the other hand.

Similarly, user defined equivalences could be taken into account, which could increase the applicability, and thus the value of the LCCP transformation, greatly. Similarly, associativity and commutativity may be supported in a straightforward way.

3.9 Related Work

Code optimization techniques have a long history. Methods like loop-invariant code motion or strength reduction have been standard techniques in the context of scalar variables for a long time [MR79, ASU86, KRS94, Mor98]. LCCP can be viewed as an extension of the original code motion technique to the case of arrays referenced in `for`-loops.

Extended Code Motion There exist several extensions of the traditional, scalar-based technique: Rüthing, Knoop, and Steffen extend partial dead code elimination on scalars by a semantic component building equivalence classes of representations of the same value [SKR90, RKS98]. These equivalence classes – which are still based on scalars – do not directly correspond to those used in the LCCP method in its present form. In their semantic code motion algorithm, assignments are analyzed explicitly in order to identify values that are equivalent, although possibly reading from a completely different set of variables. Such an analysis is not supported by the LCCP method, as described above. However, a similar effect could be achieved with an iterative application of LCCP and a subsequent dependence analysis.

Agrawal, Saltz and Das extend the traditional fixed point analysis in that they build a graph called the **Full Program Representation** (FPR) [ASD95]. The FPR connects function calls to the same function within different procedures. This leads to an *interprocedural* code motion technique, called **Interprocedural Partial Redundancy Elimination** (IPRE). IPRE is targeted at assigning very expensive function calls to different procedures: call sites of the same function in several procedures A and B may be deleted due to a new call site in some other procedure C. This is in contrast to LCCP, which is presented here only as an intraprocedural optimization. The fixed point approach of their technique handles loops and other control structures simply as unpredictable and thus yields quite coarse dependence information. In later work, Agrawal proposes another code motion strategy called **Interprocedural Balanced Code Placement** (IBCP) based on the same framework. ICBP calculates predicates which define the movement of start and end statements for asynchronous (split-phase) functions, such as checkpointing [Agr99]. Parallelizing compilers face a similar problem when generating communication statements within a procedure or for redistribution at procedure boundaries. In this context, IPRE can be used to implement message coalescing. Asynchronous communication routines can take advantage of IBCP to overlap communication with computation once they are extracted from enclosing loops. In contrast, the focus of LCCP is to extract subexpressions (that lead to communication) from as many loops as possible within a procedure.

Common Subexpressions with Arrays Dehbonei describes in his dissertation [Deh90] a generation algorithm for parallel loop code loosely based on a parallel code generation algorithm by Allen, Callahan, and Kennedy [ACK87]. Dehbonei's code generation algorithm determines syntactically equivalent common subexpressions containing arrays in loop nests and hoists these subexpressions out of a loop, if the loop index is not contained in the array subscripts. However, this work predates an exact dependence analysis on loops and thus is not able to analyze more complicated expressions. E.g., it fails to recognize `A(i1+i2)**2` as an expression that can actually be computed in a one-dimensional loop.

Polyhedron Scanning in `HPF` compilers `HPF` compilers like the dHPF compiler [MCA97, AJMCY98] have long improved their communication code through optimizations like **message coalescing** (sending the same array element only once to a target processor) and **message vectorization** (sending a sequence of array elements in one message) [HKT91]. While earlier compilers used a purely syntactic approach to communication generation, newer versions moved to polyhedron scanning techniques in order to copy data speedily into a communication buffer [ACIK95, AMC98]. The LCCP transformation can make some of these optimizations explicit in that read accesses to the same memory cell are represented by the same occurrence instance in the COIG. Thus, some message coalescing can be done via an application of LCCP: the occurrences for which communication should be generated are identified by the interesting sets computed. Communication code generation for `HPF` compilers is thus simplified and can be improved substantially, in particular for compilers that feature efficient communication code generation for simple access patterns such as subscript triplets but only less efficient methods for arbitrary linear subscripts (which is the case, e.g., for ADAPTOR and the Portland Group's `pghpf` compiler).

Subspace Model Knobe and Dally introduce a so-called **subspace model** [KD95], which is described in greater detail in Knobe's dissertation [Kno97]. This model essentially relates a subexpression with the set of indices found in this expression, ignoring loop bounds. Index space restrictions that may limit the domain of flow dependences (and thus the validity of equivalences) are modelled by converting the code to SSA form prior to applying the methods of this model. Operations are executed in a loop nest whose depth corresponds to the number of indices associated with a given subexpression, reducing the number of dimensions to be enumerated, and different subexpressions that can be proven to be associated with the same range of arrays and subscript values are combined into a single operation. The model also handles non-affine subscript functions. However, since a dedicated affine analysis is *not* employed, the equivalence of array accesses cannot be established if these subscripts differ in some respects other than a permutation of index names or a constant shift. Thus, expressions like `A(i1+i2)**2` are, again, not identified as essentially one-dimensional. Neither can a term like `C(i,i)` be identified as a special case of the term `C(i,j)`.

Common Subexpressions in Numerical Codes A very powerful common subexpression elimination (CSE) system called DICE (**D**omain-shift **I**nvariant **C**ommon subexpression **E**liminator) is part of the CTADEL compiler [vE98]. CTADEL is a language for numerical codes based on a restricted SSA form. This simplified input format almost eliminates the need for a dependence analysis and simplifies the procedure for determining equivalent expressions. Equivalence is determined by a set of rules for a term rewriting system which even exploits associativity and commutativity laws. Thus, the method is able to find a range of equivalences that escape the presented LCCP method. However, a methodical reduction of index space dimensions as for the example above (`A(i1+i2)**2`) is not within the focus of that work. CSE may increase communication overhead due to the fact that values are stored in memory at their definition sites in order to be reused later. Therefore, DICE features switches with which the user can decide how aggressively the CSE algorithm should operate.

Wonnacott's Dead Code Elimination Wonnacott extends dead code elimination to arrays by removing elements from the set of operations to be executed, if they are not in the domain of a dependence relation (i.e., in our framework in the image of an h-transformation) [Won01]. This strategy fits well into the framework presented here: there is no reason to keep track of the set $\mathfrak{O}\mathfrak{I}$ of occurrence instances to be executed – it suffices to execute the pre-image sets of the h-transformations defining dependence relations. We do not explicitly consider this case. Instead, our approach removes computations of values that are needed but already computed. However, Wonnacott's approach can be implemented as part of our code generation phase.

Program Equivalence and Template Recognition Barthou, Feautrier, and Redon examine the problem of comparing two SAREs [BFR02], which is in essence the same problem as deciding whether two occurrence instances belong to the same equivalence class in LCCP. However, the idea behind this transformation is conceptually orthogonal to the approach discussed in this chapter. In contrast to LCCP, they compare a code fragment to some template code fragment stored in a database so that the code in question can be replaced, e.g., by a simple library call to an optimized version of the same calculation. In order to do so, they create a **memory state automaton** (**MSA**) – a finite state automaton that is augmented by index space information – from the cross products of the MSAs representing the two SAREs. Equivalence of the SAREs can then be rephrased as the reachability problem in the combined MSA [BFR02, AB03]. Due to the undecidability of the equivalence of two SAREs, this is a semi-decision procedure. Christophe Alias expands this to a complete framework for template substitution [AB05, Ali05], extending the comparison algorithm to a matching algorithm so that subterms may be left unspecified in a template. The decision algorithm presented is very powerful and able to detect equivalences not covered by LCCP, such as reduction recognition. On the other hand, LCCP can be extended easily to include associative-commutative operators by standard techniques in computer algebra (introducing n-ary operators and sorting the operands accordingly) [Che86]. These operators cannot be handled as easily by the template matching algorithm. While the template matching approach recognizes known patterns in program fragments, LCCP identifies equivalent portions of arbitrary code fragments and creates directly optimized code from the given source program.

Reduction Optimization Data reuse may lead to the computation of a series of similar values that may be obtained from each other. For example, in the code

```
DO i=1,n
  Y(i)=0
  DO j=1,i
    Y(i)=Y(i)+X(j)
  END DO
END DO
```

X is reused along the i-axis. This code can also be expressed as the following SARE:

$$Y(i) \quad = \quad (\textstyle\sum j : j \in \{1,\ldots,i\} : \; X(j)) \tag{3.36}$$

The reuse of X in this case implies that the elements of Y can be obtained from each other by adding or subtracting corresponding values of X:

$$Y(i+1) \quad = \quad Y(i-1) + X(i+1) \tag{3.37}$$

Equation (3.37) can then be computed in a single scan. Following this idea, which is profoundly similar to the approach of LCCP, Gupta and Rajopadhye propose a method for optimizing reductions [GR06]. In contrast to our method, their approach is solely based on the simplifications of reductions, which have to be defined in a form like the following:

$$Y \quad = \quad \mathrm{reduce}(+, (\begin{pmatrix} i \\ j \end{pmatrix} \mapsto i), X) \tag{3.38}$$

This form can be produced from the SARE of a program (which, in turn, is based on the program's SSA form) and is rewritten into equivalent definitions in this method. Problems due to reassignments etc. do not come into play, here. LCCP works directly on the arrays of the program and simplifies general expressions based on Herbrand equivalence; however, it is problematic to extend to the case of recursion and thus reductions. On the other hand the method by Gupta and Rajopadhye depends on a representation that already identifies reductions and specifically optimizes them – which does not, in general, include the expressions optimized by LCCP.

Chapter 4

Placement Computation and Replication

The previous chapter introduced the LCCP transformation, which can be used to reduce the number of occurrence instances to be executed by removing index space dimensions along which the same value is computed. With the method for representative selection presented in Section 3.5.2, which we employ here, the resulting reduced COIG needs to be scheduled in order to obtain a legal execution order. This is completely independent of the fact that one may want to execute independent occurrence instances in parallel: we just need to have some legal execution order of occurrence instances in order to assure that the target program is semantically equivalent to the input program, as discussed in Section 3.6. On the other hand, a schedule already suffices to produce a legal output program with reduced occurrence instance sets: a schedule can be extended easily to a complete injective mapping (which is needed for program transformation in the polyhedron model) by appending an identity mapping on dimensions that are linearly independent on the schedule and among each other. However, especially in the context of parallel execution, it is important to produce also a placement function that distributes work in an beneficial way. In particular, on modern architectures, the communication cost is very high compared to the computation cost, so that communication should be avoided as much as possible. As already hinted in Section 3.7, this applies to on several levels:

1. Communication between nodes is usually done using communication libraries like `MPI` [Mes97] or `PVM` [Sun90] and incurs the largest time penalty – one may divide this cost by discerning on-site communication from off-site communication.
2. Communication between processes on the same node is not as expensive as communication between nodes but still involves possibly time consuming copy operations within the node's memory. It is sometimes also managed by communication libraries as above.
3. Threads share the same memory within a common process. However, they can be distributed on different CPUs, which incurs communication cost in form of cache invalidations and misses. This is the level of parallelization usually employed by `OpenMP` (notwithstanding cluster implementations of `OpenMP` [Hoe07]).
4. Within a single thread, instruction level parallelism (ILP), dedicated graphics processing units (GPUs) or other coprocessors may be used to exploit further parallelism. This may again lead to communication between registers, cache and possibly main memory (for communication between processing units).

At all the levels enumerated above, locality, both between source and targets of a dependence (which roughly translates to temporal locality), and between different accesses to consecutively stored (computed) values (spatial locality) is an important performance factor. Especially the latter two points of the enumeration above gain from locality of consecutive accesses, since current cache architectures usually fetch data in contiguous chunks.

A comparison between the INDEPENDENT loops of the loop nest of Figure 3.15 in Section 3.7 with the above points suggests to implement the former two points with the outer INDEPENDENT loops – enumerating p_1,..., p_#pdims – and the latter two points with the inner loops – enumerating ilp_1,...,ilp_$n_{\mathtt{depth}}$.

The connection between an occurrence instance α in the original (or LCCP-transformed) program and the iteration of INDEPENDENT loops that is used to enumerate α is given by a placement. We propose to create a placement for the loops corresponding to each of the points above by algorithms especially taylored for the task, starting from point 1 onward, with each placement algorithm possibly using the scheduling information and placement information of previous modules, since that order reflects the influence on execution time. This may possibly make some or all of these placements non-invertible, i.e., they may represent replicated placements.[1] With the usual HPF and MPI implementations that hide the mapping between virtual and physical processors, it is not even possible to draw a line between the first two points with a pure source-to-source transformation. Therefore, we will only consider the first point – communication and placement between nodes in this thesis. In our approach, the other placement dimensions do not necessarily *have* to be defined by an explicit placement. However, one might also choose to use the placement method we will now examine in closer detail with adjusted parameters to obtain placements of the lower levels. The placement method presented in the following sections has been introduced in previous work [FGL03, FGL04]. Wondrak implemented and evaluated this approach in his diploma thesis [Won05].

4.1 Computation Placement Based on Initial Data Placement

Although the use of replicated placements as described above is not limited to this scenario, we are primarily concerned with creating a placement for a COIG, i.e., for an LCCP generated loop program. In particular, the nodes in the reduced COIG that correspond to interesting sets need to be associated with a placement, since these occurrence instances represent computations whose results have to be stored somewhere. In contrast, the occurrence instances of non-interesting sets are trivially best executed where their respective (unique) targets are executed.

Example 4.1 *Let us consider the following program fragments:*

```
!HPF$ INDEPENDENT
      DO i1=2,n-2
!HPF$ INDEPENDENT
        DO i2=1,n
          A(i2,i1)=(B(i1)  *2)*i2 &
&                 -(B(i1+2)*2)/i2 &
&                 +(B(i1+1)*2)+i2 &
&                 +(B(i1-1)*2)
        END DO
      END DO
```

```
!HPF$ INDEPENDENT
      DO i1=1,n
        TMP(i1)=B(i1)*2
      END DO
!HPF$ INDEPENDENT
      DO i1=2,n-2
!HPF$ INDEPENDENT
        DO i2=1,n
          A(i2,i1)=TMP(i1)   *i2 &
&                 -TMP(i1+2)/i2 &
&                 +TMP(i1+1)+i2 &
&                 +TMP(i1-1)
        END DO
      END DO
```

[1] Only at the lowest level (which does not necessarily have to be the level of source code), the space-time mapping – and thus the placement – has to be a function, since different tasks cannot be handled by the same unit within a CPU at the same time.

The program fragment on the left contains four computations of $B(i) \cdot 2$. The same value is computed repeatedly for different iterations of both the i_2-loop and the i_1-loop. LCCP can transform this program fragment into the one on the right (the actual transformation depends on the exact space-time mapping used). In the transformed code, $B(i) \cdot 2$ is calculated once and then assigned to a temporary. As discussed in Section 3.7 (page 130), this assignment is not *explicit in the reduced COIG and does not correspond to a write access in the reduced COIG, but is implemented as a write access in the target code.*

The least expensive placement in terms of communication cost is, of course, to place all occurrence instances at the same place. However, this trivial placement should be avoided in most cases, since it does not leave any room for performance improvement through parallelization. Assuming that the user has specified non-trivial data placements, one way to avoid this, which we adopt here, is to use placements derived from the user-given ones as follows. We suppose that basic interesting sets, i.e., interesting sets that represent write accesses, already have a placement. Instances of write accesses have to be executed at the place of the memory cell to which to write is located. This information is given by the placement defined for the dummy loops introduced in Section 2.3.2. Since parallelism in HPF is completely defined by data placements, this is a safe assumption for HPF programs. But, if we handle programs in some language other than HPF, in which placement directives may be not quite as explicit, this is no unrealistic supposition either: the only additional assumption is that a placement of input and result data has been decided upon before computations and placement of intermediate results is tackled. And we can assume with good reason that input and output data may be somehow restricted in their possible placement, since, e.g., in a cluster system, this data is often already distributed for larger problems.

Depending on the distribution of A and B in Example 4.1, different distributions of *TMP* may be beneficial: most probably, an alignment with B is a good approach. But if B is replicated, it is probably more useful to align with A.

However, whether or not a certain data distribution actually improves performance may depend on several factors such as the ability of the underlying compiler to produce efficient (communication) code and the target machine. Therefore, our aim is to create a flexible placement method for occurrence instances. Note that we do not distinguish between occurrence instances that represent calculations and occurrence instances that represent memory accesses or even memory cells: an occurrence instance α is just the execution of an operation whose result value may be needed sometime at some place, possibly several times and at several places – in which case we have to store the result in memory. Since α is executed only at a certain location, the result has to be stored there first (even if it is only used somewhere else, we do need some buffer). The basic idea behind our placement method is thus to propagate the placements of all sources and targets of an occurrence instance set $\mathfrak{O}$ (along the transitive closure of flow dependences and structural output dependences) to $\mathfrak{O}$ itself. For a sequence $\mathfrak{O}_1, \ldots, \mathfrak{O}_{n_\mathfrak{O}}$, this results in a set of possible placement relations $\mathfrak{A}_{\mathfrak{O}_i}$ for each such set $\mathfrak{O}_i$. The method then estimates the communication cost for each combination of placements for the different occurrence instance sets $\mathfrak{O}_1, \ldots, \mathfrak{O}_{n_\mathfrak{O}}$. I.e., for each element of $\prod_{i=1}^{n_\mathfrak{O}} \mathfrak{A}_{\mathfrak{O}_i}$ a graph representing the communication structure in the target program is constructed and an associated cost is computed. The combination producing the lowest cost is then selected as the set of placement relations to be used for the loop program.

LCCP identifies those points in the code whose result values should be stored in memory, because they are needed more than once during program execution. For the same reason, these are also the occurrence instance sets where redistribution of data may be appropriate. It is therefore important to select the right placement relation for these interesting sets. Note that the points in the code need not consist of complicated computations, but may well represent plain read accesses, as the following example shows.

Example 4.2 *Let us reconsider the code from Example 3.23, this time with explicit placements for the arrays used:*

```
!HPF$ DISTRIBUTE PROCS1(BLOCK)
!HPF$ DISTRIBUTE PROCS2(BLOCK,BLOCK)
```

```
!HPF$ ALIGN B WITH PROCS2
!HPF$ ALIGN A WITH PROCS1
!HPF$ ALIGN C WITH PROCS1
!HPF$ INDEPENDENT
      DO i1=1,6
!HPF$ INDEPENDENT
        DO i2=(i1-1)/3+1,(i1+1)/3
          B(i2,i1)=A(i1+i2)**3+C(3)
        END DO
      END DO
```

The placements for these arrays are quite intuitive: assuming all the elements of C (there should be more than just $C(3)$ – otherwise one would not store it in an array) can be computed independently, a placement along a 1-dimensional processor mesh is a good choice, since it leads to the best processor utilization for the computation of C. The same statement holds for A. For B, however, one may want to exploit as much parallelism as possible, perhaps in the context of the later use of B. However, since $C(3)$ is used for every *iteration of the loop nest, it would most certainly be best to replicate this value. The usage in every iteration is reflected by the fact that the occurrence instance set representing the read access is a singleton and in fact an interesting set. It will therefore be considered for an alternate placement, and the corresponding code might look as follows:*

```
!HPF$ ALIGN TMP WITH PROCS1(*,*)
:
      TMP=C(3)
:
```

This code prescribes the explicit communication of $C(3)$ to all processors.

As the example shows, the placement we create in this case through propagation is not necessarily a function, but a *relation*: there is only a scalar *TMP*, which is stored on *all* processors of the processor mesh $PROCS_1$. Therefore, the placement is replicated – several copies of the same occurrence instance are placed onto several different physical processors. Thus, the computation to be done to produce the corresponding value is executed on several processors, and the result is also stored on this set of physical processors.

The remainder of this section introduces our main representation of these replicated placements and how to obtain such a representation from a given HPF program. Our placement algorithm is then developed in Section 4.2. This placement algorithm is based on a general cost model. Section 4.3 presents a benchmark suite for developing a cost model for a specific system and contains a case study in which we develop such a cost model for the ADAPTOR compiler with the Passau hpcLine cluster as the target computer.

4.1.1 How to Model Replication

Our aim is to use replication (of data and computations) to reduce communication costs. Example 2.28 in Section 2.4 (page 42) shows that (possibly replicated) HPF distributions can be expressed by polyhedra. Ignoring the array bounds, we can also create a placement description that only consists of two linear mappings. In our approach, the vertices of the reduced COIG are at some point transformed into vertices representing partitions so that the actual placement description does not need to contain a domain description but can be defined by a linear relation, i.e., by a pair of unrestricted linear mappings between $\mathbb{Q}^{n_{src}+n_{blob}+2}$ and $\mathbb{Q}^{n_{tgt}+n_{blob}+2}$. Section 2.4 already shows the basic representation for such a relation. Here, we consider the specifics of representing an HPF placement relation for occurrence instances and dependences between occurrence instances.

Example 4.3 *An* HPF *placement of the form*

```
!HPF$ DISTRIBUTE T(BLOCK(B1),BLOCK(B2),BLOCK(B3)) ONTO P
!HPF$ ALIGN A(i1,i2) WITH T(a1*i1+c1,*,i2)
```

defines an alignment of the alignee `A` *onto an align target* `T`. *This alignment is represented by two mappings*

$$\Phi_L \quad : \quad \begin{pmatrix} i_1 \\ i_2 \end{pmatrix} \mapsto \begin{pmatrix} \left\lfloor \frac{a_1 \cdot i_1 + c_1}{b_1} \right\rfloor \\ 0 \\ \left\lfloor \frac{i_2}{b_3} \right\rfloor \end{pmatrix}$$

and

$$\Phi_R \quad : \quad \begin{pmatrix} p_1 \\ p_2 \\ p_3 \end{pmatrix} \rightarrow \begin{pmatrix} p_1 \\ 0 \\ p_3 \end{pmatrix}$$

We do not consider `CYCLIC` or `BLOCK-CYCLIC` distributions as non-linear, special distributions, but approximate them by a corresponding linear function: since the number of processors is unknown at compile time, we cannot make any assertion as to which elements may get aligned. The only statement we can make is that two elements will be aligned if the nominators and denominators of their alignment function are the same. Since the denominator, again, may depend on the number of processors and even the value of parameters at run time, we only consider alignments to the same template object as pertaining to the same denominator. Alignment to different template variables is always seen as an alignment to different arrangements of processors that are completely independent of each other.

In `HPF`, placements for arrays are defined while, in our model, there is no difference between occurrence instances representing data and those representing calculations. But every calculation that is executed does yield some result value. This result value has to be stored *somewhere* – otherwise the calculation would be unnecessary and could be discarded. The process of storing the result value for future use is made explicit in our method by assigning this value to an array variable. This determines the placement of the execution through the placement of the array.[2] Note that we assume a placement to be defined via an alignment to a template that is distributed across a processor array. If an array is distributed directly via a `DISTRIBUTE` directive, we can rewrite this definition so that an element $(i_1, \ldots, i_r)^T$ of an array A is aligned to element $(i_1, \ldots, i_r)^T$ of a new template of the same shape and extent as A that is distributed across the processor array just as A was before.

We will now first describe how to obtain basic placements technically from the `HPF` distributions specified by the programmer. Section 4.1.3 shows how our method handles dependences between occurrence instances that are distributed in a replicated fashion. The actual placement algorithm is then examined in the following sections.

4.1.2 Parsing `HPF` Placements

In this section, we will briefly sketch the technicalities of obtaining a placement description, as discussed above for the arrays in an `HPF` program, from the distribution directives.A possible implementation for translating an `HPF`-like description of distributions to a version using two mappings is presented in Algorithm 4.1.1.

In Step 1, Algorithm 4.1.1 (*ParseHPF*) creates two matrices:

M''_L: defines a mapping from the space spanned by the identifiers used in the `ALIGN` clause to the space defined by the subscript dimensions of the alignee.

M''_R: defines a mapping from the space spanned by the identifiers used in the `ALIGN` clause to the space defined by the subscript dimensions of the template.

[2]Actually, there is one exception since, in contrast to our framework, `HPF` supports the notion of a subroutine call without any return value. Therefore, in the case of subroutine calls, a placement has to be implemented by creating a template T that "stores" the (non-existent) result value and places the calculation by determinig the home of the enclosing loops with respect to the coordinates of T using the `HOME` directive.

We suppose that a function LinearExpr is defined that is used here to create the matrix rows representing these linear expressions.

In Step 2, the matrices M''_L and M''_R are inverted in order to obtain mappings to the common space of identifiers used. M''_L is *per definitionem* of full rank and thus invertible. But M''_R may define a relation so that several elements of the alignee are aligned with only one specific coordinate in a template dimension, i.e., there is a condition on template elements to store any element of the alignee at all. In this case, this condition is defined by a matrix M''^C_R, as discussed in the context of Algorithm 2.2.1 (*GeneralizedInverseMin*) in Section 2.2.

Step 3 transforms these matrices so that their source spaces are the index space of the original program – i.e., array dimensions are represented by the index dimensions enumerated by the corresponding dummy loops, and template dimensions are represented by parallel indices of the target program. We suppose that these embeddings are defined by a call to a function CreateProjections. In addition, this step transforms the function mapping the (parallel) target indices to the common vector space into an endomorphism by prepending and appending sufficiently many zero-rows. This means that the common subspace is actually a virtual processor mesh, as shown in Section 2.4.

Algorithm 4.1.1 [*ParseHPF*]:
Input:
alignDirective :
: an `ALIGN` directive of the form
`ALIGN` *alignee* `(`*alignDummyList*`)` *alignTarget* `(`*alignSubscriptList*`)`.

templateNameToIndex :
: mapping from the template name to the number of the first dimension in the target program that represents a dimension of that processor array.

identifierList :
: list of all identifiers used in *alignDummyList* or *alignSubscriptListorasparameters*.

parameterList :
: list of all structural parameters in the (source and target) program; the i-th element in the list contains the identifier of the parameter dimension i.

Output:
$\Phi_{alignee} = (\Phi_L, \Phi_R)$:
: placement for *alignee* as a relation between the index space of the dummy loop for *alignee* and the processor mesh *alignee* is placed onto.

Procedure:

/* STEP 1: initialize matrices M''_L, M''_R defining mappings from a common identifier space */
/* to the data space of the alignee on the one hand and */
/* to the data space of the template on the other hand */
let $M''_L \in \mathbb{Q}^{(\text{size}(alignDummyList)+\text{size}(parameterList))\times\text{size}(identifierList)}$;
`for` $i = 1$ `to` size(*alignDummyList*)
  set the i-th row of M''_L to the vector
  LinearExpr(*alignDummyList*`[`i`]`, *identifierList*) $\in \mathbb{Z}^{\text{size}(identifierList)}$
  that represents the linear expression *alignDummyList*`[`i`]`;
`endfor`
/* Parameters are mapped to themselves */
`for` $i = 1$ `to` size(*parameterList*)
  set the size(*alignDummyList*) + i-th row of M''_L to the vector
  LinearExpr(*parameterList*`[`i`]`, *identifierList*) $\in \mathbb{Z}^{\text{size}(identifierList)}$ that represents
  the i-th parameter in *parameterList*;
`endfor`
let $M''_R \in \mathbb{Q}^{(\text{size}(alignSubscriptList)+\text{size}(parameterList))\times\text{size}(identifierList)}$;
`for` $i = 1$ `to` size(*alignSubscriptList*)

```
   set the i-th row of M''_R to the vector
   LinearExpr(alignSubscriptList[i], identifierList) ∈ Z^size(identifierList) that represents
   the linear expression alignSubscriptList[i];
endfor
for i = 1 to size(parameterList)
   set the size(alignSubscriptList) + i-th row of M''_R to the vector
   LinearExpr(parameterList[i], identifierList) ∈ Z^size(identifierList) that represents
    the i-th parameter in parameterList;
endfor
/* STEP 2: invert the mappings above to create mappings from the alignee's data space and */
/* from the template's data space back top the space of common identifiers */
/* The placement relation can basically be expressed by the inverses of */
/* M''_L and M''_R now. */
/* What remains are restrictions of the relation, e.g., to a single template coordinate. */
/* Only copy rows of M''_L/M''_R into M'_L/M'_R, */
/* if they define actual conditions... */
let M'_L ∈ Q^(size(alignSubscriptList)+size(parameterList))×(size(alignDummyList)+size(parameterList));
let M'_R ∈ Q^(size(alignSubscriptList)+size(parameterList))×(size(alignSubscriptList)+size(parameterList));
currRow := 1;
for i = 1 to size(identifierList)
   /* If row i of M''go_L or M''go_R is zero, */
   /* neither that row nor the corresponding row of the other matrix */
   /* have to be considered for constraints; */
   /* otherwise, those two rows together define a constraint for the placement relation. */
   if M''go_R[i] ≠ 0 then
      M'_L[currRow] := M''go_L[i];
      M'_R[currRow] := M''go_R[i];
      currRow := currRow + 1;
   endif
endfor
/* There may have been additional contitions for the processors that store a given element. */
/* These are represented by the rows of M''C_R. */
NumberOfConditions := size(arraySubscriptList) + size(parameterList) − rk(M''_R);
for i = 1 to NumberOfConditions
      M'_L[currRow] := 0;
      M'_R[currRow] := M''C_R[i];
      currRow := currRow + 1;
endfor
/* STEP 3: produce a relation based on the actual indices of dummy loops */
/* for the arrays to be allocated as target space for the placement */
CreateProjections(alignDirective, templateNameToIndex, M_{P_L}, M_{P_R});
M_L := M'_L · M_{P_L};
M_R := M'_R · M_{P_R};
/* currRow is less or equal size(alignSubscriptList); */
/* i.e., we can now extend M_R so that the corresponding function is an endomorphism */
prepend templateNameToIndex(alignee) rows containing only zero to M_R and M_L;
append enough zero-rows to M_R and M_L so that M_R becomes quadratic;
let Φ_L : Q^(n_src+n_blob) → Q^(#pdims+n_blob) : ν ↦ M_L · ν;
let Φ_R : Q^(#pdims+n_blob) → Q^(#pdims+n_blob) : ν ↦ M_R · ν;
return Φ = (Φ_L, Φ_R);
```

The following example shall give a short impression on the structure of such a placement relation.

Example 4.4 *The following code fragment defines two templates – a 2-dimensional and a 3-dimensional one – that together build a 5-dimensional target space (ignoring time dimensions for now). The 3-dimensional array A is then aligned with the 3-dimensional template.*

```
      REAL A(1:50,1:100)
!HPF$ TEMPLATE T1(1:100,1:100)
!HPF$ DISTRIBUTE T1(BLOCK,BLOCK)
!HPF$ TEMPLATE T2(1:100,1:100,1:100)
!HPF$ DISTRIBUTE T2(BLOCK,BLOCK,BLOCK)
!HPF$ ALIGN A(i,j,k) WITH T2(2*i,3,1)
```

The only actually aligned dimension of A is the first one, which is aligned to the first dimension of T_2; but only template elements with second coordinate 3 *and third coordinate* 1 *store any elements at all. This is sketched in Figure 4.1. With parameterList* $= (m_\infty, m_c)$ *and identifierList* $= (i, j, k, m_\infty, m_c)$*, the initial matrices produced are:*

$$M''_L = \begin{array}{c} \begin{matrix} i & j & k & m_\infty & m_c \end{matrix} \\ \begin{pmatrix} 1 & 0 & 0 & 0 & 0 \\ 0 & 1 & 0 & 0 & 0 \\ 0 & 0 & 1 & 0 & 0 \\ 0 & 0 & 0 & 1 & 0 \\ 0 & 0 & 0 & 0 & 1 \end{pmatrix} \end{array} \qquad M''_R = \begin{array}{c} \begin{matrix} i & j & k & m_\infty & m_c \end{matrix} \\ \begin{pmatrix} 2 & 0 & 0 & 0 & 0 \\ 0 & 0 & 0 & 0 & 3 \\ 0 & 0 & 0 & 0 & 1 \\ 0 & 0 & 0 & 1 & 0 \\ 0 & 0 & 0 & 0 & 1 \end{pmatrix} \end{array}$$

The generalized inverses and the corresponding conditional matrices produced by Algorithm 2.2.1 are then:

$$(M''_L)^{g_0} = \begin{pmatrix} 1 & 0 & 0 & 0 & 0 \\ 0 & 1 & 0 & 0 & 0 \\ 0 & 0 & 1 & 0 & 0 \\ 0 & 0 & 0 & 1 & 0 \\ 0 & 0 & 0 & 0 & 1 \end{pmatrix} \qquad (M''_R)^{g_0} = \begin{pmatrix} \frac{1}{2} & 0 & 0 & 0 & 0 \\ 0 & 0 & 0 & 0 & 0 \\ 0 & 0 & 0 & 0 & 0 \\ 0 & 0 & 0 & 1 & 0 \\ 0 & 0 & 0 & 0 & 1 \end{pmatrix}$$

$$(M''_L)^C \in \mathbb{Q}^{0\times 5} \qquad (M''_R)^C = \begin{pmatrix} 0 & 0 & 1 & 0 & -1 \\ 0 & 1 & 0 & 0 & -3 \end{pmatrix}$$

As expected, there is no condition defined by $(M''_L)^C$. $(M''_R)^C$*, on the other hand, defines the condition that the second coordinate of the template is* 3 *and the third is* 1. *The algorithm then replaces the rows from the mappings for which* $M''^{g_0}_R$ *contains only zero-entries by the conditions defined in* M''^C_R.[3] *This results in the following matrices:*

$$M'_L = \begin{pmatrix} 1 & 0 & 0 & 0 & 0 \\ 0 & 0 & 0 & 0 & 0 \\ 0 & 0 & 0 & 0 & 0 \\ 0 & 0 & 0 & 1 & 0 \\ 0 & 0 & 0 & 0 & 1 \end{pmatrix} \qquad M'_R = \begin{pmatrix} \frac{1}{2} & 0 & 0 & 0 & 0 \\ 0 & 0 & 1 & 0 & -1 \\ 0 & 1 & 0 & 0 & -3 \\ 0 & 0 & 0 & 1 & 0 \\ 0 & 0 & 0 & 0 & 1 \end{pmatrix}$$

A placement relation only exists between array elements and template elements that are mapped to the same element by M'_L *and* M'_R*, i.e., the coordinate in the first array dimension has to be half as large as the coordinate in the first template dimension. This relation, too, can be gleaned from Figure 4.1.*

[3] These zero-rows come from the fact, that the corresponding coordinates in the template dimension was free to choose during inversion, i.e., a placement is defined for all these coordinates, which means that the array elements are replicated along these dimensions.

The final step of Algorithm 4.1.1 introduces zero-columns and rows so that the domains of the mappings are the indices of dummy loops and the target indices, respectively. In particular, the first two dimensions are reserved for the unused template T_1. We suppose that the program does not contain any additional loops and also ignore the operand and occurrence dimensions. This results in the matrices:

$$M'_L = \begin{pmatrix} 0 & 0 & 0 & 0 & 0 \\ 0 & 0 & 0 & 0 & 0 \\ 2 & 0 & 0 & 0 & 0 \\ 0 & 0 & 0 & 0 & 0 \\ 0 & 0 & 0 & 0 & 0 \\ 0 & 0 & 0 & 1 & 0 \\ 0 & 0 & 0 & 0 & 1 \end{pmatrix} \qquad M'_R = \begin{pmatrix} 0 & 0 & 0 & 0 & 0 & 0 & 0 \\ 0 & 0 & 0 & 0 & 0 & 0 & 0 \\ 0 & 0 & 1 & 0 & 0 & 0 & 0 \\ 0 & 0 & 0 & 0 & 1 & 0 & -1 \\ 0 & 0 & 0 & 1 & 0 & 0 & -3 \\ 0 & 0 & 0 & 0 & 0 & 1 & 0 \\ 0 & 0 & 0 & 0 & 0 & 0 & 1 \end{pmatrix}$$

These matrices finally define a placement relation whose associated function of M_R is an endomorphism in $\mathbb{Q}^{n_{tgt}+n_{blob}+2}$ mapping from physical processors to virtual processors as discussed in Section 2.4.

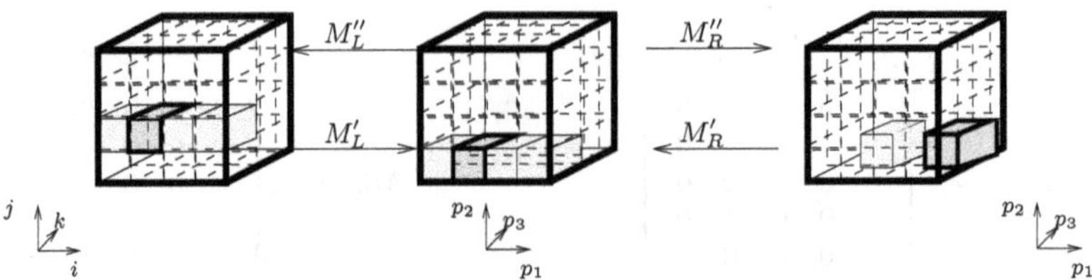

Figure 4.1: Placement relation of Example 4.4. The array elements $A(\cdot, 2, 1)$ and their placements are marked with boxes. In particular, $A(2, 2, 1)$ is marked with dark grey boxes and fat borders.

Given a placement relation defined by two mappings (Φ_L, Φ_R), the set of processors on which an occurrence instance α is to be stored is:

$$\Phi(\alpha) = \Phi_R^{-1} \circ \Phi_L(\alpha)$$

A consistency condition is that the image of Φ_L is a subset of the image of Φ_R. For the placements given as input, we can safely assume this; the placements we compute in Section 4.2 satisfy this property by construction.

This approach corresponds to the description of replicated data in HPF, as sketched in Sections 2.1 and 2.4. Algorithm 4.1.1 describes a way to obtain a linear relation for a placement of an array in an HPF program. In order to obtain placements for interesting sets that are newly created by the LCCP transformation, we will view the COIG of the program fragment to be considered with output and anti dependences removed. Our method is based on following these dependences and propagating the placement relation from the final read and write access instances. The placement selection is then done based on the communication taking place in the transformed program – which is modelled by the flow dependences and structural output dependences in the target index space of the program. The following section will therefore focus on how to treat dependences to occurrence instances that might be distributed according to a replicated placement.

4.1.3 Dependences in the Presence of Replication

The transformation in the polyhedron model is represented by a space-time mapping. This mapping is usually a piecewise linear function that can be represented by a family of linear functions $(T_i)_{i \in \{1,\ldots,n_T\}}$ where each function T_i is defined on a polyhedron of occurrence instances and has the form $T_i = \begin{pmatrix} \Theta_i \\ \Phi_i \end{pmatrix}$. In this function definition, Θ_i is a schedule (i.e., the time component of the space-time mapping) and Φ_i a placement (the space component). LCCP returns a transformed dependence graph in the form of the reduced COIG. A space-time mapping is then computed for the vertices of the reduced COIG that define interesting sets – possibly after an additional partitioning of the sets represented by these vertices. For the computation of a placement, we view the expected dependence graph for the currently considered space-time mapping in order to estimate communication cost. Since at this point we may have replicated placements, the space-time mapping is not defined by a mapping but by a linear relation.

In order to estimate the costs for communication induced by the target program, we view the dependences with respect to space coordinates – i.e., after application of the placement relation (Φ_L, Φ_R). These space coordinates determine in our case the nodes in a cluster on which to execute some computation and store the result. Since neither of the two functions defining the placement relation is necessarily invertible, it is not immediately clear how to represent dependences in the target program.

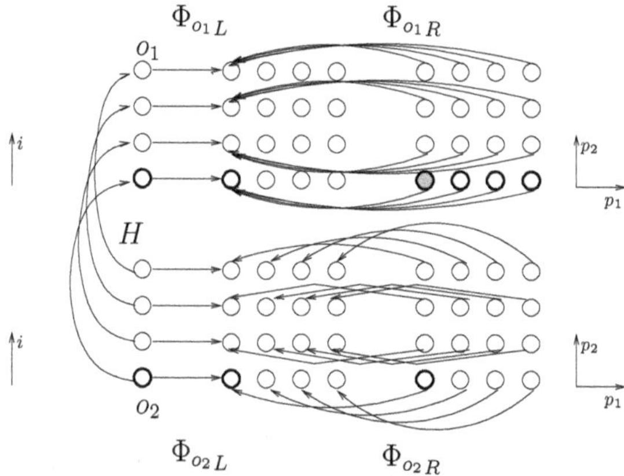

Figure 4.2: Dependence of several different instances of occurrence o_2 from the same instance of occurrence o_1.

Example 4.5 *Consider the following code fragment:*

```
       INTEGER i,j
       REAL A(1:4)
       REAL B(1:4,1:4)
!HPF$ TEMPLATE PROCS(1:4,1:4)
!HPF$ DISTRIBUTE PROCS(BLOCK,BLOCK)
!HPF$ ALIGN B(j,i) WITH PROCS(j,i)
!HPF$ ALIGN A(i)   WITH PROCS(j,i)
!HPF$ INDEPENDENT
       DO i=1,4
!HPF$ INDEPENDENT
         DO j=1,4
           B(j,i)=B(j,i)+A(i)**3
         END DO
       END DO
```

An application of LCCP yields the common subexpression A(i)**3 *which is assigned to an auxiliary variable in the transformed code and needs a placement for computation (and assign-*

ment to this auxiliary variable). Let o_1 be the occurrence of the write access in the dummy loop representing the placement of A specified by the user. This dummy loop is not present in the code above. Let o_2 be the occurrence of the exponentiation `A(i)**3`*, which indirectly depends on o_1. Since the read access instances of* `A(i)` *do not form an interesting set, this dependence becomes direct in a compressed graph that only contains interesting sets. Figure 4.2 represents a placement as defined in the following code fragment:*

```
       INTEGER i,j
       REAL A(1:4)
       REAL B(1:4,1:4)
       REAL TMP(1:4)
!HPF$ TEMPLATE PROCS(1:4,1:4)
!HPF$ DISTRIBUTE PROCS(BLOCK,BLOCK)
!HPF$ ALIGN B(j,i) WITH PROCS(j,i)
!HPF$ ALIGN A(i)   WITH PROCS(j,i)
!HPF$ ALIGN TMP(i) WITH PROCS(1,i)
!HPF$ INDEPENDENT
       DO i=1,4
         TMP(i)=A(i)**3
       END DO
!HPF$ INDEPENDENT
       DO i=1,4
!HPF$ INDEPENDENT
         DO j=1,4
           B(j,i)=B(j,i)+TMP(i)
         END DO
       END DO
```

The auxiliary array TMP storing the intermediate values $A(i)^3$ is replicated along the first dimension of the physical processor mesh and distributed along the second dimension. This is because the instances of occurrence o_1 are stored in a replicated fashion (all processors with the same p_1 coordinate own a copy of an occurrence instance). Physical processors storing the lowest respective instance of an occurrence are marked by fat circles. Just as the mappings $(\Phi_{o_1 L}, \Phi_{o_1 R})$ define a placement relation for the instances of o_1, the mappings $(\Phi_{o_2 L}, \Phi_{o_2 R})$ define a placement for the instances of o_2, as depicted in the lower part of Figure 4.2. The instances of o_2 – the exponentiations – are only allocated on the first column of the processor space by their placement.

We now view only the dependence of the exponentiation `A(i)**3` *on the write access to A. Figure 4.2 shows this dependence as an h-transfromation H. In the upper part of the figure, four instances of occurrence o_1 are mapped to a virtual processor array by $\Phi_{o_1 L}$; physical processors are mapped to the same processor array by $\Phi_{o_1 R}$. The h-transformation maps a target occurrence instance α (with $\mathrm{OccId}(\alpha) = o_2$) to its source occurrence instance $H(\alpha)$ ($\mathrm{OccId}(H(\alpha)) = o_1$).*

This means that all instances of occurrence o_2, each of which depends on a different instance of o_1, may select the source processor from which to load the data needed for computation.

Example 4.5 shows a single placement for each interesting set. To assess the cost of executing the target program, our method will iterate over different possible combinations of placements for the different interesting sets and estimate the cost (in our case the plain communication cost) of executing the program generated for this particular placement combination.[4] In the situation of the transformed code of Example 4.5, an occurrence instance that computes the value $A(i)$ for some given i can choose among several different places from where to get the value $A(i)$: since A is stored in a replicated fashion, any one of the copies in the p_1-dimension will do.

[4] Actually, the placement of the computation of $A(i)^3$ and thus of the auxiliary array *TMP* in the example follows neither the placement of the source nor the targets of the computed expression – it will therefore *not* be generated as a valid possibility in our method. This placement relation is only used for the sake of the argument, here.

Correspondingly, if *TMP* were distributed in a replicated fashion, we would have the obligation to compute the corresponding value on several processors. Let us take a closer look at what options and obligations replicated placements incur.

Which Copy to Read from The set of copies to choose from for the execution of an occurrence instance α (with $\text{OccId}(\alpha) = o_1$) in Example 4.5 is given by

$$({\Phi_{o_1}}_R^{-1} \circ {\Phi_{o_1}}_L \circ H)(\alpha)$$

With Φ_{o_1} and H represented as pairs of mappings, as introduced in Section 2.4.2, this becomes either

$$(\Phi_{o_1} \circ_- H)(\alpha)$$

or

$$(\Phi_{o_1} \circ_+ H)(\alpha)$$

Let us now take a closer look in order to decide which operator is appropriate here. Which one of the copies is actually chosen during execution may depend on various factors that are beyond our control: the final choice may depend on the compiler or even the hardware. Especially in the context of inter-node communication, which is our concern here, it is usually advantageous to choose a copy for which as many dimensions as possible feature the same coordinates as the occurrence instance α. In our example, in which the copies are layed out along the p_1-dimension, this is the one with p_1-coordinate 1, marked as a shaded fat circle in Figure 4.3 (for the case $i = 1$). If copies can only be placed along an axis of the processor mesh, as is the case for `HPF`, this is also the nearest copy. The $\circ_-$ operator from Definition 2.36 is designed to produce exactly this copy. Therefore, we suppose to choose the copy of an occurrence instance α as in the example according to the following mapping.

$$H' \quad = \quad (\Phi_{o_1} \circ_- H \circ_+ \Phi_{o_2}^{-1})(\alpha) \tag{4.1}$$

with H and H' being represented as pairs of mappings – a representation that can be used for mappings, according to Corollary 2.31. This leaves the dependence in the portion of the target index space that we consider here as the identity: each occurrence instance has a copy of the source of its dependence available on the local processor. Note that we have not yet justified the $\circ_+$ operator in the formula above. This will follow immediately.

Which Copies to Execute We have seen that we may pick any copy of a replicated dependence source. On the other hand, several target occurrence instances may have to be executed on the same processor. Considering a placement relation defined by (Φ_L, Φ_R), a non-empty kernel of Φ_R (as can be observed for ${\Phi_{o_1}}_R$ in Example 4.5) defines several places to execute/store an occurrence instance and thus represents a choice from which processor to read the source of a dependence. A non-empty kernel of Φ_L, on the other hand, represents several occurrence instances that have to be executed on the same processor. Thus, this non-empty kernel represents the obligation to consider several occurrence instances. Therefore, Equation (4.1) contains the $\circ_+$ operator which returns the complete set of occurrence instances placed on the same physical processor indicated by α. Example 4.6 shows this case.

Example 4.6 *Let us change the code of Example 4.5 a little bit. Instead of* `A(i)**3`*, we now write* `A(1)**i`*. This means that the occurrence instances of the eyponentiation operator still form an interesting set, and this interesting set is non-trivial (*`A(1)**i` *represents different values for*

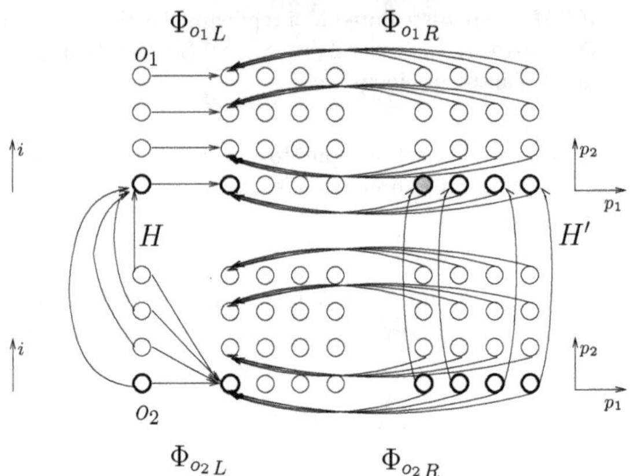

Figure 4.3: Dependence from several instances of occurrence o_2 to the same instance of occurrence o_1.

different values of i), but all these occurrence instances read the same element of A. The resulting code looks as follows.

```
      INTEGER i,j
      REAL A(1:4)
      REAL B(1:4,1:4)
      REAL TMP(1:4)
!HPF$ TEMPLATE PROCS(1:4,1:4)
!HPF$ DISTRIBUTE PROCS(BLOCK,BLOCK)
!HPF$ ALIGN B(j,i) WITH PROCS(j,i)
!HPF$ ALIGN A(i)   WITH PROCS(j,i)
!HPF$ ALIGN TMP(i) WITH PROCS(j,1)
!HPF$ INDEPENDENT
      DO i=1,4
        TMP(i)=A(1)**i
      END DO
!HPF$ INDEPENDENT
      DO i=1,4
!HPF$ INDEPENDENT
        DO j=1,4
          B(j,i)=B(j,i)+TMP(i)
        END DO
      END DO
```

Figure 4.3 shows the occurrence instances for the exponentiation (o_2) stored in the auxiliary array TMP and the write access (o_1) in the dummy loop for array A, from which o_2 effectively reads. The complete set of o_2-instances is executed on each of the processors on the lowest row of the processor mesh. All the instances of o_2 depend on the same instance of o_1 – the one that defines A(1). Therefore, the corresponding computation is placed on the same set of processors as this single element. The resulting dependence in the target index space is named H' in Figure 4.3. Again, the dependence is essentially the identity, but only because all instances of o_2 are executed on the same virtual processor.

Recapitulating, we observe that, when modelling a dependence in the target index space, there may be two sets to consider:

- Within the set of copies of the source occurrence instance of a dependence, the copy to read from may be chosen freely (this is done by the $\circ_-$ operator).
- The set of dependence targets executed on the same processor has to be enumerated completely. Therefore, we cannot just chose a single element, but have to take into account the complete set (this set is returned by the $\circ_+$ operator).

We have now established how to handle dependences between processors in the target index space. In the next section, we will develop the algorithms that use these transformed dependences in a selection criterion for good placements of occurrence instances.

4.2 The Placement Algorithm

This section discusses the placement algorithm in detail. The description is based on Wondrak's diploma thesis [Won05], in particular, most of the algorithms can also be found there in a slightly different formulation.

Our placement method follows a simple approach: in HPF, the placement of arrays is always defined by the user – and can be obtained using Algorithm 4.1.1. It only remains to introduce placement relations for the interesting sets created by LCCP. These interesting sets mark the points in the computation which lend themselves naturally to storing in main memory, since the result of these operations are needed at several different places of the program. It is now our aim to produce placement relations for these interesting sets.

4.2.1 Restriction to Interesting Sets

As already observed in Section 3.2, it is not necessary to compute a space-time mapping for vertices of the reduced COIG (*RCOIG*) that do not represent interesting sets. Therefore, we assume the reduced COIG to be given in a compacted form that only consists of vertices representing interesting sets. An algorithm for producing such a reduced COIG that only consists of interesting sets is presented in Wondrak's diploma thesis [Won05, Algorithm 3]. This transformation consists in removing each vertex v representing a non-interesting set and replacing the edges originating and ending in v with the combination of each edge ending in v with each edge originating in v. Example 4.7 shows this transformation.

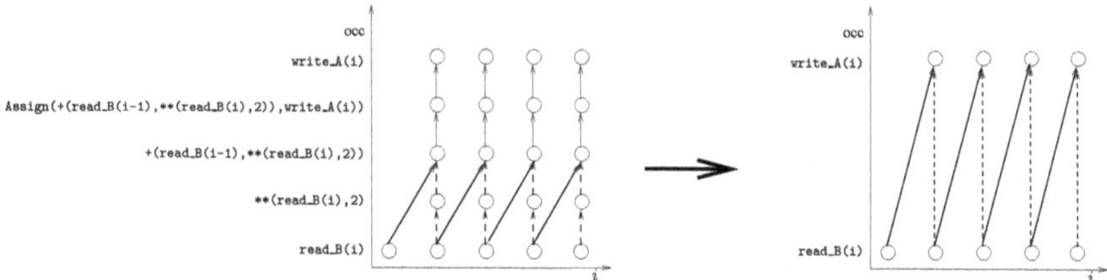

Figure 4.4: COIG and compacted COIG for Example 4.7.

Example 4.7 *Figure 4.4 shows a simplified COIG of the following code fragment (occurrence instances representing different operands are folded into a single instance, linear expressions are omitted).*

```
!HPF$ INDEPENDENT
      DO i=2,5
        A(i)=B(i-1)+B(i)**2
      END DO
```

The right hand side of Figure 4.4 shows a COIG of the same program fragment with elements of non-interesting sets removed. Only the instances of the write access and the read accesses represent interesting sets. There are two possible ways to reach an instance of the write access from a read access instance and vice versa: either through the fat edge or through the fat dashed edges. Figure 4.5 shows the reduced forms of the graphs from Figure 4.4.

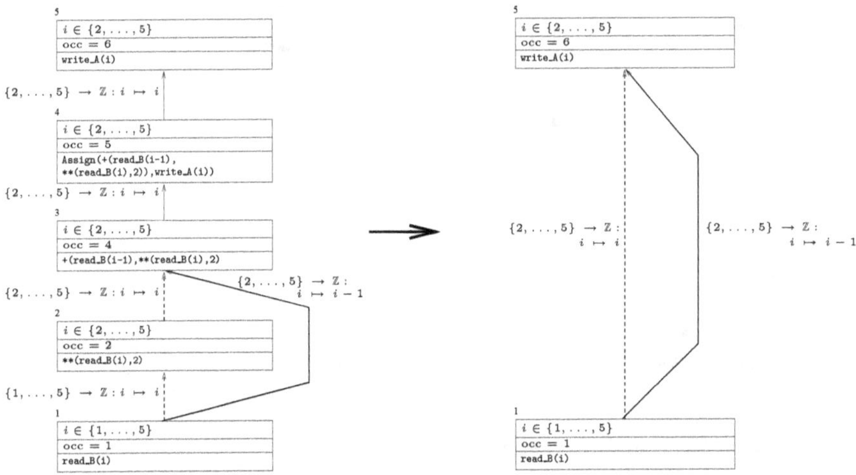

Figure 4.5: Reduced COIG and compacted reduced COIG for Example 4.7.

Considering Example 4.7, we observe that the number of vertices in the COIG is greatly reduced (and thus in the reduced COIG, too). Both possible ways to reach a write access from a read access have to be retained for the placement selection phase. These dependence relations are constructed from the composition of the appropriate h-transformations. In the following, we suppose that the reduced COIG processed by the placement method presented only contains vertices representing interesting sets.

4.2.2 Initial Placement Relations

We assume that the best placement for some given occurrence instance α is to align it either with one of the values that are necessary to calculate the value of α (i.e., align α with one of its input arguments) or with one of the occurrence instances that need output from α in order for their values to be computed (i.e., align α with one of its output arguments). In this way, at least the communication for the input argument or for the output of α can be done in an inexpensive manner. We will therefore propagate the placement information from source to target and from target to source for each flow dependence (and each structural output dependence). This propagation yields a set of possible placement relations for each interesting set. Each combination of these possible placement relations incurs different communication costs, which we will estimate by examining a communication graph based on the dependences in the target index space of the loop program, as discussed in Section 4.1.3. The placement combination with the least cost will be selected as the set of placement relations for the program. One may argue that, in order to align α effectively with one of its dependence targets, from the point of view of efficient communication, it suffices to distribute α to a *superset* of processors that need the output value(s) of α. However, we will only take a superficial glance at this topic.[5]

[5]See the discussion of placement combination on page 160.

The only occurrence instances that feature an *a priori* placement are occurrence instances of write accesses – note that interesting sets, which *lead* to write accesses in the target program, do *not* necessarily represent write accesses. These occurrence instances need to have a definite placement, because they *have* to be executed at the same places at which the array element to be written to is stored (i.e., at its homes) – any other convention would only introduce a placeholder for some messaging scheme that finally introduces an instance of a write access to be executed at the home of the array element considered. However, since dummy loops also enumerate write accesses, every read access in the program is dependent on *some* write access. Therefore, a read access may also propagate a possible placement to an interesting set through this dependence on some write access. Algorithm 4.1.1 creates a representation of placement relations for array elements. In order to obtain the placement relation for an occurrence instance set representing write accesses, it is necessary to consider its subscript function and the actual placement of the array written to. Algorithm 4.2.1 produces such an initial placement for all write accesses of the program. The algorithm just applies the placement relation of the array written to to the subscript function of the write access. The resulting composition defines the single possible placement relation for the occurrence instance set containing instances of this write access and thus for the corresponding vertex v of the reduced COIG. The set of possible placement relations for v, $\mathfrak{A}_v$, is thus a singleton.

Algorithm 4.2.1 [*InitPlacement*]:
Input:
$\mathfrak{G}_{RCOIG} = (\mathfrak{V}_{RCOIG}, \mathfrak{E}_{RCOIG})$:
the reduced COIG of a loop program, with only flow dependences represented as edges in $\mathfrak{E}_{RCOIG}$.

$\widetilde{\Phi} = \left\{ \Phi_{A_1}, \ldots, \Phi_{A_{n_{\mathtt{srcArrays}}}} \right\}$:
for each array A_i in the source program the placement relation Φ_{A_i}.

Output:
$\mathfrak{V}_{\mathrm{write}}$: the set of vertices representing write accesses.

$\mathfrak{A} = \{\mathfrak{A}_v \mid v \in \mathfrak{V}_{\mathrm{write}}\}$:
for each vertex v representing write access instances the corresponding set of possible placements $\mathfrak{A}_v$.

Procedure:
$\mathfrak{V}_{\mathtt{write}} := \left\{ v \,\middle|\, v \in \mathfrak{V}_{RCOIG} \wedge \mathtt{write\_A}_j\mathtt{(\ldots)} = \mathrm{Occ}^{-1}(\mathrm{OccId}(v)), j \in \{1, \ldots, n_{\mathtt{srcArrays}}\} \right\}$;
`forall` $v \in \mathfrak{V}_{\mathtt{write}}$
  Let $\mathrm{Occ}^{-1}(\mathrm{OccId}(v)) = \mathtt{write\_A}_j\mathtt{(}\varphi_v(i_1, \ldots, i_{n_{src}}, n_1, \ldots, n_{n_{blob}})\mathtt{)}$;
  /* i.e., φ_v is the subscript function to A_j in the dummy loop */
  $\Phi_v := \Phi_{A_j} \circ \varphi_v$;
  $\mathfrak{A}_v := \{\Phi_v\}$;
`end forall`
$\mathfrak{A} := \bigcup_{v \in \mathfrak{V}_{\mathtt{write}}} \mathfrak{A}_v$;
`return` $(\mathfrak{V}_{\mathtt{write}}, \mathfrak{A})$;

Applying Algorithm 4.2.1 to a reduced COIG yields a graph with each of the vertices representing write accesses augmented by a set containing the single placement relation eligible for the occurrence instance set represented by that vertex (we recall that write access instances have to be executed on exactly the processors that hold the array element written to). The following deduction of possible placements for the other vertices proceeds roughly according to these steps:

1. Partition the graph so that a placement relation propagated from a write access to a vertex v (and from there further on) is well defined for *all* the instances represented by v.

2. Compute the transitive closure of dependence relations in the partitioned graph.
3. Create new possible placements by building the composition of the transitive dependence relation with the placement relation of the corresponding write access.
4. Postprocess the created placement relations to obtain only placement relations compatible with the (HPF) compiler.
5. Evaluate the possible placement relations for each vertex in the graph according to a parameterized cost model depending on the compiler and target machine.

In order to create additional placement possibilities, we may then reiterate these steps in reverse (from dependence target to source or vice versa). The following sections will elaborate on these steps.

4.2.3 Partitioning

The polyhedron model solves the problem of representing infinite dependence graphs by representing a whole set of instances of dependence sources or targets by a vertex in the dependence graph. This also holds for the input to our placement method, which is given by a compact form of the reduced COIG discussed in the previous section. The question then arises which occurrence instances to include in one of the sets to be represented as a vertex.

Our next step will be to propagate placement relations along the edges of the reduced COIG $RCOIG = (\mathfrak{V}_{RCOIG}, \mathfrak{E}_{RCOIG})$. We can ignore the domains and images of the dependences represented by the edges of the graph, if for each edge $\mathfrak{E}_{RCOIG} \ni e = (v_s, v_t, H_e)$ the following holds:

$$\mathrm{dom}(H_e) \supseteq \mathrm{OccInst}(v_t) \tag{4.2}$$

$$\mathrm{im}(H_e) \supseteq \mathrm{OccInst}(v_s) \tag{4.3}$$

Where $\mathrm{OccInst}(v)$ yields the set of occurrence instances for vertex v. If Equation (4.2) holds, it is possible to propagate a relation from dependence source to target without considering domain and target of the respective relations, because it guarantees that the relation defined by the edge e actually holds for the complete target set represented by v_t. Thus, a relation computed as $H_e \circ H_n \cdots \circ H_1$ holds for the complete set represented by v_t. Correspondingly, if Equation (4.3) holds, it follows that a relation computed as $H_{e_n}^{-1} \circ \cdots \circ H_1^{-1} \circ H_e^{-1}$ holds for the complete set represented by v_s. If both Equations (4.2) and (4.3) held, placement relations could be propagated from source to target and from target to source of a dependence without problems, i.e., there would be no need to consider the domain of the dependence relation along which to propagate the original placement relation to obtain a new one.

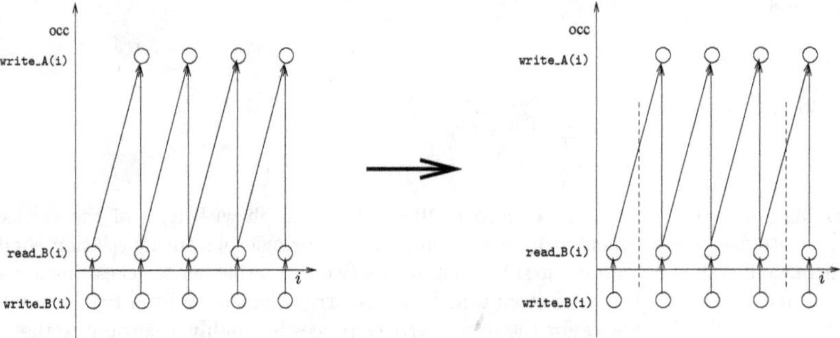

Figure 4.6: COIG for Example 4.8 with partitioning for Equation (4.3).

Example 4.8 *Reconsider the code fragment from Example 4.7, augmented by distribution directives, here:*

```
      REAL A(2:5)
      REAL B(1:5)
!HPF$ TEMPLATE T1(2:5)
!HPF$ TEMPLATE T2(1:5)
!HPF$ DISTRIBUTE T1(BLOCK)
!HPF$ DISTRIBUTE T2(BLOCK)
!HPF$ ALIGN A(i) WITH T1(i)
!HPF$ ALIGN B(i) WITH T2(i)
!HPF$ INDEPENDENT
      DO i=2,5
        A(i)=B(i-1)+B(i)**2
      END DO
```

The left hand side of Figure 4.6 shows the COIG of this code fragment, with the write access instances of the dummy loop defining B also represented. After LCCP, the read accesses `B(i)` *and* `B(i-1)` *can be represented by the same occurrence instance set, as shown on the left hand side of Figure 4.7.*

As can be gleaned from this figure, the graph on the left hand side has Property (4.2). It is thus possible to propagate a placement from source to target, i.e., from the write accesses to B to the write accesses to A along the dependences.

On the other hand, the graph on the right side does not *meet Property (4.2). In the code above, array A is aligned with a 4-element template T_1. This does not suffice for all 5 elements of B, should we try to distribute them according to the same placement of A. In order to satisfy Equation (4.2), a partitioning of the write accesses to B is necessary that discerns which combination of targets depend on a specific subset of occurrence instances. This partitioning is shown on the right hand side of Figure 4.6 as the separating dashed lines. Figure 4.7 shows the partitioned reduced COIG as resulting from the application of Algorithm 4.2.2 (Partition) with direction =* target_to_source. *Note, however, that the graph on the right hand side of Figure 4.7 has now gained Property (4.3), but also lost Property (4.2).*

Algorithm 4.2.2 (*Partition*) generates a partitioned reduced COIG as shown on the right hand side of Figure 4.7 for Example 4.8. The input is here expected to be in the form of sorted arrays *occurrenceLevels* and *restDeps* as produced by Algorithm 3.2.1 (*GroupOccurrenceInstances*), which is also used by Algorithm 3.4.1 (*LCCPLowLevel*). The first step initializes the data structures. In particular, the direction in which relations are to be propagated defines the order in which the vertices are visited – just as for the computation of $\equiv_{(\Delta^{(s,f)} \cup \Delta^{f}, \mathrm{id})}$. In the second step, each of the vertices v on the same level i of *occurrenceLevels* is partitioned into vertices $v'_1, \ldots, v'_n$ with $\mathrm{OccInst}(v) = \bigcup_{j \in \{1,\ldots,n\}} \mathrm{OccInst}(v_j)$ so that all elements of v_j read from the same combination of source vertices, if *direction* = source_to_target. This partitioning from dependence sources to targets guarantees that Equation (4.2) is satisfied. On the other hand, if *direction* = target_to_source, the partitioning is done in the opposite direction. In this case, each of the vertices v on the same level i of *occurrenceLevels* is partitioned so that into vertices $v'_1, \ldots, v'_n$ with $\mathrm{OccInst}(v) = \bigcup_{j \in \{1,\ldots,n\}} \mathrm{OccInst}(v_j)$ so that all elements of v_j *write* to the same combination of *target* vertices. This procedure then guarantees that the resulting graph satisfies Equation (4.2).

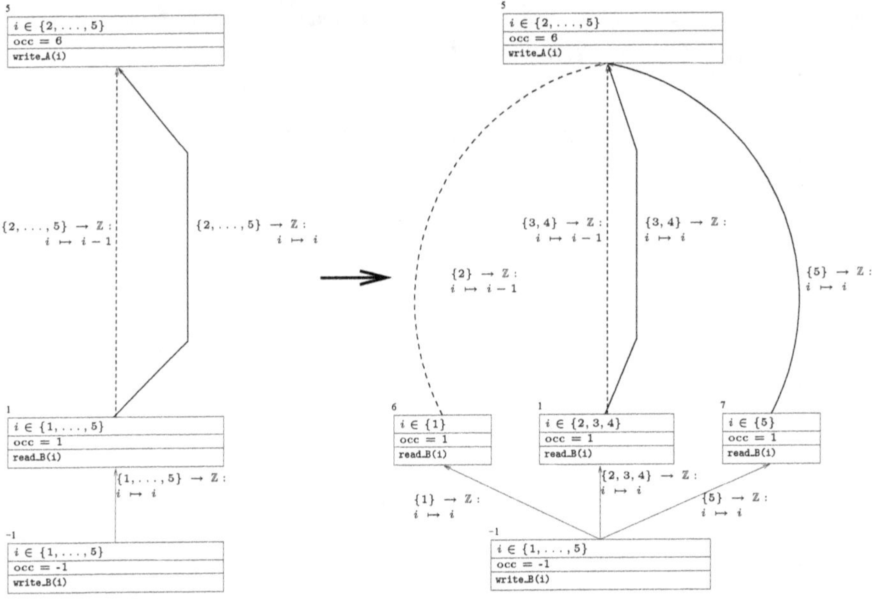

Figure 4.7: Reduced COIG for Example 4.8 with partitioning for Equation (4.3).

Algorithm 4.2.2 [*Partition*]:
Input:
occurrenceLevels :

array of sets of occurrence instances, each set *occurrenceLevels* [i] representing a union of vertices of an reduced COIG *RCOIG* = $(\mathfrak{V}_{RCOIG}, \mathfrak{E}_{RCOIG})$. I.e., each set *occurrenceLevels* [i] corrsponds to one level produced by Algorithm 3.2.1 (*GroupOccurrenceInstances*).

restDeps :

array of flow dependences and structural output dependences between the elements of *occurrenceLevels*.

$\mathfrak{A} = \{\mathfrak{A}_v \mid v \in \mathfrak{V}_{RCOIG}\}$:

for each vertex $v \in \mathfrak{V}_{RCOIG}$ the corresponding set of possible placements $\mathfrak{A}_v$.

direction ∈ {source_to_target, target_to_source} :

direction in which to partition $\mathfrak{G}_{RCOIG}$.

Output:
occurrenceLevelsNew :

array of sets of occurrence instances, each set *occurrenceLevelsNew* [i] representing a union of vertices of a partitioned reduced COIG $\mathfrak{G}'_{RCOIG} = (\mathfrak{V}_{RCOIG}{}', \mathfrak{E}_{RCOIG}{}')$.

restDepsNew :

array of flow dependences and structural output dependences between the elements of *occurrenceLevelsNew*. *restDepsNew* represents the edges of the partitioned reduced COIG $\mathfrak{G}'_{RCOIG}$. $\mathfrak{G}'_{RCOIG}$ has the following property:
if *direction* = source_to_target:

$$\left(\forall e : e = (v_s, v_t) \in \mathfrak{E}_{RCOIG} : \begin{array}{l} \mathrm{dom}(H_e) = \mathrm{OccInst}(v_s) \lor \\ \mathtt{write\_\cdot} = \mathrm{Occ}^{-1}(\mathrm{OccId}(v_t)) \end{array}\right)$$

if *direction* = target_to_source:

$$\left(\forall e : e = (v_s, v_t) \in \mathfrak{E}_{RCOIG} : \begin{array}{l} \mathrm{im}(H_e) = \mathrm{OccInst}(v_t) \lor \\ \mathtt{write\_\cdot} = \mathrm{Occ}^{-1}(\mathrm{OccId}(v_t)) \end{array}\right)$$

$\mathfrak{A}' = \{\mathfrak{A}'_v \mid v \in \mathfrak{V}_{RCOIG}'\}$:

for each vertex $v \in \mathfrak{V}_{RCOIG}'$ the corresponding set of possible placements $\mathfrak{A}'_v$.

Procedure:

```
/* STEP 1: initialization */
Let restDepsSource be a function that takes a node v of RCOIG
and returns a set of edges 𝔈 from restDeps with
(∀e = (v_s, v_t) : e ∈ 𝔈 : v_s = v);
Let restDepsTarget be a function that takes a node v of RCOIG
and returns a set of edges 𝔈 from restDeps with
(∀e = (v_s, v_t) : e ∈ 𝔈 : v_t = v);
/* if we partition from source to target, step up (forward) */
/* in occurrenceLevels, otherwise down (backward) */
if direction = source_to_target then
  start := 2;
  end := size(occurrenceLevels);
  stride := 1;
else
  start := size(occurrenceLevels);
  end := 2;
  stride := − 1;
endif
/* the lowest level of occurrenceLevels represents write accesses – which are not partitioned */
occurrenceLevelsNew[1] := occurrenceLevels[1];
/* STEP 2: for each level, partition each set of occurrence instances according to the */
/* comnbination of source (or targets, depending on direction) */
/* for this occurrence and create appropriate dependence relations */
for i = start to end, step stride
  /* for each set of occurrence instances */
  forall v ∈ occurrenceLevels[i]
    if direction = source_to_target then
      currInEdges := restDepsTarget(v);
      currOutEdges := restDepsSource(v);
    else
      currInEdges := restDepsSource(v);
      currOutEdges := restDepsTarget(v);
    endif
    toBePartitioned := OccInst(v);
    /* for each combination of sources (or targets)... */
    for i = #(currInEdges) to 1, step −1
      forall {e_1, ..., e_i} ∈ {𝔈 ∈ 𝒫(currInEdges) | #(𝔈) = i}
        /* ... partition the set according to the possible sources/targets */
```

```
        vNew := toBePartitioned ∩ ⋂_{j∈{1,...,i}} dom(H_{e_j});
        toBePartitioned := toBePartitioned \ OccInst(vNew);
        /* any possible placement already found for v may also be used for vNew */
        if OccInst(vNew) ≠ ∅ then
          𝔄'_vNew := 𝔄'_v;
          𝔄' := 𝔄' ∪ {𝔄'_vNew};
          occurrenceLevelsNew[i] := append(occurrenceLevelsNew[i], vNew);
          /* adapt the dependence relation to the created partition */
          if direction = source_to_target then
            forall e_o = (v, v_o, H_{e_o}) ∈ currOutEdges
              H_{e_n} := H_{e_o}|_{H_{e_o}^{-1}(OccInst(vNew))};
              if dom(H_{e_n}) ≠ ∅ then
                restDepsNew := append(restDepsNew, (vNew, v_o, H_{e_n}));
              endif
            end forall
          else
            forall e_o = (v_o, v, H_{e_o}) ∈ currOutEdges
              H_{e_n} := H_{e_o}|_{OccInst(vNew)};
              restDepsNew := append(restDepsNew, (v_o, vNew, H_{e_n}));
            end forall
          endif
        endif
      end forall
    endfor
  end forall
endfor
/* as an optimization, one could add an additional Step 3:   */
/* optimize graph by compacting it once more */
return (occurrenceLevelsNew, restDepsNew, 𝔄');
```

The reduced COIG to which the placement method should be applied is now either in a form in which placement relations can safely be propagated in the direction from the dependence source to the dependence target or vice versa. The actual propagation is based on the computation of the transitive closure of the dependence relations discussed in the next section.

4.2.4 Transitive Closure

The transitive closure of relations defined by polyhedra, as discussed in Section 2.4.1, is not in general computable [KPRS94]. In contrast, the transitive closure of a finite graph *is* computable – and ignoring the exact h-transformations (the weights of the edges), the reduced COIG is such a finite graph. A straightforward solution of this problem is given by the Floyd-Warshall Algorithm [Flo62, War62]. The derivation given by Algorithm 4.2.3 (*TransitiveClosure*) computes the transitive closure of the reduced COIG seen as a finite graph $RCOIG = (\mathfrak{V}_{RCOIG}, \mathfrak{E}_{RCOIG})$ with an edge $\mathfrak{E}_{RCOIG} \ni e = (v_s, v_t, H_e)$ consisting of a source and a target vertex and an associated linear function H_e (represented as linear relation) as a weight.

Algorithm 4.2.3 [*TransitiveClosure*]:
Input:
$RCOIG = (\mathfrak{V}_{RCOIG}, \mathfrak{E}_{RCOIG})$:
reduced COIG representing the COIG $(\mathfrak{O}\mathfrak{I}/{\equiv_{(\Delta^{(s,f)} \cup \Delta^f,\, \mathrm{id})}}, \Delta_{\equiv_{(\Delta^{(s,f)} \cup \Delta^f,\, \mathrm{id})}})$.

Output:

$RCOIG = (\mathfrak{V}_{RCOIG}, \mathfrak{E}_{RCOIG})$:
approximation $RCOIG' = (\mathfrak{V}_{RCOIG}{}', \mathfrak{E}_{RCOIG}{}')$ (representing a COIG $(\mathfrak{DI}/{\equiv_{(\Delta^{(s,f)} \cup \Delta^{f}, \mathrm{id})}}', \Delta_{\equiv_{(\Delta^{(s,f)} \cup \Delta^{f}, \mathrm{id})}}'))$ of the transitive closure of $RCOIG$: $\mathfrak{E}_{RCOIG} \subseteq \mathfrak{E}_{RCOIG}{}' \subseteq \mathfrak{E}_{RCOIG}{}^{+}$

Procedure:

```
𝔙_RCOIG' := 𝔙_RCOIG;
𝔈_RCOIG' := 𝔈_RCOIG;
forall v1 ∈ 𝔙_RCOIG'
  forall v2 ∈ 𝔙_RCOIG'
    forall v3 ∈ 𝔙_RCOIG'
      forall e1 ∈ {e' = (v1', v2', H1) ∈ 𝔈_RCOIG | v1' = v1 ∧ v2' = v2}
        forall e2 ∈ {e' = (v2', v3', H2) ∈ 𝔈_RCOIG | v2' = v2 ∧ v3' = v3}
        /* view H1, H2 as linear relations, ignoring */
        /* the domains; these are irrelevant here */
        𝔈_RCOIG' := 𝔈_RCOIG' ∪ {(v1, v3, H1 ∘+ H2)};
      end forall
    end forall
  end forall
end forall
Remove cycles from (𝔙_RCOIG, 𝔈_RCOIG);
return (𝔙_RCOIG, 𝔈_RCOIG);
```

The result of Algorithm 4.2.3 represents a composition of finitely many applications of linear relations that are obtained from the weights of the edges in $\mathfrak{E}_{RCOIG}$ by removing the domain constraints – these constraints are not relevant after partitioning with Algorithm 4.2.2. The plain Floyd-Warshall Algorithm does not handle cycles in any special way. The resulting graph may contain paths with one or two cycles. This is where the finiteness of the result graph comes into play: in a single run, the algorithm cannot introduce paths with an arbitrary number of cycles. A cycle in this graph has to contain a write access. We have also established that write accesses have a fixed placement relation. Any path from some occurrence over a write access to – which has a fixed placement – to another occurrence (or back to the original occurrence, as is the case for a simple cycle) does not introduce a good placement choice, since the communication to (or from) the fixed write access will be the same.

The only point in which a cycle can be useful for determining a placement relation is the combination discussed in Section 2.4.2. Definition 2.37 defines the combination of linear relations that behaves like both the input relations as long as they yield the same home; if they differ in the home for some elements, the result placement defines a replication of copies of these elements along the dimensions in which the original placements differ. Thus, the combination of two placements defines a placement on a superset of both original placement relations. The combination of several possible placements Φ_v for a given node v can therefore yield a new placement relation that only distributes data along axes, if this is advantegous for several dependences. The result could very well be a good placement candidate for v. This is true especially for the combination of an already generated placement Φ_v and the placement relation $\Phi_v \circ_{+} \Psi$ obtained from Φ_v by traversing a cycle ending in v, since this cycle then represents the reuse of a value generated with the same term. However, Considering these placements greatly increases the number of combinations to be assessed by our placement method. We therefore decide not to include such placements in the set $\mathfrak{A}_v$ of possible placements for a node v. In particular, cycles are completely removed from the result graph of Algorithm 4.2.3 before returning the graph.

4.2.5 Placement Propagation

After the computation of the transitive closure, the set of possible placements is easily computed by applying the placement of a set of write access instances $\mathfrak{D}$ to the composition of the linear relations representing the path to the vertex representing $\mathfrak{D}$. This propagation is done by the following algorithm.

Algorithm 4.2.4 [*Propagate*]:
Input:
$\mathfrak{G}_{RCOIG} = (\mathfrak{V}_{RCOIG}, \mathfrak{E}_{RCOIG})$:
reduced COIG with $\mathfrak{V}_{RCOIG} = \left\{v_1, \ldots, v_{n_{\mathfrak{V}_{RCOIG}}}\right\}$.

$\mathfrak{A} = \left\{\mathfrak{A}_{v_1}, \ldots, \mathfrak{A}_{v_{n_{\mathfrak{V}_{RCOIG}}}}\right\}$:

for each $v \in \mathfrak{V}_{RCOIG}$ the set of possible placement relations $\mathfrak{A}_v$.

direction $\in$ {source_to_target, target_to_source} :
direction in which to propagate placement relations.

Output:
$\mathfrak{A}' = \left\{\mathfrak{A}'_{v_1}, \ldots, \mathfrak{A}'_{v_{n_{\mathfrak{V}_{RCOIG}}}}\right\}$:

for each $v \in \mathfrak{V}_{RCOIG}$ the set of possible placement relations $\mathfrak{A}'_v$ including those propagated from dependence sources/targets.

Procedure:

```
/* STEP 1: compute the transitive closure of the graph using Algorithm 4.2.3 */
𝔈_RCOIG' := TransitiveClosure((𝔙_RCOIG, 𝔈_RCOIG));
/* STEP 2: propagate placement information in the given direction */
if direction = source_to_target then
  forall e = (v_s, v_t, H_e) ∈ 𝔈_RCOIG'
    Φ_{v_t} := Φ_{v_s} ∘+ H_e;
    𝔄'_{v_t} := 𝔄'_{v_t} ∪ {Φ_{v_t}};
  end forall
else
  forall e = (v_s, v_t, H_e) ∈ 𝔈_RCOIG'
    Φ_{v_s} := Φ_{v_t} ∘+ H_e^{-1};
    𝔄'_{v_s} := 𝔄'_{v_s} ∪ {Φ_{v_s}};
  end forall
endif
return 𝔄' = {𝔄'_{v_1}, ..., 𝔄'_{v_{n_𝔙_RCOIG}}};
```

Algorithm 4.2.4 computes the set of possible placement relations from the approximated transitive closure applied in Step 1. Depending on the direction in which to propagate the placement relations (from source to target or from target to source), the placement relations for a node v are computed as a compositions of the placements of write accesses reachable from v with the composition of the h-transformations or the inverse h-transformations building the weights of the corresponding paths.

Example 4.9 *Let us return to the code from Examples 4.7 and 4.8. This time, we use a replicated placement for array A than in Example 4.8:*

```
      REAL A(2:5)
      REAL B(1:5)
!HPF$ TEMPLATE T1(2:5)
```

```
!HPF$ TEMPLATE T2(1:5)
!HPF$ DISTRIBUTE T1(BLOCK)
!HPF$ DISTRIBUTE T2(BLOCK)
!HPF$ ALIGN A(i) WITH T1(*)
!HPF$ ALIGN B(i) WITH T2(i)
!HPF$ INDEPENDENT
      DO i=2,5
        A(i)=B(i-1)+B(i)**2
      END DO
```

Let us now determine the possible placements for the instances of the read access to B that we have determined to constitute an interesting set. For the sake of simplicity, we will only consider the dimensions of the index space of the corresponding occurrence instances here. For the read access to B, this is $\mathfrak{O}_B := \{i\}\, 1 \leq i \leq 5$*, because we have read access to all elements from element number* 1 *to element number* 5 *in this program fragment.*
When propagating placements from source to target in this case, the read access may obtain its placement from the corresponding write access in the dummy loop. The h-transformation in this case is a linear function $H_1 : \mathrm{i} \mapsto i$*. B – and thus the write access to B in the dummy loop – is distributed to a template according to a placement relation*

$$\Phi_B = \left(\begin{matrix} i & 1 \\ \end{matrix} \begin{pmatrix} 1 & 0 \\ 0 & 0 \\ 0 & 1 \end{pmatrix}, \begin{matrix} p_1 & p_2 & 1 \end{matrix} \begin{pmatrix} 1 & 0 & 0 \\ 0 & 0 & 0 \\ 0 & 1 & 0 \end{pmatrix} \right)$$

with p_1 *being the target index for template* T_1 *and* p_2 *the target index for* T_2*. The resulting possible placement for the read access is then (due to the simple h-transformation) the same distribution*

$$\Phi_{\mathtt{read\_B}} \circ_+ H_1 = \left(\begin{matrix} i & 1 \end{matrix} \begin{pmatrix} 1 & 0 \\ 0 & 0 \\ 0 & 1 \end{pmatrix}, \begin{matrix} p_1 & p_2 & 1 \end{matrix} \begin{pmatrix} 1 & 0 & 0 \\ 0 & 0 & 0 \\ 0 & 0 & 1 \end{pmatrix} \right)$$

For the propagation from target to source, the graph has to be repartitioned as shown in Figure 4.8. This results in three sets representing the read access (again, we only represent the index space without occurrence number etc.): $\mathfrak{O}_{B,1} := \{i\}\, i = 1$, $\mathfrak{O}_{B,2} := \{i\}\, 2 \leq i \leq 4$, $\mathfrak{O}_{B,3} := \{i\}\, i = 5$*. Figure 4.7 shows the reduced COIG partitioned for placement propagation from source to target. For each of these three sets, we not only have the possible placement computed above from the write access in the dummy loop, but also one computed from the placement of the write access to A, which is distributed in a replicated fashion according to the placement relation*

$$\Phi_{\mathtt{write\_A}} = \left(\begin{matrix} i & 1 \end{matrix} \begin{pmatrix} 0 & 0 \\ 0 & 0 \\ 0 & 1 \end{pmatrix}, \begin{matrix} p_1 & p_2 & 1 \end{matrix} \begin{pmatrix} 0 & 0 & 0 \\ 0 & 0 & 0 \\ 0 & 0 & 1 \end{pmatrix} \right)$$

Let us now calculate the respective placements for each of the sets $\mathfrak{O}_{B,1}$, $\mathfrak{O}_{B,2}$ *and* $\mathfrak{O}_{B,3}$ *that are propagated from the write access to A.*

$\mathfrak{O}_{B,1}$: *The only h-transformation along which we may propagate has the form* $H_2 : i \mapsto i - 1$*. The result is therefore the placement relation*

$$\Phi_{\mathtt{write\_A}} \circ_+ H_2^{-1} =$$

$$\left(\overset{\begin{matrix} i & 1 \end{matrix}}{\begin{pmatrix} 0 & 0 \\ 0 & 0 \\ 0 & 1 \end{pmatrix}}, \overset{\begin{matrix} p_1 & p_2 & 1 \end{matrix}}{\begin{pmatrix} 0 & 0 & 0 \\ 0 & 0 & 0 \\ 0 & 0 & 1 \end{pmatrix}} \right) \circ_+ \left(\overset{\begin{matrix} i & 1 \end{matrix}}{\begin{pmatrix} 1 & 0 \\ 0 & 1 \end{pmatrix}}, \overset{\begin{matrix} i & 1 \end{matrix}}{\begin{pmatrix} 1 & -1 \\ 0 & 1 \end{pmatrix}} \right)$$

$$= \left(\overset{\begin{matrix} i & 1 \end{matrix}}{\begin{pmatrix} 0 & 0 \\ 0 & 0 \\ 0 & 1 \end{pmatrix}}, \overset{\begin{matrix} p_1 & p_2 & 1 \end{matrix}}{\begin{pmatrix} 0 & 0 & 0 \\ 0 & 0 & 0 \\ 0 & 0 & 1 \end{pmatrix}} \right)$$

This placement generates a copy of part of the array B on all processors.

$\mathfrak{D}_{B,2}$: *In this case, we have two h-transformations according to which we can propagate a placement: in addition to* H_2*, we also have* $H_3 : i \mapsto i$*. However, since A is stored in a replicated fashion, the result of the composition with the* $\circ_+$ *operator is the same:*

$$\Phi_{\texttt{write\_A}} \circ_+ H_3^{-1} \quad =$$

$$\left(\overset{\begin{matrix} i & 1 \end{matrix}}{\begin{pmatrix} 0 & 0 \\ 0 & 0 \\ 0 & 1 \end{pmatrix}}, \overset{\begin{matrix} p_1 & p_2 & 1 \end{matrix}}{\begin{pmatrix} 0 & 0 & 0 \\ 0 & 0 & 0 \\ 0 & 0 & 1 \end{pmatrix}} \right) \circ_+ \left(\overset{\begin{matrix} i & 1 \end{matrix}}{\begin{pmatrix} 1 & 0 \\ 0 & 1 \end{pmatrix}}, \overset{\begin{matrix} i & 1 \end{matrix}}{\begin{pmatrix} 1 & 0 \\ 0 & 1 \end{pmatrix}} \right)$$

$$= \left(\overset{\begin{matrix} i & 1 \end{matrix}}{\begin{pmatrix} 0 & 0 \\ 0 & 0 \\ 0 & 1 \end{pmatrix}}, \overset{\begin{matrix} p_1 & p_2 & 1 \end{matrix}}{\begin{pmatrix} 0 & 0 & 0 \\ 0 & 0 & 0 \\ 0 & 0 & 1 \end{pmatrix}} \right)$$

In this case, we have actually generated only one new placement relation when propagating from target to source (i.e., we have only two possible placements for this set). In order to notice this fact, we compare the resulting placement relations as discussed in Section 2.4.2 and only add placement relations as new possibilities that we have not encountered before.

$\mathfrak{D}_{B,3}$: *As Figure 4.7 shows, only the h-transformation* H_3 *connects* $\mathfrak{D}_{B,3}$ *with the write access. Therefore, we have the same possible placement relations for this set as for* $\mathfrak{D}_{B,1}$ *and* $\mathfrak{D}_{B,2}$*.*

The propagation algorithm described above may produce placement relations that are not expressable in HPF. The next section will handle this case.

4.2.6 Postprocessing Placement Relations

Not every placement relation computed by Algorithm 4.2.4 represents a placement that we can actually use in the target program. In our case, the restriction is that we only allow legal placements according to the HPF standard [Hig97]. As already discussed in Section 2.1.1, there are placements representable by linear relations that – for good reason – do not define a legal HPF placement. Example 2.7 on page 11 shows an illegal placement in HPF. HPF is not capable of processing general linear relations as placements. Example 4.10 shows why we may have to deal with the illegal placement of Example 2.7 at this point.

Example 4.10 *Consider the following legal* HPF *code:*

```
      REAL A(10)
      REAL B(10,10)
!HPF$ TEMPLATE T(10,10)
```

```
!HPF$ DISTRIBUTE T(BLOCK,BLOCK)
!HPF$ ALIGN A(i,j) WITH T(i,j)
!HPF$ ALIGN B(i,j) WITH T(i,j)
!HPF$ INDEPENDENT, NEW(j)
      DO i=1,10
!HPF$ INDEPENDENT
        DO j=1,10
          A(j,i)=2*B(i,i)+A(j,i)
        END DO
      END DO
```

LCCP will extract the repeated computation of the 1-dimensional expression `2*B(i,i)` *to produce code like the following:*

```
      REAL A(10)
      REAL B(10,10)
      REAL TMP(10)
!HPF$ TEMPLATE T(10,10)
!HPF$ DISTRIBUTE T(BLOCK,BLOCK)
!HPF$ ALIGN A(i,j) WITH T(i,j)
!HPF$ INDEPENDENT
      DO i=1,10
        TMP(i)=2*B(i,i)
      END DO
!HPF$ INDEPENDENT, NEW(j)
      DO i=1,10
!HPF$ INDEPENDENT
        DO j=1,10
          A(j,i)=TMP(i)+A(j,i)
        END DO
      END DO
```

The only thing missing in this transformed code is the placement for TMP. A straightforward placement would be to propagate the placement of B to its target TMP according to the flow dependence $H : i \mapsto (i,i)^T$*. Leading to a placement defined by:*

```
!HPF$ ALIGN TMP(i) WITH T(i,i)
```

Unfortunately, this is an illegal alignment, since the align-subscript `(i,i)` *contains the align dummy* `i` *twice. We can circumvent this problem by introducing replicated placements. The two placements*

```
!HPF$ ALIGN TMP(i) WITH T(i,*)
```

and

```
!HPF$ ALIGN TMP(i) WITH T(*,i)
```

are both legal HPF *placements. They define a replication in either the first or the second dimension of* T*. Thus, TMP is always stored on a superset of the processors specified by the original placement. Thus, copies will be available* at least *on the processors specified in the original placement.*

The following theorem, taken from Wondrak's diploma thesis [Won05], formalizes the restrictions a placement relation has to satisfy in order to be expressible in HPF. In the following, we will convert between the representation of a linear relation by two functions and its defining matrix by a function RelMat:

$$\text{RelMat} : (\Phi_L, \Phi_R) \mapsto d \cdot \left(\begin{array}{cc} M_{\Phi_L} & -M_{\Phi_R} \end{array} \right)$$

where $\Phi_L : \alpha \rightarrow M_{\Phi_L} \cdot \alpha$ and $\Phi_R : \beta \rightarrow M_{\Phi_R} \cdot \beta$ (and d is the lcm of all entries of M_{Φ_R} and M_{Φ_L}).

Theorem 4.11 *Let (Φ_L, Φ_R) be two mappings representing a linear relation defining a placement and let $M_\Phi = \text{RelMat}((\Phi_L, \Phi_R))$ be the corresponding matrix. Let $M_\Phi{}'$ be the reduced echelon form of $M_\Phi \in \mathbb{Q}^{q \times n_{src}+n_{blob}+2+n_{tgt}+n_{blob}+2}$. Then the placement defined by (Φ_L, Φ_R) is a legal* `HPF` *placement, iff each of the following conditions hold*

1. *No row of M_Φ contains non-zero entries in more than one source index dimension:*
$$(\forall r : r \in \{1, \ldots, q\} : \#(\{c \in \{1, \ldots, n_{src}\} \mid M_\Phi[r, c] \neq 0\}) \leq 1) \tag{4.4}$$

2. *No row of M_Φ contains non-zero entries in more than one target index dimension:*
$$(\forall r : r \in \{1, \ldots, q\} : \#(\{c \in \{n_{start}, \ldots, n_{end}\} \mid M_\Phi[r, c] \neq 0\}) \leq 1) \tag{4.5}$$
with $n_{start} = n_{src} + n_{blob} + 2 + n_{tgt}$ and $n_{end} = n_{src} + n_{blob} + 2 + n_{tgt} + n_{blob}$.

3. *No column of M_Φ representing a source index features more than one non-zero entry:*
$$(\forall c : c \in \{1, \ldots, n_{src}\} : \#(\{r \in \{1, \ldots, q\} \mid M_\Phi[r, c] \neq 0\}) \leq 1) \tag{4.6}$$

4. *No column of M_Φ representing a target index features more than one non-zero entry:*
$$(\forall c : c \in \{n_{start}, \ldots, n_{end}\} : \#(\{r \in \{1, \ldots, q\} \mid M_\Phi[r, c] \neq 0\}) \leq 1) \tag{4.7}$$
with $n_{start} = n_{src} + n_{blob} + 2 + n_{tgt}$ and $n_{end} = n_{src} + n_{blob} + 2 + n_{tgt} + n_{blob}$.

Proof:

The theorem is an obvious reformulation of the constraints to the `ALIGN` directive in the `HPF` standard [Hig97, Section 3.4]. Conditions for columns in the above theorem come from overall statements about the complete directive; a conditions on a row comes from the relation between an element of a `<align-dummy-list>` and an element of an `<align-subscript-list>`. The gentle reader is referred to Wondrak's work [Won05] for a detailed discussion. ✓

Note that the restrictions imposed by `HPF` are not arbitrary, but guarantee that arrays can be defined statically with a rectangular shape. Otherwise, a processor might have to test whether its coordinates in the processor mesh satisfy some constraint in order to determine, whether it has to hold any data at all.

The following algorithm (*MakeHPFConform*) takes a set of possible placements for a vertex of a reduced COIG, as produced by Algorithm 4.2.4, and returns a set of placements that all conform to the restrictions imposed by `HPF`. Each of the generated placements defines a distribution of array elements to a superset of the processor mesh elements onto which the original placement distributed the same array elements. This is guaranteed by the fact that a new placement is generated from an old one by zeroing out one or more of the rows in the matrix defining the placement. Thus, one or more of the conditions defining the placement relation is deleted. The naïve approach of Algorithm 4.2.5 is to delete every possible combination of conditions from the original in turn, each time testing, whether the result defines a legal `HPF` placement. Note that a row may also contradict the rules of Theorem 4.11, if one of its columns contradicts those rules. Another approach may be to heuristically delete some preferred set of rows.

Algorithm 4.2.5 [*MakeHPFConform*]:
Input:
$\mathfrak{A}_v$: the set of possible placements for a vertex v.
Output:
$\mathfrak{A}'_v$: a transformed set of possible placements, each of which is representable in HPF and defines a superset of homes defined by a placement from $\mathfrak{A}_v$.

Procedure:
```
forall Φ ∈ 𝔄_v
  if Φ is not representable in HPF according to Theorem 4.11 then
    M_Φ := RelMat(Φ);
    Determine the set 𝔋 of rows of M_Φ that do not conform to HPF according to Theorem 4.11;
    /* 𝔘 is the set of all combinations of numbers of illegal rows */
    𝔘 := P(𝔋) \ ∅;
    forall 𝔑 ∈ 𝔘
      M_Φ' := M_Φ;
      forall r ∈ 𝔑
        M_Φ'[r,·] := 0;
        if M_Φ' is HPF conform according to Theorem 4.11 then
          𝔄'_v := 𝔄'_v ∪ RelMat^{-1}(M_Φ');
        endif
      end forall
    end forall
  endif
end forall
```

Algorithms 4.2.4 and 4.2.5 generate a large number of placements for each node v of the reduced COIG. These placement relations – represented by (unscoped) linear relations – are stored in a set $\mathfrak{A}_v$. In order to implement such a set, it is necessary to eliminate duplicate entries, which are identified as discussed in Section 2.4.2. Thus, placement relations that define the same placement are handled only once. We have already encountered this problem in Example 4.9. Practical experience showed that storing these reduced sets of placement can be a challenge for even quite a simple program fragment. This is because a very large number of quite similar placement relations is generated. Our implementation in LooPo therefore stores the placement relations in the form of sparse matrices. With all the possible placements generated for each vertex of the reduced COIG now, it is time to evaluate these placements.

4.2.7 Placement Evaluation

The last step proposed for placement generation on page 154 is the evaluation of the generated placements. Algorithm 4.2.6 (*Evaluate*) performs this evaluation simply by generating each possible combination of placements, generating a reduced COIG transformed by the space-time mapping corresponding to that placement relation and returning the least expensive placement combination according to a cost function that assigns a cost to the transformed reduced COIG.

Algorithm 4.2.6 [*Evaluate*]:
Input:
$RCOIG = (\mathfrak{V}_{RCOIG}, \mathfrak{E}_{RCOIG})$:
reduced COIG with $\mathfrak{V}_{RCOIG} = \left\{v_1, \ldots, v_{n_{\mathfrak{V}_{RCOIG}}}\right\}$.

$\mathfrak{A} = \left\{\mathfrak{A}_{v_1}, \ldots, \mathfrak{A}_{v_{n_{\mathfrak{V}_{RCOIG}}}}\right\}$:
for each vertex $v \in \mathfrak{V}_{RCOIG}$ a set $\mathfrak{A}_v$ of possible placements.

CostFunc :
a cost function for the transformed reduced COIG $CostFunc : RCOIG' \to \mathbb{N}$.
Output:
$\Phi = \bigcup_{v \in \mathfrak{V}_{RCOIG}} \Phi_v$:
placement relation consisting of a linear relation Φ_v for each vertex v.

Procedure:
```
/* STEP 1: initialization */
cost_best := ∞;
𝔎 := 𝔄_v1 × ... × 𝔄_v_n𝔙RCOIG ;
/* STEP 2: evaluate the costs for each combination of placements */
forall (Φ_v1, ..., Φ_v_n𝔙RCOIG) ∈ 𝔎
  /* compute the transformed reduced COIG */
  𝔙_RCOIG' := 𝔙_RCOIG;
  𝔈_RCOIG' := ∅;
  forall e = (v_s, v_t, H_e) ∈ 𝔈_RCOIG
  /* additional information such as placement relations from */
  /* other levels and schedule may be taken into account here */
  (Ψ_L, Ψ_R) := Φ_vs ∘₋ H_e ∘₊ Φ_vt;
  𝔈_RCOIG' := 𝔈_RCOIG' ∪ {(v_s, v_t, (Ψ_L, Ψ_R))};
  end forall
  /* estimate the cost for the transformed graph */
  cost := CostFunc((𝔙_RCOIG', 𝔈_RCOIG'));
  if cost_best > cost then
    cost_best := cost;
    Φ := {Φ_v1, ..., Φ_n𝔙RCOIG};
  endif
end forall
return Φ;
```

Any cost function can be used here to determine the expected cost of executing the transformed program fragment. Thus, the cost model can be tuned to the compiler and the hardware. Section 4.3 will examine how to obtain a cost model for a given compiler (and machine). But first, let us put the single parts together to obtain the complete algorithm.

4.2.8 The Final Algorithm

The algorithms presented in the previous sections can, by and large, be plugged into each other. Only Algorithm 4.2.2 (*Partition*) depends on the representation of the set of vertices in the reduced COIG by an array. With this representation, it is possible to partition the graph for propagation in one dependence direction just as with Algorithm 3.4.1 (*LCCPLowLevel*), since the set of flow dependences (both structural and non-structural) represents a strict partial order in its transitive closure. And since structural output dependences always go from an operator instance to a write access instance, the same holds for the transitive closure of structural flow dependences together with structural output dependences. The other algorithms use a simpler representation so that they are easier to follow.

Thus, the overall algorithm in principle consists of a loop that performs the steps:

1. Partition the reduced COIG for relation propagation from source to target of flow dependences.

2. Propagate placement relations from source to target.

3. Partition the reduced COIG for relation propagation from target to source of structural flow dependences and structural output dependences to their sources.

4. Propagate placement relations from these targets back to their sources.

After completion of the loop, the algorithm still has to accomplish the following tasks:

5. Filter out illegal placements (note that the restrictions HPF raises here actually do have some merit in general). This results in a set $\left\{\mathfrak{A}_{v_1}, \ldots, \mathfrak{A}v_{n_{\mathfrak{V}_{RCOIG'}}}\right\}$ containing a set of possible placements for each vertex of the partitioned reduced COIG.

6. For each combination of these placements, i.e., each element of $\mathfrak{A}_{v_1} \times \ldots \times \mathfrak{A}v_{n_{\mathfrak{V}_{RCOIG'}}}$, compute a transformed reduced COIG and from that a (communication) cost, which is used to identify the optimal placement in this search space.

These steps are performed by the following algortihm:

Algorithm 4.2.7 [*OIAllocator*]:
Input:
$RCOIG = (\mathfrak{V}_{RCOIG}, \mathfrak{E}_{RCOIG})$:
a reduced COIG.

$\mathfrak{A} = \{\mathfrak{A}_v \mid v \in \mathfrak{V}_{\text{write}}\}$:
for each vertex v representing write access instances the corresponding set of possible placements $\mathfrak{A}_v$.

CostFunc :
a cost function for the transformed reduced COIG $CostFunc : RCOIG' \rightarrow \mathbb{N}$.

Output:
$RCOIG' = (\mathfrak{V}_{RCOIG}{}', \mathfrak{E}_{RCOIG}{}')$:
a newly partitioned reduced COIG.

$\Phi = \bigcup_{v \in \mathfrak{V}_{RCOIG'}} \Phi_v$:
a placement relation Φ_v for each vertex $v \in \mathfrak{V}_{RCOIG}{}'$.

Procedure:

```
Let crit be some criterion indicating when to stop placement propagation;
Let 𝔈_RCOIG denote the flow dependences of 𝔈_RCOIG (𝔈_RCOIG = Δ^f ∪ Δ^(s,f));
/* call Algorithm 3.2.1 to obtain a sorted array of occurrence instance sets */
occurrenceLevels := GroupOccurrenceInstances(𝔙_RCOIG, 𝔈_RCOIG);
Let restDeps be an array representing the h-transformations of 𝔈_RCOIG;
while crit
   /* propagate from source to target */
   (occurrenceLevels, restDeps, 𝔄) := Partition(occurrenceLevels, restDeps, 𝔄, source_to_target);
   Let RCOIG' = (𝔙_RCOIG', 𝔈_RCOIG') be the representation of occurrenceLevels, restDeps as sets;
   Let depsSrcToTgt be an array representing the h-transformations of flow dependences of restDeps;
   𝔄 := Propagate(RCOIG, 𝔄, source_to_target);
   /* propagate from target to source */
   (occurrenceLevels, restDeps, 𝔄) := Partition(occurrenceLevels, restDeps, 𝔄, target_to_source);
```

Let $RCOIG' = (\mathfrak{V}_{RCOIG}', \mathfrak{E}_{RCOIG}')$ be the representation of *occurrenceLevels*, *restDeps* as sets;
Let *depsTgtToSrc* be an array representing the h-transformations of structural flow dependences and structural output dependences of *restDeps*;
$\mathfrak{A} := Propagate(RCOIG, \mathfrak{A}, \text{target\_to\_source})$;
`endwhile`
/* restrict search space to legal placements */
$\mathfrak{A} := MakeHPFConform(\mathfrak{A})$;
/* get the placement combination with the least cost */
$\Phi := Evaluate(RCOIG, \mathfrak{A}, costFunc)$;
`return` $(RCOIG', \Phi)$;

Practical experience shows that it is in general not worthwhile iterating through the `while`-loop more than once, if cycles are not handled in some special way – as proposed above in Section 4.2.4 (page 160). This is true even though, as the following example shows, cycles are not the only way to obtain additional placement candidates. However, the repeated partitioning simply results in such fine a granularity that the additional computation needed to evaluate even more possible placement relations does not pay off, since the resulting additional placements usually only differ marginally from those that can be determined in a single sweep. Therefore, we propose to use only a single loop iteration or a bound on the number of generated possible placements as abort criterion *crit* above. Another possibility for the abort criterion would be to check, whether each vertex v in the graph has already obtained a placement relation from each other vertex v' which is connected to v via a path that does not contain any write access.

Yet, the only point at which the strategy of iterating through the loop only once loses placement information that might be interesting is where different points in a larger expression influence each other. This is the case, for example, if we need to generate placements for terms `B(i)`, `C(i)` and `B(i)+C(i)` so that the read access to C may determine where to place the read access to B. However, these cases rarely reveal new good choices for a placement relation.

Example 4.12 shows the basic procedure behind Algorithm 4.2.7 and the placement strategy.

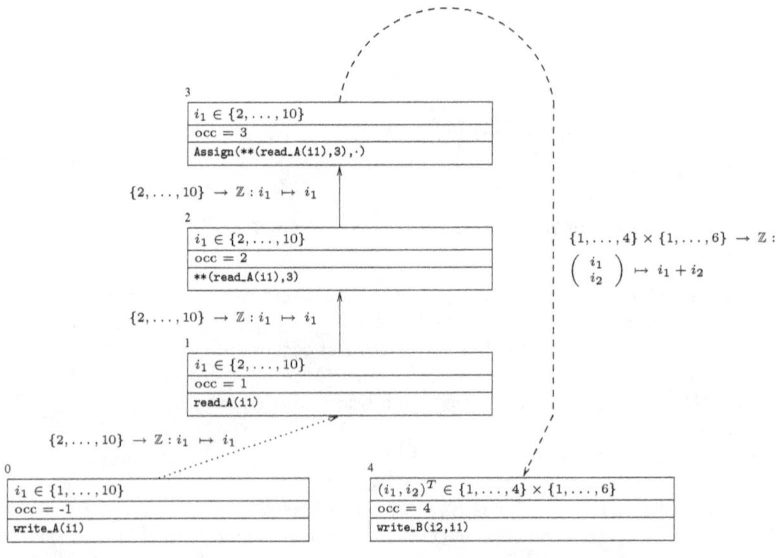

Figure 4.8: Reduced COIG of the code fragment of Example 4.12.

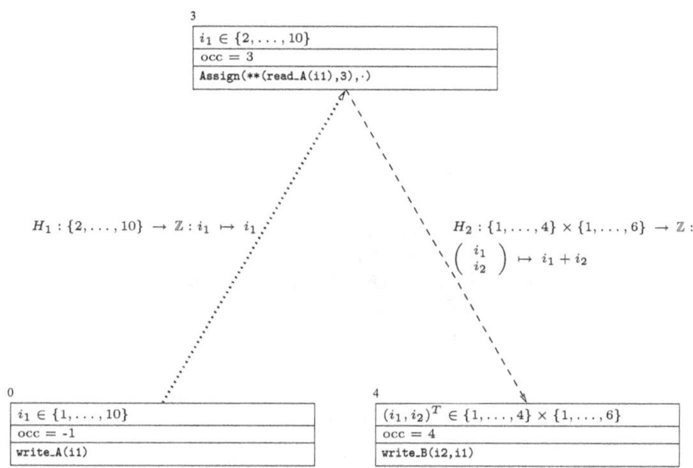

Figure 4.9: Compacted reduced COIG of the code fragment of Example 4.12.

Example 4.12 *Let us revisit the code of Example 3.18, this time augmented with an* HPF *placement:*

```
      DOUBLE PRECISION A(1:10)
      DOUBLE PRECISION B(1:4,1:6)
!HPF$ TEMPLATE T1(1:10)
!HPF$ TEMPLATE T2(1:4,1:6)
!HPF$ DISTRIBUTE T1(BLOCK)
!HPF$ DISTRIBUTE T2(BLOCK,BLOCK)
!HPF$ INDEPENDENT
      DO i1=1,4
!HPF$ INDEPENDENT
        DO i2=1,6
          B(i2,i1)=A(i1+i2)**3
        END DO
      END DO
```

Let us suppose that LCCP created a new, 1-dimensional, occurrence instance set for the computation of $A(i)^3$ resulting in the following code:

```
      DOUBLE PRECISION A(1:10)
      DOUBLE PRECISION B(1:4,1:6)
!HPF$ TEMPLATE T1(1:10)
!HPF$ TEMPLATE T2(1:4,1:6)
!HPF$ DISTRIBUTE T1(BLOCK)
!HPF$ DISTRIBUTE T2(BLOCK,BLOCK)

!HPF$ INDEPENDENT
      DO i1=2,10
          TMP(i1)=A(i1)**3
      END DO
!HPF$ INDEPENDENT
      DO i1=1,4
!HPF$ INDEPENDENT
```

```
    DO i2=1,6
      B(i2,i1)=TMP(i1+i2)
    END DO
  END DO
```

In fact, here the complete assignment – viewed as a function application – is a 1-dimensional expression. Figure 4.8 shows the reduced COIG for this example. Figure 4.9 shows only the interesting sets of this reduced COIG. The fat dotted arrows indicate non-structural flow dependences, and the fat dashed arrows indicate structural output dependences. The graphs mirror the sorting of Algorithm 3.2.1 (GroupOccurrenceInstances): the lowest levels produced by the algorithm are depicted at the bottom of the figure.
With the parallel index p_1 corresponding to template T_1 and the parallel indices p_2 and p_3 corresponding to template T_2, we can use Algorithm 4.1.1 to obtain the following placement relations:

A	$\Phi_A = \left(i_1 \mapsto \begin{pmatrix} i_1 \\ 0 \\ 0 \end{pmatrix}, \begin{pmatrix} p_1 \\ p_2 \\ p_3 \end{pmatrix} \mapsto \begin{pmatrix} p_1 \\ 0 \\ 0 \end{pmatrix} \right)$
B	$\Phi_B = \left(\begin{pmatrix} i_1 \\ i_2 \end{pmatrix} \mapsto \begin{pmatrix} 0 \\ i_1 \\ i_2 \end{pmatrix}, \begin{pmatrix} p_1 \\ p_2 \\ p_3 \end{pmatrix} \mapsto \begin{pmatrix} 0 \\ p_2 \\ p_3 \end{pmatrix} \right)$

A partitioning in direction from source to target concerns the vertices 0 and 3 of the figures (the only place that involves as flow dependence). However, since $\mathrm{dom}(H_1)$ *covers the complete set represented by vertex 3, Equation 4.2 holds; i.e., it is immediately possible to propagate the placement* Φ_A *that holds for vertex 0 can be propagated to vertex 3. The corresponding placement is*

$$\Phi_1 = \Phi_A \circ_+ H_1 \quad = \quad \left(i_1 \mapsto \begin{pmatrix} i_1 \\ 0 \\ 0 \end{pmatrix}, \begin{pmatrix} p_1 \\ p_2 \\ p_3 \end{pmatrix} \mapsto \begin{pmatrix} p_1 \\ 0 \\ 0 \end{pmatrix} \right)$$

In the next step, placements are propagated from target to source. The propagation in this direction concerns vertices 3 and 4, which are connected by a structural output dependence. Again, a paritioning is not necessary, since $\mathrm{im}(H_2)$ *covers the complete set represented by vertex 3. We obtain*

$$\Phi_2 = \Phi_B \circ_+ H_2^{-1} \quad = \quad \left(i_1 \mapsto \begin{pmatrix} 0 \\ 0 \\ 0 \end{pmatrix}, \begin{pmatrix} p_1 \\ p_2 \\ p_3 \end{pmatrix} \mapsto \begin{pmatrix} 0 \\ -p_2 + p_3 \\ 0 \end{pmatrix} \right)$$

As discussed above, we choose to iterate only once through the `while`*-loop, which means that we have now found all possible placement relations for the one vertex that did not have an associated placement from the beginning on. This leaves us with the following sets of placements for the interesting sets represented by vertices 0, 3 and 4 in Figure 4.9:*

$$\begin{aligned} \mathfrak{A}_0 &= \{\Phi_A\} \\ \mathfrak{A}_3 &= \{\Phi_1, \Phi_2\} \\ \mathfrak{A}_4 &= \{\Phi_B\} \end{aligned}$$

With a set of placements created for each node, we have to check all the placements in turn for legality. For Φ_A *and* Φ_B*, this is clear.* Φ_1 *is represented by matrices*

$$\begin{array}{c} \begin{matrix} i_1 & i_2 \end{matrix} \\ \begin{pmatrix} 1 & 0 \\ 0 & 0 \\ 0 & 0 \end{pmatrix} \end{array}, \quad \begin{array}{c} \begin{matrix} p_1 & p_2 & p_3 \end{matrix} \\ \begin{pmatrix} 1 & 0 & 0 \\ 0 & 0 & 0 \\ 0 & 0 & 0 \end{pmatrix} \end{array} \tag{4.8}$$

Not surprisingly, this placement is representable in HPF.
Φ_2 *is represented as*

$$\begin{array}{c} \begin{matrix} i_1 & i_2 \end{matrix} \\ \begin{pmatrix} 0 & 0 \\ 0 & 0 \\ 0 & 0 \end{pmatrix} \end{array}, \quad \begin{array}{c} \begin{matrix} p_1 & p_2 & p_3 \end{matrix} \\ \begin{pmatrix} 0 & 0 & 0 \\ 0 & -1 & 1 \\ 0 & 0 & 0 \end{pmatrix} \end{array} \tag{4.9}$$

which does *represent an illegal placement according to Theorem 4.11, since several target dimensions are present in a single row of the matrix on the right. The only condition we can remove from this placement is* $-p_2 + p3 = 0$. *This results in a single alternative placement*

$$\begin{array}{c} \begin{matrix} i_1 & i_2 \end{matrix} \\ \begin{pmatrix} 0 & 0 \\ 0 & 0 \\ 0 & 0 \end{pmatrix} \end{array}, \quad \begin{array}{c} \begin{matrix} p_1 & p_2 & p_3 \end{matrix} \\ \begin{pmatrix} 0 & 0 & 0 \\ 0 & 0 & 0 \\ 0 & 0 & 0 \end{pmatrix} \end{array} \tag{4.10}$$

which means to replicate all occurrence instances onto all processors. Let us denote this resulting placement with Φ_3.

We now proceed to the evaluation. For this, we compute the transformed h-transformations H_1' *and* H_2', *which correspond to the original h-transformations* H_1 *and* H_2, *respectively. This is a trivial case again, since we have only a single vertex with more than one associated possible placements. The following tables associate the different combinations with the corresponding dependences in the target space:*

vertex 0 Φ_{v_0}	*vertex 3* Φ_{v_3}	*vertex 4* Φ_{v_4}	$H_1' = \Phi_{v_0} \circ_- H_1 \circ_+ \Phi_{v_3}^{-1}$	*cost*
Φ_A	Φ_1	Φ_B	$\left(\begin{pmatrix} p_1 \\ p_2 \\ p_3 \end{pmatrix} \mapsto \begin{pmatrix} p_1 + p2 \\ 0 \\ 0 \end{pmatrix}, \begin{pmatrix} p_1 \\ p_2 \\ p_3 \end{pmatrix} \mapsto \begin{pmatrix} p_1 \\ 0 \\ 0 \end{pmatrix}\right)$	100
Φ_A	Φ_3	Φ_B	$\left(\begin{pmatrix} p_1 \\ p_2 \\ p_3 \end{pmatrix} \mapsto \begin{pmatrix} 0 \\ 0 \\ 0 \end{pmatrix}, \begin{pmatrix} p_1 \\ p_2 \\ p_3 \end{pmatrix} \mapsto \begin{pmatrix} 0 \\ 0 \\ 0 \end{pmatrix}\right)$	10

vertex 0 Φ_{v_0}	*vertex 3* Φ_{v_3}	*vertex 4* Φ_{v_4}	$H_2' = \Phi_{v_3} \circ_- H_2 \circ_+ \Phi_{v_4}^{-1}$	*cost*
Φ_A	Φ_1	Φ_B	$\left(\begin{pmatrix} p_1 \\ p_2 \\ p_3 \end{pmatrix} \mapsto \begin{pmatrix} 0 \\ 0 \\ 0 \end{pmatrix}, \begin{pmatrix} p_1 \\ p_2 \\ p_3 \end{pmatrix} \mapsto \begin{pmatrix} 0 \\ 0 \\ 0 \end{pmatrix}\right)$	10
Φ_A	Φ_3	Φ_B	$\left(\begin{pmatrix} p_1 \\ p_2 \\ p_3 \end{pmatrix} \mapsto \begin{pmatrix} p_2 \\ p_3 \\ 0 \end{pmatrix}, \begin{pmatrix} p_1 \\ p_2 \\ p_3 \end{pmatrix} \mapsto \begin{pmatrix} p_2 \\ p_3 \\ 0 \end{pmatrix}\right)$	1

The final columns in the tables represent some (communication) costs that we assign to the corresponding h-transformation somewhat arbitrarily. We choose these costs roughly from what can be expected to be handled well by HPF *compilers. For estimating the cost of a specific transformed reduced COIG, we just sum up the costs of the edges. We thus obtain* 110 *for* Φ_1 *and* 11 *for* Φ_3. *With this estimation, we would choose the second placement,* Φ_3 *for the assignment – the replication. This may not seem to be a good choice at first glance. But the replication actually*

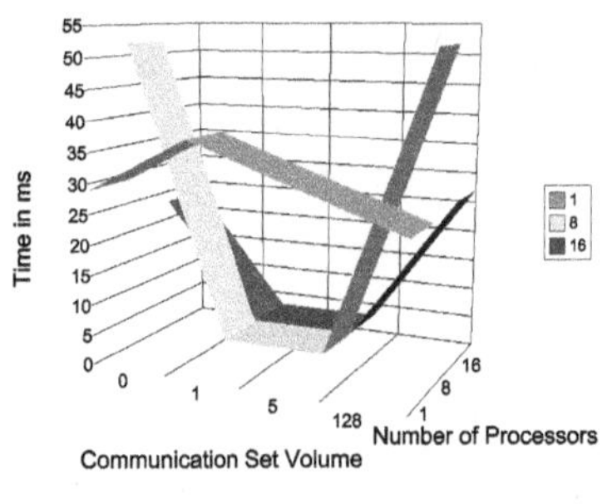

(a) Regular accesses

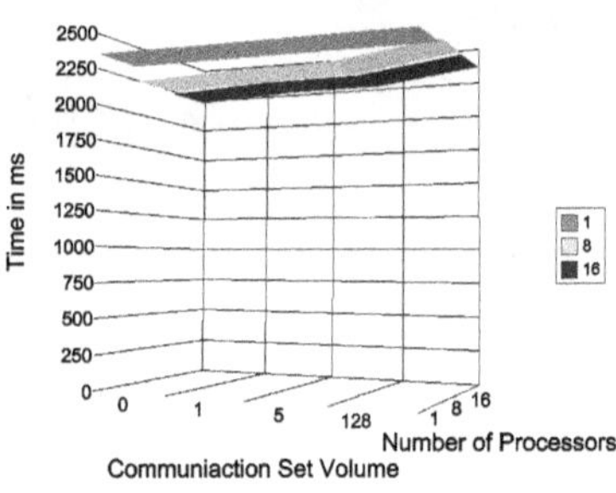

(b) Irregular accesses

Figure 4.10: Run time for a 1D-shift in dependence of communication set value and number of processors

allows the compiler to broadcast the complete array A to all processors, which may indeed be less time consuming than finding the correct owner of the array element to read from.

In the example above we introduced some more or less random cost function to evaluate the placement relation for an occurrence instance set. In the next section, we will examine how to obtain a cost function that actually models the behaviour of an HPF compiler.

4.3 Cost Models

We have now introduced a placement method that is guided by a function (*CostFunc*) in selecting a promising placement relation. It is therefore important to choose this cost function appropriately for the system at hand.

When developing a cost model, one may view different factors that may influence the performance (or other cost). The *Bulk Synchronous Parallelism* programming model (BSP) [Val90, SHM97] comes with its own coarse but very effective cost model. Program execution consists of a series of so-called *supersteps.* Between these supersteps, the processors are synchronized and data is exchanged among the distributed memory banks. The time for program execution is thus modelled as the actual computation time plus for each superstep the communication cost, which conists itself of a startup-time l and a communication volume dependent time:

$$\text{cost of superstep} \quad = \quad \max(\{w_i \mid i \in \textit{processes}\}) + \max(\{h_i \mid i \in \textit{processes}\}) \cdot g + l \qquad (4.11)$$

where w_i is the workload of process i and g is the normalized bandwidth of the system and h_i a so-called *h-relation* – which is not to be confused with the h-transformation. An h-relation is usually modelled as the maximum of the `send` and `receive` operations to be performed by the processes (although variations are possible). The machine-dependent parameters w_i, g and l are inferred from a run of a benchmark suite.

We also take the approach of a global time step in that the programs produces here are all synchronous programs – with outer loops enumerating these global time steps. Asynchronous loop programs are not well (easily and efficiently) representable in languages like HPF (or language extensions like OpenMP). With each time step of the synchronous program generated viewed as a superstep, we will take a similar approach. However, as the following example shows, the BSP model does not supply enough information to select an appropriate placement.

Example 4.13 *The following code fragment is an implementation of a shift:*

```
!HPF$ DISTRIBUTE(BLOCK,BLOCK)::S,T
!HPF$ INDEPENDENT
      DO i1=l1,u1
!HPF$ INDEPENDENT
        DO i2=l2,u2
          T(i2,i1)=S(i2-(l2-1),i1)
        END DO
      END DO
```

This loop nest shifts the array S by $l_2 - 1$ *elements in the first dimension (the second dimension is left unchanged). An equivalent formulation of the same communication pattern is given by the following code – assuming that* $idx(1, i_2, i_1) = i_2 - (l_2 - 1)$ *and* $idx(2, i_2 i_1) = i_1$*:*

```
!HPF$ DISTRIBUTE(BLOCK,BLOCK)::S,T
!HPF$ INDEPENDENT
      DO i1=l1,u1
!HPF$ INDEPENDENT
        DO i2=l2,u2
          T(i2,i1)=S(idx(1,i2,i1),idx(2,i2,i1))
        END DO
      END DO
```

with idx replicated on all processors. Since the communication pattern is the same in both codes – and so is the size of the sets to be communicated – one would not expect too large a difference in the resulting performance according to a BSP-like cost model. Actual run times of these two codes for the ADAPTOR-7.1 compiler on the Passau hpcLine computer (SCI-connected 1 GHz Pentium III nodes) without any further optimization switches are depicted in Figure 4.10(a) for the first code and in Figure 4.10(b) for the second one. We see that the communication set volume per node (corresponding to an h_i *in Equation 4.11 of the BSP cost model) actually has no significant impact on the time spent for this communication. In the first version, there is some variation to be observed, however, we cannot deduce any direct dependence from either the communication set volume or the number of participating nodes. A difference of more than two orders of magnitude, however, exists between the execution times of the first and the second program versions: see the different scales of the time axis in Figure 4.10(a) and Figure 4.10(b).*

The different run times of Example 4.13 are explained by the different communication code introduced by ADAPTOR for the two access patterns (we observed similar behavious with the Portland Group's HPF compiler). A short description of how syntactically driven HPF compilers (and in particular ADAPTOR) create communication code can be found in Wondrak's diploma thesis [Won05, Chapter 5]. As soon as the compiler is not able to use an optimized library call, performance drops dramatically. Therefore, the first of our goals is to avoid this worst-case scenario. A similar observation can be made for the shape of the index space: a non-rectangular index space often leads to run times comparable with those in the presence of indirect array adressing. When developing our cost model, these factors should be taken into account. Different kernels varying in this respect will therefore be a part of the benchmark suite on which we will base our cost estimations. However, most non-trivial transformations in the polyhedron model will result in a non-rectangular parallel loop nest. The restrictions resulting for the target programs appeared just too restrictive to us. Therefore, we implemented the work-around for the ADAPTOR compiler discussed in Section 3.7 (page 125) instead of penalizing placement relations on the grounds of index set shapes.

The benchmark suite we use to develop a cost model for the target system will be the subject of the next section. Section 4.3.2 shows the application of that benchmark suite to the ADAPTOR compiler in combination with the hpcLine PC-Cluster at the University of Passau.

4.3.1 The Benchmark Suite

In order to design a benchmark suite that is able to measure the influence of several possible aspects that may have an impact on the overall performance of the parallel program, we start from a simple copy operation of the form

```
!HPF$ DISTRIBUTE(BLOCK)::S,T
!HPF$ INDEPENDENT
      DO i1=1+shift1,N1
! Statement S1:
        T(i1)=S(i1-shift1)
      END DO
```

which implements a 1-dimensional shift of S. Then we vary the following factors:

1. Array dimensionality.
2. Index space size.
3. Index space form.
4. Communication pattern.
5. Communication set size (if possible).

In the following, we review each of these points in closer detail.

Array Dimensionality Varying array dimensionality is implemented by using multidimensional distributed arrays assigned to in a multiply nested loop of the form

```
!HPF$ DISTRIBUTE(BLOCK,BLOCK)::S,T
!HPF$ INDEPENDENT
      DO i1=1+shift1,N1
!HPF$ INDEPENDENT
        DO i2=1+shift2,N2
! Statement S1:
          T(i1)=S(i1-shift1,i2-shift2)
        END DO
      END DO
```

Index Space Size The index space size is tuned simply by setting $shift_1, shift_2$ and N_1, N_2 appropriately. Note that the communication set volume also changes with these variables in the 2-dimensional case (in the 1-dimensional case, this value only changes with $shift_1$). In order to keep the array alignment constant relative to the enlarged index space, the array shapes are also changed accordingly.

Index Space Form As discussed above, non-rectangular parallel loop nests may not be handled well by an HPF compiler. A more complicated index space is generated from the code fragment above simply by splitting the index space into two triangular index spaces:

```
!HPF$ DISTRIBUTE(BLOCK,BLOCK)::S,T
!HPF$ INDEPENDENT
      DO i1=1+shift1,N1
!HPF$ INDEPENDENT
        DO i2=1+shift2,i1-1
```

```
! Statement S1:
          T(i1)=S(i1-shift1,i2-shift2)
        END DO
      END DO
!HPF$ INDEPENDENT
      DO i1=1+shift1,N1
!HPF$ INDEPENDENT
        DO i2=MAX(i1,1+shift2),N2
! Statement S2:
          T(i1)=S(i1-shift1,i2-shift2)
        END DO
      END DO
```

As reported earlier, ADAPTOR-7.1 generates very inefficient code in this case. For this compiler we have decided to work around this problem. However, a comprehensive benchmark suite for a general cost model needs to include this variant because it represents a common result of target codes automatically generated from polyhedra and at the same time may have a large impact on the overall run time of the program.

Communication Pattern The communication pattern is actually always a completely local communiation or a shift. This may be made explicit to the compiler or hidden, depending on the subscript expression used. Syntactically, the subscript expression of the right hand sides of S_1 and S_2 above varies from the local access `i1` and the shift `i1-shift1` through a permutation and a broadcast to the most general indirect addressing via `idx(i1)`.

Communication Set Size Finally, the communication set size is changed by varying the value of *shift*. This determines the amount of data to be exchanged between processors.

By default, the benchmark suite works with the following array sizes and shift values:

Dimensionality	Array Size	Shift Values
1	$N_1 \in \{128, 256, 512, 1024, 2048, 8192, 16384, 32768, 65536\}$	$shift_1 \in \{0, 1, 5, 128, 1024\}$
2	$N_1 = N_2 \in \{128, 256, 512, 1024, 2048\}$	$shift_1 = 0$ $shift_2 \in \{0, 1, 5, 128, 1024\}$

The benchmark kernels are developed with the functionality of the shift operator in mind. This is actually a quite natural approach: copying data can be viewed as special case of shift, and the shift operator is usually already an optimized communication primitive in `HPF` compilers (`Fortran` even defines intrinsics for array shifts), and any communication pattern can be represented as a composition of a copy between templates and a series of shifts.

Wondrak divides these benchmark kernels into four groups:

1. Kernels that copy data between arrays that are aligned to each other.
2. Kernels that copy data between arrays aligned to different templates; these arrays are thus placed onto different processor meshes.
3. Kernels that measure the cost of executing floating point operators.
4. Kernels that identify the influence of the index space form (here, this is a single kernel).

The following table gives an overview of these kernels, the actual assignment statement S_1 used in the respective kernel, and the associated group.

Kernel	Central Code	Group
local1D	T(i1)=S(i1)	1
local2D	T(i2,i1)=S(i2,i1)	1
shift1D	T(i1)=S(i1-shift1)	1
shift2D	T(i2,i1)=S(i2-shift2,shift1)	1
permute2D	T(i2,i1)=S(i1,i2)	1
broadcast	T(i2,i1)=S(shift2,1)	1
general	T(i2,i1)=S(idx(i2),idx(i1))	1
copy1D2D	T(i1,1)=S(i1)	2
copy1D2Ddiag	T(i1,i1)=S(i1)	2
copy2D1D	T(i1)=S(i1,1)	2
copy2D1Ddiag	T(i1)=S(i1,i1)	2
calcPlus	T(i2,i1)=S(i2,i1)+5.87328344	3
calcMinus	T(i2,i1)=S(i2,i1)-5.87328344	3
calcMult	T(i2,i1)=S(i2,i1)*5.87328344	3
calcDiv	T(i2,i1)=S(i2,i1)/5.87328344	3
calcExp	T(i2,i1)=S(i2,i1)**5.87328344	3
triangle	DO i1=low1,up1 DO i2=low2,i1-1 T(i2,i1)=S(i2-shift2,i1-shift1) END DO END DO ...	4

Each of the kernels is run repreatedly in order to yield accurate time measurements, and each of these measurements is again repeated several times during a run of the benchmark suite. The minimal time measured is then used for comparing kernels. All results are normalized wrt. to the simplest kernel – `local1D` for 1-dimensional kernels and `local2D` for 2-dimensional kernels. With this strategy, we will now put our benchmarks suite to work.

4.3.2 A Case Study: ADAPTOR

We have run the benchmark suite developed in the previous section with the `HPF` compiler ADAPTOR-7.1. Our platform was the Passau hpcLine PC-cluster, which consists of SCI-connected dual-processor Pentium III nodes running at 1 GHz with 512 MB RAM per node. We used a single CPU per node in all tests. The tests were conducted by Wondrak [Won05], who also developed a cost model for the ADAPTOR compiler based on these results, which we shall describe here.

As we have already seen above, the main contributing factor in the overall performance of an `HPF` program compiled with ADAPTOR is the subscript function: a shift with a constant element count is implemented with an optimized communciation function in the compiler's run time library. A slightly more complicated subscript function already forces the compilerto use run time resolution. The following table lists the factors of the time taken for each of the kernels wrt. the time for the corrsponding run of `local` for a representative sample of array sizes:

Kernel	Array Size (N_1)	Number of Processors	Factor wrt. `local`·
shift1D	65536	4	10
general1D	65536	4	259
shift1D	2048	4	13
general1D	2048	4	551
shift2D	2048	4	2
permute2D	2048	4	3
general2D	2048	4	477
broadcast	2048	4	1
shift1D	65536	8	13
general1D	65536	8	691
shift1D	2048	8	18
general1D	2048	8	870
shift2D	2048	8	2
permute2D	2048	8	5
general2D	2048	8	759
broadcast	2048	8	1
shift1D	65536	16	13
general1D	65536	16	1383
shift1D	2048	16	25
general1D	2048	16	1528
shift2D	2048	16	3
permute2D	2048	16	3
general2D	2048	16	1611
broadcast	2048	16	1
triangle	2048	16	15

We observe that, indeed, the different subscript functions have a crucial impact on performance which is even almost constant wrt. array size and number of processors, with the exception of the general communication pattern, whose run time increases roughly linearly with the number of participating processors. We can therefore use Algorithm 4.3.1 to determine the actual cost of a target program by adding up the average factors obtained from the benchmark suite for a given class of subscript functions.[6] This class is determined by Algorithm 4.3.2 (*Classify*).

Algorithm 4.3.1 [*CostFuncADAPTOR*]:

Input:

$RCOIG' = (\mathfrak{V}_{RCOIG}', \mathfrak{E}_{RCOIG}')$: transformed reduced COIG.

Output:

c : estimated communication cost for the target program.

Procedure:

$c := 0$;

`forall` $e = (v_s, v_t, (H'_L, H'_R)) \in \mathfrak{E}_{RCOIG}'$

/* Algorithm 4.3.2 yields the class of the communication introduced by this edge in the graph */

$class := Classify((H'_L, H'_R))$;

$$c := c + \begin{cases} 1 & \text{if } class = \texttt{local} \\ 2 & \text{if } class = \texttt{shift} \\ 3 & \text{if } class = \texttt{permutation} \\ 146 & \text{if } class = \texttt{diffTemplates} \\ 700 & \text{if } class = \texttt{general} \end{cases};$$

`end forall`

`return` c;

[6] These factors are taken as the average over all runs of the benchmark kernels (not just from the table above).

Algorithm 4.3.2 (*Classify*) determines the class of a communication defined by an h-transformation in the target space. This h-transformation is obtained as decribed in Section 4.1.3 and thus defined by two mappings, one of which can be thought of defining the processor mesh from which to read the value. If source and target space of this mapping differ, we have to deal with a redistribution across processor meshes, which is indicated by the classification `diffTemplates`. The other mapping defines the subscript expression with which to read that value:

- If it is defined by the unit matrix, the communication is a local.
- If the mapping of the indices is the unit matrix, but there is also a non-zero entry on the part of the parameters, the communication is a shift.
- If the mapping of the indices is a permutation (and there is no parameter in the resulting expression), the communication is a permutation (which can be handled by an optimized function in the ADAPTOR run time library).
- Otherwise, the communication represents some general pattern that ADAPTOR can only implement using run time resolution.

Algorithm 4.3.2 [*Classify*]:
Input:
(H'_L, H'_R) :
dependence relation in the target space.
Output:
classification $\in$ {`local`, `shift`, `permutation`, `diffTemplates`, `general`} :
classification of the communication generated by (H'_L, H'_R).

Procedure:
/* STEP 1: initialize and determine the templates to which to be aligned */
$c_1 := \min(\{c \in \{1,\ldots,\#pdims\} \mid M_{H'_L}[\cdot,c] \neq 0\})$;
$c_2 := \min(\{c \in \{1,\ldots,\#pdims\} \mid M_{H'_R}[\cdot,c] \neq 0\})$;
Let t_1 be the dimension of the outermost index of the nest of dummy loops also containing index c_1;
Let j_1 be such that $\#pdim[j_1]$ is the depth of the nest of dummy loops containing index c_1;
Let t_2 be the dimension of the outermost index of the nest of dummy loops also containing index c_2;
Let j_2 be such that $\#pdim[j_2]$ be the depth of the nest of dummy loops containing index c_2;
/* STEP 2: classify the accesses */
/* STEP 2.1: communication between processor meshes */
if $t_1 \neq t_2$ **then**
  return `diffTemplates`;
else
  /* 2.2: detailed consideration of function matrices */
  Let $M_2 = L \cdot M_{H'_R}$ be the reduced echelon form of $M_{H'_R}$;
  $M_1 := L \cdot M_{H'_L}$;
  $t'_1 := t_1 + \#pdim[j_1] - 1$;
  $t'_2 := t_1 + \#pdim[j_2] - 1$;
  /* only the dimensions $t_1,\ldots,t'_1$ – from which the subscript expression for */
  /* the fetch is built – are of interest from now on */
  Let $\mathfrak{R}=\{r_1,\ldots,r_{n_{\text{eqs}}}\}$, $r_1 < \cdots < r_{n_{\text{eqs}}}$ be the set of rows such that:
  $\left(\forall r : r \in \{1,\ldots,\#pdims\} : \begin{array}{c}(\exists c : c \in \{t_1,\ldots,t'_1\} : M_1[r,c] \neq 0) \\ \Leftrightarrow r \in \mathfrak{R}\end{array}\right)$;
  /* expressions in dimensions of the dependence target space */

$$M_T := \begin{pmatrix} M_1\,[r_1, t_1] & \dots & M_1\,[r_1, t_1'] \\ \vdots & & \vdots \\ M_1\,[r_{n_{eqs}}, t_1] & \dots & M_1\,[r_{n_{eqs}}, t_1'] \end{pmatrix};$$

/* expressions in dimensions of the dependence source space */

$$M_{T'} := \begin{pmatrix} M_2\,[r_1, t_2] & \dots & M_2\,[r_1, t_2'] \\ \vdots & & \vdots \\ M_2\,[r_{n_{eqs}}, t_2] & \dots & M_2\,[r_{n_{eqs}}, t_2'] \end{pmatrix};$$

/* expressions in dimensions of constants */

$$M_K := \begin{pmatrix} M_2\,[r_1, 1 + \#pdims] & \dots & M_2\,[r_1, n_{blob} + \#pdims] \\ \vdots & & \vdots \\ M_2\,[r_{n_{eqs}}, 1 + \#pdims] & \dots & M_2\,[r_{n_{eqs}}, n_{blob} + \#pdims] \end{pmatrix};$$

```
if M_T = I_{#pdim[t'_1+1],#pdim[t'_1+1]} then
  if M_T' = I_{#pdim[t'_2+1],#pdim[t'_2+1]} then
    if M_K = 0 then
      return local;
    else
      return shift;
    endif
  else
    if M_K = 0 ∧ (M_T' is a permutation of I_{#pdim[t'_2],#pdim[t'_2]}) then
      return permutation;
    else
      return general;
    endif
  endif
else
  return general;
endif
endif
```

Recapitulating Example 4.12, the assessment of the different placements roughly corresponds to the cost model presented here. The placement we choose would actually be chosen for the ADAPTOR compiler. Note that, although this is a replicated placement, which basically annihilates the effect of minimizing the dimensionality of an expression that is gained through LCCP, there is still a second order effect that could improve run time: the reduced dimensionality effectively removed the computation of $A(i)^3$ from a loop that is distributed in a blocked fashion over processors. Even if the same value is now computed on different processors, it is not computed inside the corresponding loop enumerating a block of operations to be executed on each single processor.

4.4 Related Work

Determining a placement for a loop program is quite naturally an important step in the code generation of programs represented in the polyhedron model, since it completes the space-time mapping of which the schedule is only a part: to obtain a parallel program, we also have to specify on which processor to execute the operations of the program. On the one hand, one should try to utilize as many processors as possible in order to reduce the wall-clock time needed to perform a sequential step of the program. On the other hand, communication between different processors may well be more expensive than the time saved by parallelism. Therefore, it is a non-trivial task to produce a placement that leads to good performance results. Here, we want to highlight just some methods of placement computation that are relevant to this work.

Feautrier's Placement Method The placement method introduced by Feautrier [Fea93] views local accesses as essentially no-cost operations and any non-local access as some expensive constant-cost operation. The placement method is based on the granularity of statements. The counterpart of the occurrence instance there is therefore an operation, i.e., a statement instance that combines write access and read access of the respective statement. The strategy is then to define a linear function

$$\Phi_{\mathfrak{O}} : \mathbb{Q}^{n_{src}+n_{blob}} \to \mathbb{Q}^{\#pdim}$$

for each set of occurrence instances $\mathfrak{O}$ with

$$\Phi_{\mathfrak{O}'} \circ H_e = \Phi_{\mathfrak{O}} \tag{4.12}$$

if the reduced dependence graph features a dependence of $\mathfrak{O}$ from $\mathfrak{O}'$ defined by the h-transformation H_e. Equation (4.12) then defines a linear equation system. The placement functions $\Phi_{\mathfrak{O}_i}$ solving all the program's constraints of the form of Equation (4.12) may lead to a trivial placement, i.e., *all* operations (or occurrence instances) are placed on the same processor. In order to avoid this situation, the h-transformations to be taken into account – and thus the dependences to be *cut* – are sorted according to the expected communication volume, i.e., according to the dimensionality of the image $\text{im}(H_e)$ of the corresponding h-transformation. These h-transformations are only taken into account as long as they guarantee non-trivial placements. The heuristic employed here to obtain non-trivial placements does not take into account any particular system characteristics, nor does it consider replication in any way. The approach is therefore quite different from ours, which basically just enumerates placement candidates (which may well define replicated placements) but evaluates each placement candidate with a cost function that is taylored for the target system.

The Placement Method by Dion and Robert Dion and Robert propose a placement method based on a so-called **access graph** [DR95]. This is a bipartite graph with one set of vertices representing arrays and the other set representing statements reading or writing access these arrays. Depending on the rank of the subscript function F that defines the weight of an edge (v_1, v_2, F) between the vertices v_1, v_2, either a placement for v_1 or v_2 is computed from a placement of the other vertex (for full rank mappings, both directions are possible). This means that placements are obtained for both data and statements (i.e., computations). Both methods, the one by Dion and Robert and ours, allow the placement of both computations and data. Our approach does not distinguish between computations and data at all: we assign placement relations to computations that represent interesting sets, which are then also associated with arrays, since the resulting values have to be stored somewhere, and thus arrive at data placements. The result of the placement method of Dion and Robert – as of Feautrier's – is always a family of linear functions defining the placements. Again, not every edge of the graph can be honored for placement computation. Therefore, a maximum branching of the graph is calculated and only the edges in this maximum branching are used for placement computation. The maximum branching is based on the rank of the functions associated with the edges as their priorities. The placements for inner nodes of the graph are derived from those for **root vertices** for which placements are found by heuristically trying to minimize direct communication. In comparison to our method, this approach is quite similar to Feautrier's placement algorithm. While Dion and Robert use families of functions, which are calculated according to one hard-coded heuristic, our method may yield placement relations defining replication and uses a quite simple enumeration scheme to evaluate different placement candidates using a cost function that can be adapted to the target system. In principle, our method can be extended to compute data placements (starting from some default placements), as the method by Dion and Robert already does, but it is actually targeted at propagating placement relations from already given data placements to computations (and auxiliary variables associated with those computations).

Beletskyy's Placement Method Beletskyy [Bel03] proposes a modification of an algorithm by Lim and Lam [LL97]. The method by Lim and Lam classifies different program fragments according to their communication behaviour. Beletskyy's extension considers the case of synchronization-free parallelism. The basic idea behind this modification is remotely similar to the basic idea behind LCCP: as a first step, statements are enforced to be placed on the same processor by linear constraints corresponding to Equation (4.12) – the difference here being that, while LCCP combines equivalent occurrence instances, the operations combined in Beletskyy's method are dependence sources with their targets (and vice versa). The actual placements are then computed from a kernel base of the resulting equation

$$(\Phi_{\mathfrak{D}'} \circ H_e) - \Phi_{\mathfrak{D}} = 0 \tag{4.13}$$

This is more efficient than the original method proposed by Lim and Lam, which uses the Farkas Lemma to linearize the placement constraints and obtain systems of linear equations. As with the previous two methods, the result is again a family of linear functions, based on general placements heuristics, so that it compares to our placement method roughly the same as the previous methods did.

Placement Propagation in the dHPF Compiler The technique closest to our work is the placement propagation of the HPF compiler dHPF which has been developed at Rice University [MCA97, AJMCY98, AMC98, AMC01]. The dHPF compiler is not only able to implement a replicated placement (as defined in the HPF standard), it is also capable of implementing the union of different HPF representable placements as a placement for a statement (which even exceeds the set placements examined in this chapter) and to use the placement for any of the arrays referenced in a statement S as the placement for S. Thus, the compiler may avoid the owner computes rule. An additional aim is to find placements for statements that write to privatizable arrays and auxiliary (control) statements. The dHPF compiler propagates the HOME directive along the edges of the dependence graph of the program in static single assignment form from source to target and determines the most promising placement description from a set of candidates according to an internal cost model. The basic idea is thus similar to our approach. In addition to the dHPF approach, our method also propagates from target to source; this propagation is made possible by the additional partitioning phase discussed in Section 4.2.3. Moreover, the dHPF approach chooses placements employing heuristics for this one specific compiler, whose characteristics are already well known, while our approach can be adapted to different compilers.

Communication Visualization with HPF-Builder HPF-Builder was developed at the University of Lille in order to help the HPF programmer to find good data placements [DL97]. HPF data placements are represented with the same approach using pairs of mappings as described in Section 2.4.2 that is also used by our placement method (albeit without using the operators $\circ_-$ and $\circ_+$). HPF-Builder creates a visualization of the amount of data that has to be communicated between processors due to a given HPF placement. This amount is computed by counting the integer points in a polytope via *Ehrhart polynomials*, a technique rediscovered and introduced to linear loop transformations by Clauss [Cla96]. In contrast to our work and other related work cited here, HPF-Builder is a pure visualization tool. The actual selection of placements is left to the user.

Loop Transformations for Cache Improvement Ehrhart polynomials are polynomials whose coefficients are not necessarily plain numbers of the underlying ring $\mathfrak{K}$, but **periodic numbers**. I.e., depending on some (parameter) value n, the coefficient is actually defined as some $c_{n\%p} \in \{c_0, \ldots, c_{p-1}\} \subseteq \mathfrak{K}$. Clauss and others have been using Ehrhart polynomials for determining the (integer) volume of a polytopes in different areas. In particular, Clauss determines the number of accesses to distributed array elements [Cla97]. Clauss and Meister also use Erhart polynomials to determine a mapping of array elements to actual memory cells in RAM in order to improve spatial cache locality [CM00]. In later work, Loechner, Meister, and Clauss suggest a loop transformation based on a subspace creating temporal reuse and a completion to the full space in

order to improve cache locality [LMC02]. Again, not *all* accesses can be optimized. Instead, the optimization selects iteratively as large as possible a set of accesses that have been least optimized up to the current iteration.
Slowik also creates a loop transformation that can be seen as a placement [Slo99]; again, the aim is improving cache locality. Slowik constructs transformations by enumerating a range of values at predefined places in the transformation matrix and evaluating the influence on cache performance by calculating the corresponding Ehrhart polynomials.
Bastoul and Feautrier compute **chunking functions** that define a chunk of operations to be executed together in order to improve data locality. As with the approach of Loechner, Meister and Clauss, the basic step here is to compute bases for data reuse within the index space; in contrast to this approach, Bastoul uses a heuristic to prioritize sets in order to avoid to compute a parameter dependent and possibly unknown value. This method has been implemented in the optimizer `chunky`.
All these transformations can be viewed as additional placement relations refining replicated placements, as proposed in the introduction of this chapter. However interactions between these approaches have not been studied yet.

4.5 Summary and Future Work

In this chapter, we have introduced a method for computing possibly replicated placement relations for occurrence instances. Placements are propagated from dependence sources and targets, which ultimately have to be associated with a user-defined placement.[7] All combinations of placements are then compared to each other by means of a cost function, which can be tuned for the target system. In order to design and adjust an appropriate cost model, we have developed a benchmark suite that systematically tests the performance of programs generated by the compiler. We have also established a cost model for the ADAPTOR compiler in combination with the Passau hpcLine cluster. The costs are here mainly influenced by the communication pattern and thus by the subscript functions in the target code. Apart from recalibrating the numerical values associated with a placement combination (which can simply be read from the results of the benchmark suite), this cost model should also be applicable to other `HPF` compilers.

However, the cost model as it is favours replicated placements, since there is no provision for estimating the amount of work placed on a single processor. In addition, computation of placements in practice is very time consuming. This is because *all* combinations of placement relations are evaluated. Other placement methods use heuristics at this point in order to restrict the search space. These restrictions usually make assumptions about the target system. Our goal here was to make as few such assumptions as possible. So, another possibility is to use combinatorial methods such as genetic algorithms to examine only part of the search space while still delivering useful results.

On the other hand, we already restricted the search space by ignoring paths containing cycles in the dependence graph (cf. Section 4.2.4, page 160). Including these paths into our considerations increases the search space, but could add valuable placement relations. Even without including cycles, the combination operator defined in Section 2.4.2 may yield further placement relations that may be worthwhile to include in our search space.

In addition, further placement relations may be obtained by repeatedly propagating placement relations – recall that we decided to iterate only once through the main loop of Algorithm 4.2.7. Our decision prohibits for example the propagation of a placement from one read access (which may get a placement from the array it reads) to another read access within the same expression. Placements like these may well be worth considering.

The main point for future work, however, is to extend our current simplistic strategy of computing only crude placements for processes running on different nodes to a truly hierarchical placement strategy. Ideally, each level representing a finer granularity should be able to take into account transformations already defined on the levels of coarser granularity.

[7] Note that one could also use some predefined default placement instead of a user-defined one.

Chapter 5

Putting It All Together

We have seen the theory behind the LCCP transformation in Chapter 3. This transformation ejects so-called interesting sets. These are sets of occurrence instances that each need to be associated with space-time mappings, i.e., with a schedule and a placement. We decided to stay with well-known methods for the time part of this transformation, and Chapter 4 introduced a method for obtaining a possibly replicated placement for the *space* part. It is now time to regard the results of applying all these methods together.

For this purpose, we have assembled a series of synthetic examples in Section 5.2 and an example based on a real application in Section 5.3. We used an implementation of the LCCP and *OIAllocator* placement algorithms presented in Chapters 3 and 4 that has been integrated into the LooPo source-to-source compiler to perform the transformations of these examples automatically. Section 5.1 gives a short overview over this implementation.

The automatically transformed code presented in Sections 5.2 and 5.3 was scheduled using the LooPo implementation of Feautrier's scheduler [Fea92a, Fea92b, Wie95]. The resulting polyhedra were scanned using `CLooG` [BCG$^+$03] in combination with a postprocessing step that identifies LCCP pseudo parameters and produces actual code fragments following the strategy described in Section 3.7 to produce annotated `HPF` code. We used the `HPF` compiler ADAPTOR-7.1 and ran the resulting parallel program on a varying number of processors on the Passau hpcLine cluster (consisting of 32 SCI-connected dual Pentium III-PCs running at 1 GHz with 512 MB per node), with only one process allocated to a node (i.e., one CPU per node was kept idle). The ADAPTOR compiler produces `Fortran` code with calls to its run time library, DALIB, which in turn uses (in our case) `MPI` for communication. For the compilation of the transformed `Fortran` code, the `g77` compiler was used without any further optimizations turned on in order to evaluate only the changes caused directly by our transformations. The ScaMPI `MPI` library for SCI was used as the underlying communication library. This is the same testbed as the one used for the computation of placement relations in Section 4.3.2.

5.1 Implementation in LooPo

In order to test the algorithms presented in the previous chapters, they have been implemented and integrated as modules into the LooPo system. LooPo is a modular system that consists of several independent programs which communicate with each other via files that are placed in the `/tmp` directory. Section 5.1.1 discuss the transformation steps applied to the original polyhedra that define the input program. Section 5.1.2 then describes the generation of target code from these transformed polyhedra after LCCP transformation and the subsequent application of a space-time mapping. Finally, Section 5.1.3 sketches a wrapper that calls all LooPo modules necessary to transform a complete `HPF` program with appropriate options (including `lccp`, the implementation of the loop-carried code placement technique).

5.1.1 Transformation of Polyhedra

The algorithms for the LCCP transformation have been added as a module that can be run optionally after completion of a dependence test. The transformed polyhedra are then passed to the usual scheduling, placement and code generation algorithms. This section is devoted to the processing steps before the code generation phase.

For our implementation of LCCP to work correctly, the dependence test needs to yield *exact* results – i.e., the dependence information must contain neither over- nor underapproximations. This is because, as described in Section 3.7, the target code is constructed in the code generation phase directly from the transformed dependence graph. If the input program satisfies the conditions for applying the polyhedron model discussed in Section 2.3, different dependence analysis methods can be used to obtain such exact dependence results, in particular the method by Feautrier [Fea91]. Therefore, we choose to run `lccp` only in conjunction with the LooPo implementation of the Feautrier dependence test.

Let us now consider the general implementation of the LCCP algorithm. There are some points in which the actual implementation in LooPo differs from the model discussed in Chapter 3, which we will address on the way. We will then take a look at the interactions between LCCP and scheduling and placement techniques.

General Procedure

Basically, the implementation of LCCP, realized in the program `lccp`, follows the same steps as presented in Algorithm 3.4.1. However, in order to apply a transformation in our finer grained model presented in Section 2.3.1, we need a program representation based on the two additional dimensions: occurrence number and operand number. Since these additional dimensions are not used by any other module, `lccp` first inserts them into the program's symbol table – along with entries for LCCP pseudo parameters and the parameter m_∞. For historic reasons, the order of the dimensions for operand numbers and occurrence numbers in the occurrence instance vectors are reversed in the actual LCCP implementation, i.e., the dimension of the occurrence numbers is the last index dimension in an occurrence instance vector. This has no adverse effect on the dependence analysis phase. Then, the structural dependences are computed and the OIG is determined.

Linear Expressions During the generation of occurrence instance sets (i.e., the vertices of the COIG), linear expressions do not have to be decomposed and analyzed any further. Therefore, they do not need to be associated with an occurrence. However, the largest (uppermost) linear expression is represented by an occurrence in the model presented in Section 2.3 (c.f. Example 2.19). The first reason for this is that, if there were only a single representative for all linear expressions, terms like `a(i)+2` and `a(i)+3` would be assumed equivalent and one would be substituted with the other by `lccp`.[1] In contrast to this decision in the model, linear expressions are not represented by any occurrence instance in the LCCP implementation. The only exception is the set of instances of occurrence 0 (i.e., `num(i)`, introduced in Section 2.3.2), which represents the abstract concept of an integer number. The reason is as follows. As noted on page 27 in Section 2.3.1, we may omit the other occurrence instances, since every occurrence instance α representing a linear expression is dependent on a certain instance of occurrence 0. Therefore, we can just replace a dependence of β on α with a dependence of β on the corresponding instance of occurrence 0. Since we assume linear expressions to be essentially zero cost operations, it does not have any particular merit to represent them in our algorithms. However, in order to avoid performance penalties due to representing even more occurrence instances than needed, we omit them completely in the implementation.

[1] On systems on which calculations on integers are expensive, this behaviour might be desirable, since it would enable the removal of computations of integer expressions. However, since transformations in the polyhedron model usually change the integer expressions occurring in a program without considering different integer expressions as incurring different costs, such a strategy does not integrate well into the framework employed here.

Non-Linear Expressions Definition 2.18 prescribes an enumeration scheme for occurrences that asserts that the execution order of the program corresponds to the lexicographic order of occurrence instances. In particular, output arguments of a procedure call have to be executed after input arguments. In our implementation, arguments are always visited from left to right so that the lexicographic order no longer defines the original execution order. As discussed in Section 3.5.1 (page 85), this is not a problem because the occurrence instances of the `lccp` output are always scheduled according to the exact dependence information.

Groups of Occurrence Instances In the next step, the occurrence instance sets representing the program are sorted according to their respective levels in the operator tree by an implementation of Algorithm 3.2.1 (*GroupOccurrenceInstances*). After these preprocessing steps, the LCCP algorithm is invoked.

Equivalence Classes of Occurrence Instances As decribed in Section 3.2, the LCCP algortihm consists of a single sweep from the bottom of the operator tree upwards with a condensation phase on each level that creates a new set of representatives for each equivalence class of occurrence instances on this level. The vertices of the operator tree are represented by the structural dependences in the OIG. This walk through the operator tree, and the condensation phases, employ set operations that are implemented by calls to the `Omega` library. Practical experience has shown that the algorithms often produce an excessive number of empty sets. As an example, such an empty set in the implementation using `Omega` may be of the form

$$\left\{ \left(\begin{array}{c} i \\ 1 \\ 0 \end{array} \right) \middle| \; 1 \leq i \leq -m_\infty \right\}$$

which is empty because, by definition, no integer is smaller than $-m_\infty$. These sets should be immediately identified and removed. However, the current implementation removes them late in the process, before they are stored and forwarded to other LooPo modules, leading to extremely high memory consumption and slow program execution. Performance is further degraded by the fact that occurrence instances representing different operand positions of the same occurrence are stored in separate sets in our implementation, although it would suffice to use a single set.

Selection of Representatives The core algorithm of the condensation procedure and, thus, of the complete LCCP algorithm is the selection of representatives discussed in Section 3.5. Our implementation lets the user choose between the selection methods presented in Section 3.5.1 (lexicographic minimum) and Section 3.5.2 (change of basis). As discussed earlier, the former method leads to very complicated set representations. The method employing a change of basis performs much better on our test examples. Therefore, this is the default in our implementation (and the development of the `lccp` program was done mainly with this approach in mind). The generation of representative occurrence instances with this method partly ignores the extents of occurrence instance sets, in particular during the generation of a matrix representing the change of basis for a single occurrence with equivalent instances. In addition, the `Omega` library does not directly support generation of a function matrix (and an associated function) as defined by Algorithms 3.5.4 to 3.5.6. Therefore, equivalence relations computed by the *LiftRel* operator are transformed into a matrix representation (with each matrix expressing the equality constraints holding in a given relation) and transformed back into an `Omega` relation after Algorithm 3.5.4 has computed a function matrix that can be used to map instances of the same occurrence to their respective representatives. The implementation of Algorithm 3.5.4 allows for a more complex result than described in Section 3.5.3: the input may be a disjunctive normal form of `Omega` relations and the result may consist of several separate `Omega` relations (again in disjunctive normal form). In this case, the implementation may create new occurrences in order to avoid mapping occurrence instances that are not equivalent to the same representative. In theory, this could happen for general representative mappings, since non-equivalent occurrence instances may be mapped to their respective representatives by completely unrelated mappings. However, with the representative mappings we use in the method described in Section 3.5.2, this case cannot occur. This is because equivalences between instances of the same occurrence are reused along the same directions in the index space and thus lead to the same change of basis. The only exception is the case that one of the values needed to compute the expression at hand is reassigned in the meantime. This may only lead to different transformations insofar as the reuse space is the span of the original reuse space (without taking onto account other occurrence instances) minus the space spanned by a set of unit vectors. I.e., if we have two different representative mappings for two subsets of the original index space, these mappings differ only in the set of index variables that are eliminated in the corresponding images: e.g., one mapping may replace indices i and j by an LCCP pseudo parameter, whereas another may only replace index i. All condensed occurrence instance sets of the same occurrence that represent polyhedra in the same coordinate system (i.e., with a non-trivial extent in the same set of dimensions) are the result of the same change of basis. This implies that they contain the same integer vector only if this integer vector actually represents the same value in the original program. Therefore, this feature of the implementation is actually redundant.

Space-Time Transformation

Figure 3.1 on page 60 shows LCCP as a transformation that can be simply plugged in between the dependence test and the computation of a space-time mapping. However, there are some points that have to be considered for modules that are invoked after `lccp`.

Placement Method We have already established that it is advisable to use a placement method like the one suggested in Chapter 4, so the natural choice for a placement method is the *OIAllocator* placement detailed in that chapter. Indeed, within the LooPo framework, the implementation of the *OIAllocator* placement method is the only one that is fully compatible with `lccp`. The reason is as follows. A central design decision in the *OIAllocator* algorithm is to produce replicated placements for the results of `lccp`. Therefore, the implementation of *OIAllocator* uses a different file format than other implementations of placement algorithms. The program `cloogInterface` is called after schedules and placements have been computed in order to generate a syntax tree representing the transformed program. This module only supports reading linear functions if they define schedules. Placements have to be supplied in the file format defining linear relations. Additionally, for example, the placement algorithm suggested by Dion and Robert [DR95] is not necessarily supplied with all the information it expects. This is because this method bases its placement decisions on an access graph that contains array accesses. However, the array accesses to auxiliary variables in which the results of computations are stored that have been generated by `lccp` as new interesting sets do not yet have an associated array (actually, in our framework, this association is done by the placement method). Therefore, we always use the implementations of LCCP and *OIAllocator* together.

The *OIAllocator* placement method is based on an iterative partitioning of the reduced COIG. As discussed in Section 4.2.8 (page 169), we have decided to partition the reduced COIG only once for practical reasons. We have also noticed that partitioning in one direction wrt. the operator tree is done automatically by the LCCP method. Therefore, partitioning in the opposite direction was included in the `lccp` program rather than in the implementation of *OIAllocatior*. The *OIAllocator* algorithm has been implemented by Wondrak as part of his diploma thesis [Won05, Chapter 4].

Scheduler Method Section 3.6 showed that Lamport's hyperplane method cannot be used in conjunction with LCCP because dependences may reverse their directions in the program fragment transformed by LCCP. We have also noticed that this is no problem for algorithms like the Feautrier scheduler [Fea92a, Fea92b] or the scheduling method due to Darte and Vivien [DV94]. In practice, the implementation of the Feautrier scheduler in LooPo turned out to run more stable for our examples. Therefore, we use this method by default (in particular, this module was used for all examples in the following sections).

5.1.2 Target Code Generation

The usual procedure in the polyhedron model, after schedules and placements have been computed, is to use a scanning technique like the `targgen` module of LooPo [Wet95, GLW98] or the algorithm by Quilleré, Radjopadhye and Wilde [QRW00]. Bastoul developed and implemented an improved version of the latter algorithm in his tool `CLooG` [BCG+03, Bas03, Bas04]. We make use of his implementation for our own code generation phase by embedding a call to the `CLooG` program in a pre- and a postprocessing phase. The preprocessing phase generates a `CLooG` input file defining the interesting sets and the transformation into space and time coordinates. The second phase generates a LooPo syntax tree from the output of `CLooG`. In particular, the interesting sets representing computations (and not array accesses) have to be implemented as assignment statements assigning to auxiliary variables. Both these phases are handled by the program `cloogInterface`.

For the implementation of the preprocessing phase, it is necessary to represent the parameter m_∞. `CLooG` allows us to define relations between parameters so that we are able to define m_∞ as larger than any other parameter multiplied by a huge coefficient (which essentially represents infinity by the largest number used in the input file).

When aiming for `HPF` output, a placement relation can be represented by `DISTRIBUTE` and `ALIGN` directives alone – with loop bounds enumerating a normalized range – or by `HPF` directives that define only replicated or non-replicated storage and defining alignments by appropriate loop bounds (the traditional way in the polyhedron model). In our implementation, we decided for the latter option.

As noted in Section 3.7, `lccp` pseudo parameters may have to be enumerated in order to find a legal value. Our implementation does not generate the elaborate code presented in Figure 3.15. Instead, `lccp` pseudo parameters are presented to `CLooG` as loop indices. The loops generated for these indices by `CLooG` are then transformed to `if` statements.

The postprocessing phase of `cloogInterface` uses the original syntax tree of the program fragment as a lookup table that maps an occurrence number to the corresponding operator. Thus, the syntax tree can be said to implement the function Occ^{-1} (and, thus, $\mathrm{head} \circ \mathrm{Occ}^{-1}$). The remaining code (the relation between the different occurrences) is generated according to the reduced COIG as produced by `lccp`.

For the sake of easier coding, both the implementation of *OIAllocator* and the `cloogInterface` program ignore the dimension storing the operand number in the description of interesting sets. Therefore, the computation statements produced by `cloogInterface` only assign to at most one output variable. Procedure calls with more than one output argument are not implemented. As discussed in Section 3.7, storage mapping optimization (SMO), as introduced by Lefebvre [LF98], should be applied to the code produced by the postprocessing phase in order to reduce overhead. However, this technique was not included in our implementation due to time constraints. This does not show in any of the examples discussed below: due to the dependences in these examples, the target code with SMO applied does not differ from the one presented here. Instead, the `cloogInterface` module generates a new auxiliary array for each interesting set (i.e., a set of occurrence instances for which a schedule and a placement is defined) with the result of each instance stored in a different array element. This technique is, in general, not optimal but represents a safe choice that guarantees that each individual value computed by a occurrence instance is stored separately.

5.1.3 Embedding the Transformation in a Compiler Framework

In order to apply `lccp` to a complete `HPF` program, it is necessary to preprocess the input, since LooPo is only able to transform (small) loop nests that satisfy the properties identified in Section 2.3. In the course of his diploma thesis, Wondrak implemented a wrapper module for this task [Won05, Chapter 4].

The wrapper program consists of preprocessing programs and a script called `loopohpf`, which calls all these programs and the necessary LooPo modules with the appropriate options in the coorect order:

1. A parser for HPF directives (basicallocs) is called (which also parses additional directives prefixed with !LOOPO$).

2. For each program fragment to be processed by LooPo, as identified in the previous step, the following modules are called:

   (a) The LooPo parser (scanpars) is called.

   (b) A dependence test (in our tests the implementation of Feautrier's method, fr_depend) is called.

   (c) The LCCP main program (lccp) is called.

   (d) A scheduler (in our tests the implementation of Feautrier's scheduler) is called.

   (e) The *OIAllocator* placement algorithm (oiallocator) is called.

   (f) The script cloogTransf uses the LCCP postprocessing program cloogInterface to adapt the lccp output to the needs of the polyhedron scanning module CLooG. CLooG generates a parse tree from the reduced COIG supplied by lccp. Finally, cloogTransf calls cloogInterface again; this time, it transforms the output of CLooG back into a form readable by the LooPo core modules.

   (g) The modules genAlign and targout generate source program text representing the code produced with CLooG in the previous step.

3. All occurrences of the program fragments identified in Step 1 in the original program are replaced by the corresponding transformed code fragments.

4. The HPF compiler (ADAPTOR's source-to-source transformation module fadapt) is called.

5. As discussed in Section 3.7 on page 125, the bounds of parallel loop nests are replaced by bounds that define a rectangular loop nest in the HPF compiler input. These artificial bounds are replaced with the original loop bounds after the run of fadapt.

6. The g77 compiler is used to compile the resulting Fortran program and link it with ADAPTOR's DALIB and an MPI library (in our case ScaLi's ScaMPI library).

With these steps, loopohpf transforms an HPF input program automatically into an executable program whose loop nests have been processed by lccp. We used this wrapper for the compilation of all the examples discussed in the following sections.

5.2 Some Synthetic Examples

The examples presented in this section were specifically designed to be amenable to the LCCP transformation and yet show some limits of the transformation. Similar code can also be found in real applications and benchmark kernels, such as the multigrid solver of the NAS Parallel Benchmarks [BHS+95].

Example 5.1 lccp *transforms the following code fragment on the left to the code fragment on the right. The placement relations obtained from our placement method are also presented in the code on the right.*

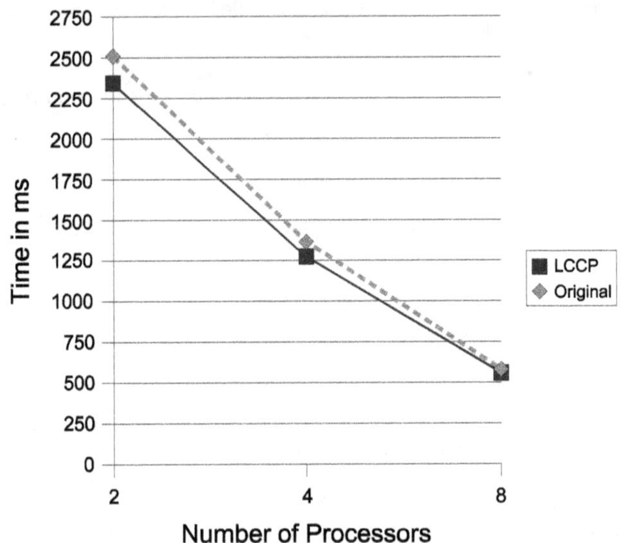

Figure 5.1: Run times for Example 5.1

```
      REAL A(n,n)
      REAL B(n)


!HPF$ TEMPLATE TEMP1(n,n)
!HPF$ DISTRIBUTE TEMP1(BLOCK,BLOCK)
!HPF$ ALIGN A(i,j) WITH TEMP1(i,j)
!HPF$ ALIGN B(i) WITH TEMP1(i,1)


!HPF$ INDEPENDENT
      DO i=1,N
!HPF$ INDEPENDENT
        DO j=1,N
         A(i,j)=A(i,j)-B(i)*B(i)+B(i)
        END DO
      END DO
```

```
      REAL A(n,n)
      REAL B(n)
      REAL ARRAY1(2:2,n)
      REAL ARRAY2(1,n)
!HPF$ TEMPLATE TEMP1(n,n)
!HPF$ DISTRIBUTE TEMP1(BLOCK,BLOCK)
!HPF$ ALIGN A(i,j) WITH TEMP1(i,j)
!HPF$ ALIGN B(i) WITH TEMP1(i,1)
!HPF$ ALIGN ARRAY1(*,j) WITH TEMP1(j,*)
!HPF$ ALIGN ARRAY2(*,j) WITH TEMP1(j,*)
      t1=1
!HPF$ INDEPENDENT
      DO p7=1,n
        ARRAY2(1,p7)=B(p7)
      END DO
      t1=2
!HPF$ INDEPENDENT
      DO p7=1,n
        ARRAY1(2,p7)=ARRAY2(1,p7)*       &
&                    ARRAY2(1,p7)
      END DO
      t1=3
!HPF$ INDEPENDENT
      DO p7=1,n
!HPF$ INDEPENDENT
        DO p8=1,n
          A(p7,p8)=ARRAY2(1,p7)+         &
&                  A(p7,p8)-             &
&                  ARRAY1(2,p7)
        END DO
      END DO
```

Figure 5.1 shows the run times of the original version and the LCCP transformed version of this code fragment for $n = 7000$ *with 2, 4, and 8 processors, respectively.*

The original code in Example 5.1 contains several instances of the occurrences representing the terms `B(i)` and `B(i)*B(i)`. The occurrence instances of `B(i)` represent an interesting set, because the values $B(i)$ are used both in the multiplication and in the subtraction *and* because the same value $B(i)$ is used for several different iterations of the j-loop. Therefore, it may pay off to generate a copy of B at another place closer to the respective use site. On the other hand, the same value $B(i) \cdot B(i)$, too, is used in several iterations of the j-loop. Therefore, the basic idea behind code placement applies: we may hoist the multiplication out of the j-loop. Then, we may use another placement for this interesting set in order to improve processor utilization. So, LCCP generates one-dimensional occurrence instance sets for each, $B(i)$ and $B(i) \cdot B(i)$ and new arrays $ARRAY_2$ and $ARRAY_1$ storing the corresponding results.
The placement method now has to select placements from the following possibilities:

Result Array	Placement from A	Placement from B
$ARRAY_1$	$\left(i \mapsto \begin{pmatrix} i \\ 0 \end{pmatrix}, \begin{pmatrix} p_7 \\ p_8 \end{pmatrix} \mapsto \begin{pmatrix} p_7 \\ 0 \end{pmatrix}\right)$	$\left(i \mapsto \begin{pmatrix} i \\ 1 \end{pmatrix}, \begin{pmatrix} p_7 \\ p_8 \end{pmatrix} \mapsto \begin{pmatrix} p_7 \\ p_8 \end{pmatrix}\right)$
$ARRAY_2$	$\left(i \mapsto \begin{pmatrix} i \\ 0 \end{pmatrix}, \begin{pmatrix} p_7 \\ p_8 \end{pmatrix} \mapsto \begin{pmatrix} p_7 \\ 0 \end{pmatrix}\right)$	$\left(i \mapsto \begin{pmatrix} i \\ 1 \end{pmatrix}, \begin{pmatrix} p_7 \\ p_8 \end{pmatrix} \mapsto \begin{pmatrix} p_7 \\ p_8 \end{pmatrix}\right)$

So, there are the same possibilities open for both arrays. However, since all these possible placement relations represent alignments to the same two-dimensional template $TEMP_1$ – corresponding to parallel indices p_7 and p_8 – there is no possibility to actually improve processor utilization due to the idenification of the corresponding occurrence instance set as one-dimensional. A better processor utilization could only be achieved through alignment to a template that contains only one distributed dimension. Unfortunately, since we only propagate already given placements and do not introduce new template dimensions in our method, this is no option here. Moreover, it turns out that B is not distributed very well: it is only aligned with the first column of $TEMP_1$, but the same element of B is used for the complete row of A – and thus $TEMP_1$. Therefore, the copy of B in $ARRAY_2$ is distributed according to the distribution of A. The same holds for $ARRAY_1$.

If an interesting set of the target program leads to the generation of a new auxiliary array, we have to take care not to overwrite any data in this auxiliary array. Therefore, all index dimensions of this interesting set that enumerate finite values are present as subscript dimensions in the generated auxiliary array. In particular, this holds for the time dimensions. Therefore, $ARRAY_1$ and $ARRAY_2$ also feature an additional dimension (which is not distributed) corresponding to the (one-dimensional) time step in which a given occurrence instance is executed.
Note that the auxiliary arrays generated by LCCP should actually be subject to storage mapping optimization as proposed by Lefebvre and Feautrier [LF98]. In the examples presented here, this is not necessary, since all occurrence instances that are to be stored in the same array can be computed in the same time step and are all referenced in the following time step, so that no further optimization can be done (using this schedule). In practice, such a strategy would be essential, although it is not implemented in the LooPo code generation module.[2]
With the additional time dimension, the above placement relations correspond to the following `HPF` placement directives:

[2] In addition, the subscripts of write accesses to auxiliary arrays in the transformed program are actually not ordered in a cache-friendly manner, as suggested in Section 3.7. For example, the assignment to $ARRAY_1$ in the transformed code of Example 5.2 features the time dimension as the first subscript dimension of $ARRAY_1$ instead of the last one (as would be appropriate for `Fortran`). This is merely due to a bug in the code generation implementation in LooPo. And yet, this flaw hardly influences the examples presented in this chapter, since we have to store only a single time step anyway, so that the behaviour of the program wrt. caches is not affected by this shortcoming.

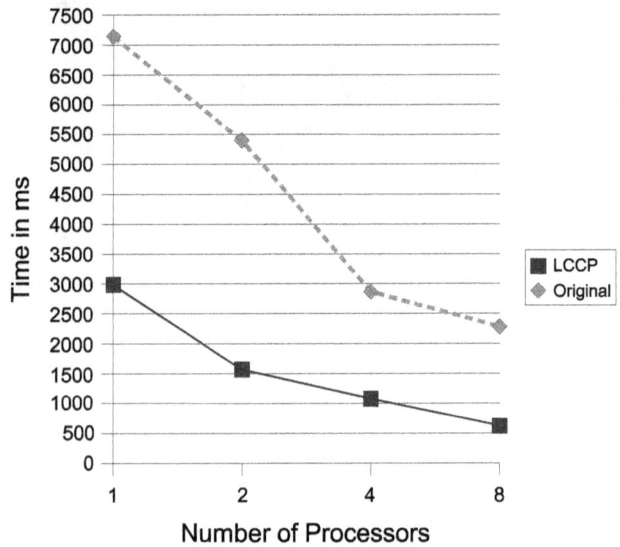

Figure 5.2: Run times for Example 5.2

Result Array	Placement from A	Placement from B (unused)
$ARRAY_1$	`!HPF$ ALIGN ARRAY1(*,j) WITH TEMP(j,*)`	`!HPF$ ALIGN ARRAY1(*,j) WITH TEMP(j,1)`
$ARRAY_2$	`!HPF$ ALIGN ARRAY2(*,j) WITH TEMP(j,*)`	`!HPF$ ALIGN ARRAY2(*,j) WITH TEMP(j,1)`

In this case, the *OIAllocator* placement method chooses the first possibility (propagate from A). Since both $ARRAY_1$ and $ARRAY_2$ are now stored in a replicated fashion, we cannot expect the LCCP transformed program to perform any better than the original. Since we have introduced copies to new arrays, we may even expect lower performance. Still, Figure 5.1 reveals an improvement of 3.5% to 6.5%. This improvement is due to the fact that the computation of $B(i) \cdot B(i)$ is still hoisted out of one of the loops. In the target program, many p_8-iterations now depend on the same value $B(p_8) \cdot B(p_8)$ that may be computed once in a single one-dimensional loop before these uses. Since there are far fewer processors available than iterations to be executed, each processor may now compute a value of $B(i) \cdot B(i)$ before using it over and over again in the following loop nest – in which the corresponding value does not need to be re-evaluated each time (it should even be present in the cache after an initial read). This effect does improve overall performance, but becomes weaker with increasing numbers of processors and thus decreasing number of iterations per processor (and decreasing reuse of the same value).

Example 5.2 *Figure 5.2 (above) shows the run times of the following original code fragment (the left one below) and the code fragment transformed by* `lccp` *(the right one below) for* $n = 4000$ *and 1 to 8 processors.*

```
      REAL A(n, n)
      REAL B(n, n)
      REAL C(n, n)
      REAL E(n, n)

!HPF$ TEMPLATE TEMP1(n,n)
!HPF$ DISTRIBUTE TEMP1(BLOCK,BLOCK)
!HPF$ ALIGN A(i,j) WITH TEMP1(i,j)
!HPF$ ALIGN B(i,j) WITH TEMP1(j,i)
!HPF$ ALIGN C(j,*) WITH TEMP1(j,*)
!HPF$ ALIGN E(i,*) WITH TEMP1(i,*)

!HPF$ INDEPENDENT
      DO i=1,n
!HPF$ INDEPENDENT
         DO j=1,n
            A(i,j)=B(j,i)*(C(i,i))
            E(i,j)=E(i,j)*(C(i,i))
         END DO
      END DO
```

```
      REAL A(n, n)
      REAL B(n, n)
      REAL C(n, n)
      REAL E(n, n)
      REAL ARRAY1(2:2,n)
!HPF$ TEMPLATE TEMP1(n,n)
!HPF$ DISTRIBUTE TEMP1(BLOCK,BLOCK)
!HPF$ ALIGN A(i,j) WITH TEMP1(i,j)
!HPF$ ALIGN B(i,j) WITH TEMP1(j,i)
!HPF$ ALIGN C(j,*) WITH TEMP1(j,*)
!HPF$ ALIGN E(i,*) WITH TEMP1(i,*)
!HPF$ ALIGN ARRAY1(*,j) WITH TEMP1(j,*)
      t1=2
!HPF$ INDEPENDENT
      DO p4=1,n
        ARRAY1(2,p4)=C(p4,p4)
      END DO
      t1=3
!HPF$ INDEPENDENT
      DO t2=1,n
!HPF$ INDEPENDENT
        DO p4=1,n
          E(p4,t2)=E(p4,t2)*ARRAY1(2,p4)
        END DO
      END DO
!HPF$ INDEPENDENT
      DO p4=1,n
!HPF$ INDEPENDENT
        DO p5=1,n
          A(p4,p5)=B(p5,p4)*ARRAY1(2,p4)
        END DO
      END DO
```

The transformed program of Example 5.2 first copies the elements needed from C into a new array $ARRAY_1$. If C is distributed in a suboptimal manner, this can lead to an improvement in run time, in particular for a term like `C(i,i)`, as in the code above, since communication code in which the subscript dimensions depend on each other (such as a condition like "the first subscript dimension equals the second one", which is the case for `C(i,i)`) can be complicated to build and induce more run time overhead. Yet, in the case of Example 5.2, C is replicated in one dimension (the same placement as chosen for $ARRAY_1$), so we were not able to find any better placement than already given. The main improvement in run time, which ranges from 58% to over 72%, is due to other effects, as we will see in a moment. In this case, where we cannot find a better distribution for an interesting set anyway (and do not gain from combining several occurrence instances into one), a further optimization step could consist of redeclaring an occurrence instance set as non-interesting, if its space-time mapping defines (essentially) the same transformation as the ones for its dependence destinations. This would eliminate time and space overhead due to the introduction of auxiliary arrays and the ensuing copy operations to these auxiliary arrays.

Although the t_2-loop in the transformed code of Example 5.2 is marked `INDEPENDENT`, it has to be enumerated by a sequential loop. This is because the second dimension of E is collapsed – i.e., all elements are stored on the same processor. Thus, we can identify the t_2-loop as an implicit placement loop (we discussed these loops in Section 3.7 on page 127). The strategy outlined in Section 3.7 suggests to place this loop insider the actual parallel (explicit placement) loop. Nevertheless, we have decided against this strategy and placed the loop *outside* the parallel loop. The reason for this decision is is that ADAPTOR does not handle this case well (it may

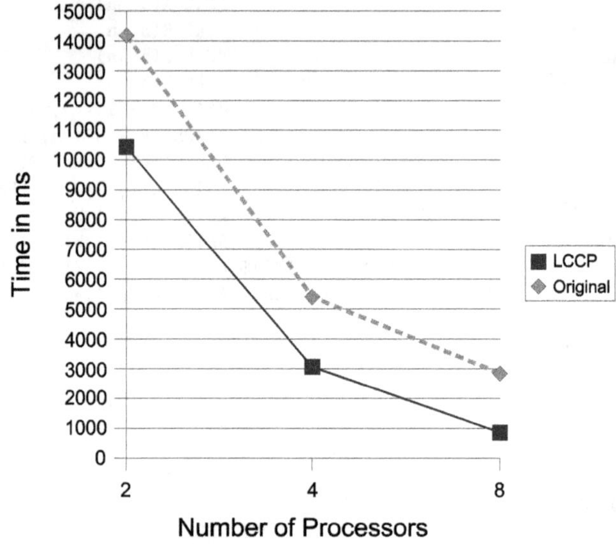

Figure 5.3: Run times for Example 5.3

not extract communication outside such loops). Therefore, we have adapted our code generation module to automatically produce this nesting order instead.

Let us now revisit the run time improvements witnessed above. These improvements actually come from the pure use of code generation in the polyhedron model. As discussed in Section 3.7, we view each target occurrence instance as a vector featuring dimensions for all parallel indices in the target program. However, we only enumerate coordinates with a finite bound. In this way, computations that are actually aligned with different template elements are not executed in the same parallel loop nest, avoiding the pitfall of using different homes in the same `INDEPENDENT` loop nest that we encountered in Section 2.1.2. This is not a result of applying LCCP, but rather of parallelization using the polyhedron model, since the loop fissioning in this case is only possible because the scheduling phase has generated a program in which all dependences are already carried by the outer loops. Note that, although, in this chapter, we have adorned the original code in the examples with `INDEPENDENT` directives, we do not *need* this information in order to produce parallel code. The usual parallelization in the polyhedron model already takes care of this. The next example shows how `lccp` performs with rather time consuming mathematical operations.

Example 5.3 *Figure 5.3 shows the run times of the following original code fragment (the left one below) and the code fragment transformed by LCCP (the right one below) for* $n = 7000$ *and 2 to 8 processors.*

```
      REAL A(n)
      REAL B(n)
      REAL C(n)
      REAL D(n,n)
      REAL E(n,n)
!HPF$ TEMPLATE TEMP1(n,n)
!HPF$ DISTRIBUTE TEMP1(BLOCK,BLOCK)
!HPF$ ALIGN A(i) WITH TEMP1(i,1)
!HPF$ ALIGN B(i) WITH TEMP1(i,1)
!HPF$ ALIGN C(i) WITH TEMP1(i,1)
!HPF$ ALIGN D(i,j) WITH TEMP1(i,j)
!HPF$ ALIGN E(i,j) WITH TEMP1(i,j)
!HPF$ INDEPENDENT
      DO i=1,n
!HPF$ INDEPENDENT
        DO j=1,n
          D(i,j)=(A(i)*B(i))+        &
&                SQRT(C(i))
          E(i,j)=SIN(A(i)*B(i))
        END DO
      END DO
```

```
      REAL A(n)
      REAL B(n)
      REAL C(n)
      REAL D(n,n)
      REAL E(n,n)
      REAL ARRAY1(3:3,n)
      REAL ARRAY2(3:3,n)
      REAL ARRAY3(2:2,n)
      REAL ARRAY4(1:1,n)
      REAL ARRAY5(1:1,n)
!HPF$ TEMPLATE TEMP1(n,n)
!HPF$ DISTRIBUTE TEMP1(BLOCK,BLOCK)
!HPF$ ALIGN A(i) WITH TEMP1(i,1)
!HPF$ ALIGN B(i) WITH TEMP1(i,1)
!HPF$ ALIGN C(i) WITH TEMP1(i,1)
!HPF$ ALIGN D(i,j) WITH TEMP1(i,j)
!HPF$ ALIGN E(i,*) WITH TEMP1(i,*)
!HPF$ ALIGN ARRAY1(*,j) WITH TEMP1(j,*)
!HPF$ ALIGN ARRAY2(*,j) WITH TEMP1(j,*)
!HPF$ ALIGN ARRAY3(*,j) WITH TEMP1(j,*)
!HPF$ ALIGN ARRAY4(*,j) WITH TEMP1(j,*)
!HPF$ ALIGN ARRAY5(*,j) WITH TEMP1(j,*)
      t1=1
!HPF$ INDEPENDENT
      DO p7=1,n
        ARRAY5(1,p7)=A(p7)
        ARRAY4(1,p7)=B(p7)
      END DO
      t2=2
!HPF$ INDEPENDENT
      DO p7=1,n
        ARRAY3(2,p7)=ARRAY5(1,p7)*          &
&                    ARRAY4(1,p7)
      END DO
      t2=3
!HPF$ INDEPENDENT
      DO p7=1,n
        ARRAY2(3,p7)=SQRT(C(p7))+          &
&                    ARRAY3(2,p7)
        ARRAY1(3,p7)=SIN(ARRAY3(2,p7))
      END DO
      t4=4
!HPF$ INDEPENDENT
      DO p7=1,n
!HPF$ INDEPENDENT
        DO p8=1,n
          D(p7,p8)=ARRAY2(3,p7)
          E(p7,p8)=ARRAY1(3,p7)
        END DO
      END DO
```

Just as with Example 5.1, array B is distributed unfavourably in Example 5.3 and copied to $ARRAY_4$ in the first step of the transformed program. Similarly, A is copied to $ARRAY_5$. Since both A and B are used several times in the following computation, this can be viewed as explicit message coalescing (we now only need to communicate A and B *once* each, whereas, previously, both arrays had to be transferred twice). C is *not* copied to an auxiliary array, because it is only needed in the term `(A(i)*B(i))+SQRT(C(i))`, whose execution itself is already represented by an interesting set. Both the value represented by this term and the expression $\sin(A(i) \cdot B(i))$ are needed on a complete row of the processor mesh and therefore get a replicated placement, just as $B(i) \cdot B(i)$ in Example 5.1. The computations of $A(i) \cdot B(i) + \sqrt{C(i)}$ are extracted from the j-loop enumerating that row of the processor mesh. Each such computation is therefore only executed once by each physical processor that is responsible for a range of that row. The same holds for the computations of $A(i) \cdot B(i)$, since the same multiplication result is needed for both adding it to $\sqrt{C(i)}$ and for taking its sine. In contrast to the previous examples, the homes of the result arrays (D and E) are identical. In this case, the fact that the relatively expensive operators `SIN` and `SQRT` do not have to be called as often as in the original program, combined with the message coalescing of A and B is responsible for a run time improvement of 26% to almost 69%.[3]

5.3 A Case Study: Binary Black Hole

This section is devoted to a small fragment of the Pittsburgh Binary Black Hole code as a real world example [GWI92]. This code was developed in the Binary Black Hole Grand Challenge project for the computation of gravitational waves [HM95]. We only consider a small loop nest of this code in Example 5.4, which is originally given in `FORTRAN 77`, but also available in an `HPF` version, which enumerates all loops that may be executed in parallel by `FORALL`-loops. We choose to use the same placement declarations, but the more appropriate `INDEPENDENT` directive for the parallel loops.[4] Another discussion of this example, primarily focussing on the application of our placement method, can be found in Wondrak's diploma thesis [Won05, Chapter 8].

Example 5.4 *Figure 5.4 shows the run times of the original program fragment for $nx = ny =$ 6000 and 2 to 16 processors, and a version with loops fissioned by hand (which is similar to the* `HPF` *version of the original). One might expect better performance from the fissioned version due to the fact that it guarantees a single home description for each loop nest. Nevertheless, the time measurements of these two versions are nearly identical so that their graphs in Figure 5.4 overlap almost completely. However, there is a third line (the lowest one) in the figure, which belongs to the LCCP transformed program fragment of the first version. The complete code for the transformed fragment is very large. Therefore, we will only review a small part of this code later on.*

Our `HPF` *program for the Binary Black Hole code, which is based on a sequential version and has been parallelized by hand, looks as follows:*

```
      REAL L2(0:ny,0:nz)
      REAL GYY(0:ny,0:nz)
      REAL GY(0:ny,0:nz)
      REAL GZ(0:ny,0:nz)
      REAL GZZ(0:ny,0:nz)
      REAL Y(0:ny)

!HPF$ TEMPLATE T1(0:Ny,0:Nz)
!HPF$ DISTRIBUTE T1(BLOCK,BLOCK)
!HPF$ ALIGN L2(p1,p2) WITH T1(p1,p2)
!HPF$ ALIGN GY(p1,p2) WITH T1(p1,p2)
!HPF$ ALIGN GZZ(p1,p2) WITH T1(p1,p2)
!HPF$ ALIGN GYY(p1,p2) WITH T1(p1,p2)
!HPF$ ALIGN Y(p1) WITH T1(p1,*)
```

[3]Note that this holds even though the computation of the more expensive operators is done in a replicated fashion.

[4]Note again that, due to the scheduling phase applied in our approach, the `INDEPENDENT` directives are actually not necessary to identify these loops as parallel.

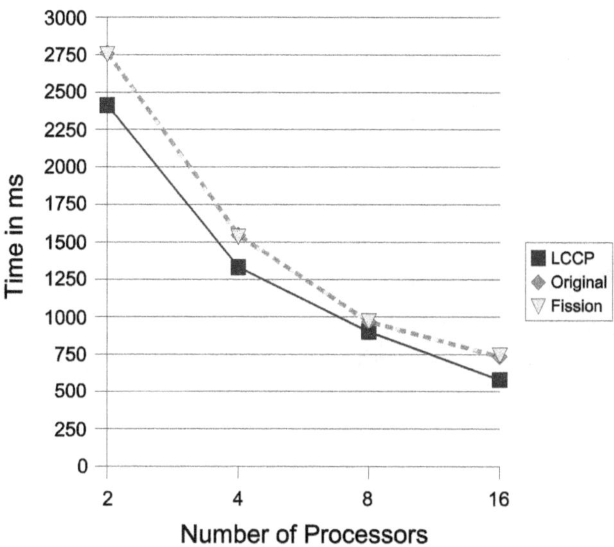

Figure 5.4: Run times for Example 5.4

```
!HPF$ INDEPENDENT
      DO k=0,nz
        L2(0,k)=-4.0                &
&              *(gm_south-g_south)  &
&              /(dz*dz)
        L2(ny,k)=-4.0               &
&              *(gm_north-g_north)  &
&              /(dz*dz)
!HPF$ INDEPENDENT
        DO j=1,ny-1
          L2(j,k)=-(((1.0-Y(j))       &
&                  *(1.0-Y(j)))       &
&                 *GYY(j,k)           &
&                 -2.0*Y(j)*GY(j,k)   &
&                 +GZZ(j,k)           &
&                 /(1.0-Y(j)*Y(j)))
         END DO
      END DO
```

The hand-fissioned – not *LCCP transformed* – *code, which is similar to the* HPF *version of the Binary Black Hole Grand Challenge project, is given below:*

```
      REAL L2(0:ny,0:nz)
      REAL GYY(0:ny,0:nz)
      REAL GY(0:ny,0:nz)
      REAL GZ(0:ny,0:nz)
      REAL GZZ(0:ny,0:nz)
```

```
      REAL Y(0:ny)

!HPF$ TEMPLATE T1(0:Ny,0:Nz)
!HPF$ DISTRIBUTE T1(BLOCK,BLOCK)
!HPF$ ALIGN L2(p1,p2) WITH T1(p1,p2)
!HPF$ ALIGN GY(p1,p2) WITH T1(p1,p2)
!HPF$ ALIGN GZZ(p1,p2) WITH T1(p1,p2)
!HPF$ ALIGN GYY(p1,p2) WITH T1(p1,p2)
!HPF$ ALIGN Y(p1) WITH T1(p1,*)

!HPF$ INDEPENDENT
      DO k=0,nz
        L2(0,k)=-4.0                   &
&                *(gm_south-g_south)   &
&                /(dz*dz)
      END DO
!HPF$ INDEPENDENT
      DO k=0,nz
!HPF$ INDEPENDENT
        DO j=1,ny-1
          L2(j,k)=-(((1.0-Y(j))          &
&                   *(1.0-Y(j)))         &
&                  *GYY(j,k)             &
&                  -2.0*Y(j)*GY(j,k)     &
&                  +GZZ(j,k)             &
&                  /(1.0-Y(j)*Y(j)))
         END DO
      END DO
!HPF$ INDEPENDENT
      DO k=0,nz
        L2(ny,k)=-4.0                    &
&                  *(gm_north-g_north)   &
&                  /(dz*dz)
      END DO
```

The original program fragment in Example 5.4 contains the following terms represented by occurrence instances that are combined into interesting sets:

Term	Reason for Applicability of LCCP	Array
Y(j)	independent of k, appears several times in body	$ARRAY_8$
1.0-Y(j)	independent of k, appears several times in body	$ARRAY_7$
Y(j)*Y(j)	independent of k, appears several times in body	$ARRAY_4$
(1.0-Y(j))*(1.0-Y(j))	independent of k	$ARRAY_6$
2.0*Y(j)	independent of k	$ARRAY_5$
1.0-Y(j)*Y(j)	independent of k	$ARRAY_3$
dz*dz	independent of k, independent of j (scalar)	–
-4.0*(gm_south-g_south)/(dz*dz)	independent of k, independent of j (scalar)	–
-4.0*(gm_north-g_north)/(dz*dz)	independent of k, independent of j (scalar)	–

Unfortunately, our prototype implementation of LCCP in LooPo was not able to combine the last three – scalar – expressions to (singleton) sets of representatives due to excessive memory usage during compilation. Therefore, equivalence recognition was turned off for scalars, and only the first 6 expressions were actually transformed into new occurrence instance sets. The table also shows the respective arrays storing the values calculated for the corresponding expressions. As already hinted, the result of our LCCP transformation is rather large, so that we only show the central part of the target code here:

```
      REAL L2(0:ny,0:nz)
      REAL GYY(0:ny,0:nz)
      REAL GY(0:ny,0:nz)
      REAL GZ(0:ny,0:nz)
      REAL GZZ(0:ny,0:nz)
      REAL Y(0:ny)
      :
      REAL ARRAY3(3:3,1:ny-1)
      REAL ARRAY4(2:2,1:ny-1)
      REAL ARRAY5(2:2,1:ny-1)
      REAL ARRAY6(3:3,1:ny-1)
      REAL ARRAY7(2:2,1:ny-1)
      REAL ARRAY8(1:1,1:ny-1)
      :
!HPF$ TEMPLATE TEMP1(0:ny,0:nz)
!HPF$ DISTRIBUTE TEMP1(BLOCK,BLOCK)
!HPF$ TEMPLATE TEMP1(0:ny,0:nz)
!HPF$ DISTRIBUTE TEMP1(BLOCK,BLOCK)
!HPF$ ALIGN L2(p1,p2) WITH temp1(p1,p2)
!HPF$ ALIGN GY(p1,p2) WITH temp1(p1,p2)
!HPF$ ALIGN GZZ(p1,p2) WITH temp1(p1,p2)
!HPF$ ALIGN GYY(p1,p2) WITH temp1(p1,p2)
!HPF$ ALIGN Y(p1) WITH temp1(p1,*)
      :
!HPF$ ALIGN ARRAY3(*,j) WITH TEMP1(j,*)
!HPF$ ALIGN ARRAY4(*,j) WITH TEMP1(j,*)
!HPF$ ALIGN ARRAY5(*,j) WITH TEMP1(j,*)
!HPF$ ALIGN ARRAY6(*,j) WITH TEMP1(j,*)
!HPF$ ALIGN ARRAY7(*,j) WITH TEMP1(j,*)
!HPF$ ALIGN ARRAY8(*,j) WITH TEMP1(j,*)
      :
      t1=1
!HPF$ INDEPENDENT
      DO p7=1,ny-1
        ARRAY8(p7)=Y(p7)
        :
      END DO
      :
      t1=2
!HPF$ INDEPENDENT
      DO p7=1,ny-1
      :
        ARRAY7(2,p7)=1.0-ARRAY8(1,p7)
      :
        ARRAY5(2,p7)=2.0*ARRAY8(1,p7)
      :
```

```
          ARRAY4(1,p7)=ARRAY8(1,p7)*ARRAY8(1,p7)
        :
        END DO
        :
        t1=3
!HPF$ INDEPENDENT
        DO 1,ny-1
          ARRAY6(3,p7)=ARRAY7(2,p7)*ARRAY7(2,p7)
          ARRAY3(3,p7)=1.0-ARRAY4(2,p7)
        END DO
        t1=4
!HPF$ INDEPENDENT
        DO p8=0,nz
!HPF$ INDEPENDENT
          DO=1,ny-1
            L2(p7,p8)=(-(GZZ(p7,p8)/ARRAY3(3,p7)+ARRAY6(3,p7)*GYY(p7,p8)&
&                     -ARRAY5(2,p7)*GY(p7,p8)))
          END DO
        END DO
```

In this case, Y is already distributed very well, so that we do not gain anything from introducing $ARRAY_8$, which is distributed exactly as Y. Since, for all interesting sets, both sources and targets suggest the same placement relation, there is only one possible placement for all the auxiliary arrays (and the corresponding computations): aligning the (single) parallel dimension with the first dimension of the processor mesh and replicating the arrays along the second dimension.

Figure 5.4 shows an improvement of the LCCP generated version of at least 6% up to over 20% wrt. the original program fragment. Further details can be gathered from the following table:

Number of Processors	Time LCCP in ms	Relative Speedup (LCCP)	Efficiency (LCCP)	Time Original in ms	Relative Speedup (Original)	Efficiency (Original)	Improvement (LCCP vs. Original)
2	2414	2.2	110%	2752	2.0	100%	12.63%
4	1331	3.99	99.8%	1535	3.56	89.0%	13.96%
8	901	5.89	73.6%	969	5.69	71.1%	6.83%
16	581	9.14	57.1%	743	7.49	46.8%	20.95%

The table also shows the relative speedup of the program run on the respective number of processors wrt. exactly the same program version run on only one processor. We observe that the relative speedup, and thus the efficiency, of the LCCP transformed program is consistently better than the one of the original program.

It should be noted that there are also examples that do not favour the LCCP transformation at all. Nevertheless, this transformation and its theoretical background, in combination with a placement method tuned to the respective target system, can be viewed as a first step towards a method for generating more efficient target code.

Chapter 6

Conclusion

Traditional code analysis does not handle loop ranges, i.e., it handles loops as completely unpredictable control structures. The polyhedron model represents an approach based on integer sets that removes this drawback. This model offers a mathematical representation of a code fragment, which has to be reconstructed into an executable program. In particular, for parallel programs running on distributed memory architectures, this has turned out to be a non-trivial task. Previous work either concentrated on elaborate scanning techniques in order to generate code that can be compiled into efficient executable code by traditional compilers, or (as in the case of loop transformations for cache optimizations) used this approach only as a coarse model of the loop nest in a program fragment.
This thesis presents two optimization techniques for loop programs that aim at improving the code on the model level. These techniques enable the code generation phase to produce source code for which a traditional (completely syntax-based) compiler generates efficient target code. This holds in particular for the case of the aforementioned parallel programs for distributed memory architectures. Both techniques are based on a common framework developed in Chapter 2. The usual granularity used in the polyhedron model is refined here further to the level of function and operator arguments. On this level, each argument position can be identified. Each of these low level operations is called an occurrence instance.

The Basic Framework In order to define intraprocedural code optimizations, it is necessary to represent the input program in great detail, close to the level of the actual program execution by the hardware. For this purpose, we have created the framework introduced in Section 2.3. This framework builds on the polyhedron model. Adding two further dimensions to the usual representation used in the polyhedron model enables us to examine not only statements or even instances of operator applications, but load and store operations (i.e., read accesses and write accesses) within call sites. Thus, not only the order of arguments in an argument list, but even actual and dummy arguments of a function call can be distinguished. This is as close as one can get to the executable code represented by the input program without knowledge of the implementation of function calls (for example, whether a given operator is to be implemented by a machine instruction or a call to some library function). The smallest entity considered in this model is the single read access, write access, or compute operation that is executed in a single well-defined loop iteration. These entities are called occurrence instances in our refined model. Each occurrence instance is associated with a function symbol that indicates whether the occurrence instance represents a compute operation (and, if so, which compute operation) or a read or write access.

Code Optimization in the Polyhedron Model Chapter 3 makes use of this approach by comparing occurrence instances based on the function symbols associated with them and on the occurrence instances on which they depend. The result is a procedure for the detection of subexpressions that compute the same value wrt. Herbrand equivalence, where terms may contain array references. This procedure, which we have dubbed loop-carried code placement (LCCP), can easily be extended to support also associativity and commutativity laws. LCCP creates a new program representation using polyhedra – as usual in the polyhedron model. In contrast to the usual transformation in the polyhedron model, LCCP introduces a new class of existentially quantified variables that have to be checked for an integer solution, but which are not used for the actual computation. The new (transformed) polyhedra are derived via a change of basis. We have observed that the safest way of obtaining valid target polyhedra (which have to consist of integer numbers), namely using unimodular transformations, is not in general possible (as Theorem 3.24 shows). Therefore, the generated polyhedra may have to be expressed in a more complicated manner than the original ones: they may contain holes even if the original ones do not. Since detecting all the redundancy in a program is not computable, we cannot expect this method to do so. In fact, it is limited to equivalences that do not follow from assignments: it cannot deduce the equivalence of A and B after an assignment of A to B. In future work, this could be improved somewhat by introducing equivalences on write accesses and defining a class of operators other than linear combination whose result can be predicted. One candidate for this is the assignment operator, which only passes on the value it reads. We could then rewrite the program representation according to the appropiate rules. However, although these improvements are straightforward extensions of the framework presented here, they also introduce new problems, whose consideration is beyond the scope of this thesis.

With its redundant code recognition features, LCCP can be used to reduce the number of occurrence instances to be executed in a given loop nest. However, this comes possibly at the price of additional assignments in the final executable. Nevertheless, in the parallel setting, this code placement technique can lead to improved processor utilization.

The method also distinguishes between plain sets of occurrence instances, which have to be represented as a computation in the target program, and interesting sets, which compute values that are needed in several places of the program fragment. Thus, the method is also useful for identifying occurrence instances for which it is worthwhile to determine an individual space-time mapping and review the possible placements in a distributed memory architecture.

Occurrence Instance Placement Determining a placement for an interesting set of occurrence instances is therefore an important aspect, not only of code generation in the polyhedron model in general, but also of postprocessing the results of LCCP transformations. With the placement method developed in Chapter 4, we aim at a universal strategy for producing placements for any target system. Of course, this means that we have to employ a highly adaptive cost model. We impose only a few assumptions on the cost model: in particular, we assume that it will predict lower costs if an occurrence instance is aligned with sources or targets of dependences pertaining to this occurrence instance. I.e., we assume that costs are decreased by data alignment and increased by the absence of alignment. The straightforward approach we follow here is to propagate placements in both directions and then evaluate all possible combinations of placements, which leads to the need of further partitioning the interesting sets so that the domain of a placement relation may be modeled accurately. This strategy produces an extremely large search space, which might better be examined using genetic algorithms or different heuristics (which might make additional assumptions about the cost model). However, this point is left for future work.

Still striving for a general placement method, we do not restrict ourselves to placement functions, but allow placement relations, i.e., a value may be computed and stored in several places. Although this replication may not utilize all the parallelism inherent in the program, it may pay off in the overall cost due to reduced communication requirements. Communication takes place in a parallel program at different levels (inter-node communication in a coarser view, communication between registers in a very fine-grained view). We propose to adopt a hierarchical approach to the placement of occurrence instances, in which placements are determined for several of these levels, moving from the coarser view to the finer one. However, for the discussion in this thesis, we restrict ourselves to placement relations on the level of inter-node communication. Placement relations on this level are easily expressed in `HPF` and, indeed, an `HPF` compiler can take advantage of these explicit placements quite well if some additional (still sometimes compiler dependent) issues are

observed (as pointed out in Sections 2.1.2, 3.7 and 5.2).
Regarding the inter-node communication using an HPF compiler, it can be observed that the complexity of array subscript expressions in the loop body has a strong impact on the performance. With the help of the benchmark suite we developed (which is described in Section 4.3.1), the cost model we have adapted to the ADAPTOR compiler as an example in Section 4.3.2 can be recalibrated to other (HPF or non-HPF) compilers that also rely heavily on the syntactic forms of subscript expressions for communication generation.

Future Work As discussed in Section 3.8 and Section 4.5, there are quite some points left for future work, some of which we want to revisit here. Let us first review possible future work on the LCCP algorithm. We will then proceed to discuss some points for improvement in the placement method.

A possible approach to eliminate performance penalties of the `lccp` transformation is to clone interesting sets into several collections of almost identical sets (with only the occurrence number changed to indicate the difference). This reverses some of the transformations achieved by LCCP, so there should be some mechanism to determine whether a set should be cloned or not.

A very important point for the general applicability of `lccp` is that with the version of `lccp` presented here, write accesses are never equivalent to each other. This restriction prevents many possible optimizations. With an equivalence on write accesses, LCCP could be used to eliminate unnecessary writes and identify all the points automatically that lend themselves to the storage of intermediate values. As discussed in Section 3.6, this restriction is not to be overlooked, since the correctnes of the transformation depends on it. However, it may be possible to circumvent these problems.

We have also observed that the transformation performed by `lccp` may give valuable hints for places where code optimizations may be perfromed in order to increase processor utilization not only for clusters but also for modern mulitcore architectures as well as stream processors and general purpose GPU (GPGPU) programming. For example, in the case of GPGPU programming, kernels may be optimized for better ressource utilization internally. Also, with the interesting sets identified by `lccp`, this technique may possibly be used to cluster code fragments into kernels automatically so as to both increase ressource utilization and to minimize kernel startups (in conjunction with a corresponding scheduler). A problem here may be to adequately model limited resources such as shared memory.

The placement algorithm presented in Chapter 4 is rather simple. It is important to choose an appropriate cost model for the problem system at hand.

The cost model we have introduced in Chapter 4 favours replicated placements, since there is no provision for estimating the amount of work placed on a single processor. In addition, computation of placements in practice is very time consuming. This is because *all* combinations of placement relations are evaluated. It remains to examine where our flexible method should employ heuristics or other search mechanisms in order to reduce the run time of the placement algorithm.

And yet, our placement method is already artificially restricted in its search space: cycles in the reduced dependence graph and the combination operator introduced in Section 2.4.2 may lead to further placement candidates. It may be worthwhile to include these placement candidates in later versions of the placement algorithm.

In addition, further placement relations may be obtained by repeatedly propagating placement relations (our placement method is parameterized with the number of propagating steps, in Chapter 4, we only use a single step).

Although the placement algorithm presented in Chapter 4 is thought to be embedded in a hierarchical placement process, we only considered a single run of the placement algorithm. Future work should take into account different levels of granularity and employ several consecutive runs of the placement algorithm with different cost models suited for the respective granularity, with each run of the placement algorithm being aware of the results of previous runs (and of the scheduler).

Final Observation We have presented a framework for performing code optimization in the polyhedron model and introduced two transformation methods based on this framework. The first transformation, LCCP, directly changes the complete program representation and may increase processor utilization as well as identify candidates for redistribution. The second method, the *OIAllocator* placement, uses replication to bring the source and target of a computation closer together. Both methods leave room for improvement but, as Section 5.3 shows, the combination of these transformation techniques, in conjunction with some heuristics in the target code generation, represents an important step towards more efficient target code by itself.

Appendix A

Options of the Program `lccp`

In Chapter 3, we were faced with several different choices on numerous occasions. For the sake of the argument, we always selected the one for which we could make the best case. Nevertheless, the program `lccp` that implements the method offers a set of options in order to change the default behaviour. Here, we present these options that can be supplied to the program `lccp` to change the default settings.[1]

`const_stat`:
: Usually, Occurrence 0 represents integer numbers. If option `const_stat` is supplied, the first statement encountered in the program is used to represent integer numbers (i.e., a dependence on a linear expression is represented as a dependence on the head node of this statement). The integer number represented is always the value of the first index variable (which may or may not lie in the actual index space of the statement).

`guess_dep_type`:
: Usually, dependences in the output are only marked as having an undefined type (as opposed to flow, anti, input or output dependence). If this option is given, the correct type of a dependence is set according to its source and target occurrences.

`highdim`:
: This option causes `lccp` to sort equivalence classes so that a set of occurrence instances is represented by a polyhedron with as many dimensions as possible. This is the method that we chose in Section 3.5.2 on page 104 for sorting the polyhedra that may represent an occurrence instance set. If the option is not given, the opposite sorting criterion is used.

`ignoredeps`:
: This option defines a set of dependence types that are ignored by `lccp` for generating equivalence classes. As discussed in Section 2.3.1, usually only flow dependences are helpful in determining equivalence classes. Therefore, the default behaviour is to use flow dependences, user given dependences (which are interpreted as flow dependences), and input dependences.

`interesting_reads`:

`interesting_writes`:

`interesting_readwrites`:

`interesting_stats`:
: Definition 3.17 in Section 3.17 defines sets of write accesses as interesting sets. With these options, different sets can be made interesting.

`keep_range`:

[1]Purely technical options that only handle format conventions between modules are omitted.

`scoped_deps`:
: With these options switched on, the exact domains, i.e., the bounds of occurrence instance sets are not ignored when searching for equivalent occurrence instances in contrast to the procedure selected in Section 3.5.2 on page 86.

`lexminmap`:

`projectmap`:
: These options are mutually exclusive. They tell `lccp` which method to use for selecting representatives. If **`lexminmap`** is given, the lexicographic minimum (as outlined in Section 3.5.1) is used. Otherwise, representatives are selected according to our method employing a change of basis (discussed in Section 3.5.2).

`memory`: The list following this option defines **dummy statements** (the statements inside dummy loops introduced in Section 2.3). Usually, dummy statements are recognized by certain reserved function names.

`norm_deps`:
: This option asserts that the dependence descriptions in the output are defined by h-transformations that represent simple expressions (as discussed in Section 3.7 on page 131). This leads to subscript expressions that can be handled well by HPF compilers.

`op_is_big`:
: Setting this option forces the operand coordinates in the complete output to be set to infinity. This marks the dimension as corresponding to a loop that does not contain any statement in its body. This is done in order to be able to call subsequent modules that work on the usual granularity of statements rather than on our refined model.

`partition_dep_stat`:
: If this option is supplied to `lccp`, an initial partitioning of occurrence instance sets from target to source is done as discussed in Section 4.2.2. Since a partitioning in the opposite direction is automatically done by the LCCP method itself, this precludes the partitioning step of the placement method discussed in Section 4.2.3.

`time`: Option `time` tells `lccp` to print out timing statistics at the end of the program run.

`v`: This option sets the verbosity level:

- `on`: Print all messages.
- `norm`: Print only warnings and errors.
- `off`: Run in silent mode.

Appendix B

Notation

Often, notations and definitions are taken for granted. But, as Wittgenstein tells us, “only facts can express a sense, a set of names cannot” [Wit18]. This work is no exception: although most of the notation encountered here is quite common, there may well be basic definitions and notations that may not be clear immediately. Therefore, we give the following small translation table in the hope that it may unveil the exact meaning of all expressions that are not immediately clear to the reader.

$\#(\mathfrak{A})$	Number of elements of a finite set $\mathfrak{A}$.
$\mathfrak{W}^C$	Complement of a linear subspace $\mathfrak{W}$.
$\mathfrak{W}^\perp$	Orthogonal complement of a linear subspace $\mathfrak{W}$.
$F : \mathfrak{A} \longrightarrow \mathfrak{B}$	Function F maps from set $\mathfrak{A}$ (its domain) to set $\mathfrak{B}$ (the image of F, some authors call this the range of F).
$F(x)$	Image of x by a function F.
$F : \mathfrak{A} \longrightarrow \mathfrak{B} : x \longmapsto y$	Function F maps each $x \in \mathfrak{A}$ (the pre-image of y) to $y \in \mathfrak{B}$ (the image of x).
$F : x \longmapsto y$	Function F maps each x to y. Domain and image of F are clear from the context, and therefore *not* stated explicitly.
$\mathrm{dom}(F)$	Domain $\mathfrak{A}$ of a function $F : \mathfrak{A} \longrightarrow \mathfrak{B}$.
$\mathrm{im}(F)$	Image of a function $F : \mathfrak{A} \longrightarrow \mathfrak{B}$; $\mathrm{im}(F) = F(\mathfrak{A}) \subseteq \mathfrak{B}$.
$\mathrm{im}(M)$	Image of matrix M, i.e., the image of the linear function $\Psi : \nu \longrightarrow M \cdot \nu$.
F^n	n applications of F: $F^n = \underbrace{F \circ \cdots \circ F}_{n \text{ times}}$.
F^{-1}	Inverse of function/relation F – possibly maps a single element to a set.
$\ker(\Psi)$	Kernel of a linear function Ψ.
$(x, y) \in F$	Relation F contains the pair (x, y). If F is a function, this means that x is mapped to y.
$x\,F\,y$	different notation for $(x, y) \in F$.
$R\vert_{\mathfrak{A}}$	Restriction of relation (or function) R to domain $\mathfrak{A}$. $R\vert_{\mathfrak{A}} = \{(a, b) \mid (a, b) \in R \wedge a \in \mathfrak{A}\}$.
$R^{=}$	Reflexive closure of relation R.
R^{-}	Symmetric closure of relation R.
R^{+}	Transitive closure of relation R.
R^{*}	Reflexive, transitive closure of relation R.
$(R^{-})^{*}$	Reflexive, symmetric transitive closure of relation R.
Ψ^g	Any generalized inverse of a linear function Ψ [Usm87, p. 84]. See Section 2.2
M^g	Generalized inverse of a matrix M.

Ψ^{g_0}	Generalized inverse of a linear function Ψ, in which undefined dimensions are mapped to 0.
M^{g_0}	Generalized inverse (as above) for matrix M.
$T^{g<}$	Generalized inverse of T that simplifies result expressions; see page 124.
Ψ^C	Conditions that have to hold for a vector to lie in the image of a singular linear function Ψ.
${M_\Psi}^C$	Matrix representing the conditions Ψ^C above for a vector α as $M_\Psi \cdot \alpha = 0^C$; (this is for a linear function $\Psi : \mu \mapsto M_\Psi \cdot \mu$).
ν^T	Vector ν, transposed.
$\mathfrak{V}/\mathfrak{W}$	Quotient space $\mathfrak{V}$ modulo $\mathfrak{W}$.
$\iota_{d,i}$	i-th unit vector in a d-dimensional space.
ι_i	i-th unit vector within a space whose dimension is clear from context.
$I_{k,k}$	k-dimensional unit matrix.
$I_{k,l}$	$k \times l$-matrix E with $E[i,j] = \begin{cases} 1 & \text{if } i = j \\ 0 & \text{otherwise} \end{cases}$.
id	The identity function.
$\text{Span}(\nu_1, \ldots, \nu_n)$	Span of the vectors $\nu_1, \ldots, \nu_n$. I.e., $\text{Span}(\nu_1, \ldots, \nu_n) = \{(\sum i : i \in \{1, \ldots, n\} : a_i \cdot \nu_i) \mid (\forall i : i \in \{1, \ldots, n\} : a_i \in \mathbb{Q})\}$. Correspondingly, $\text{Span}(\mathfrak{N}) = \bigcup_{\nu \in \mathfrak{N}} \text{Span}(\nu)$.
$\mathfrak{V} \odot \mathfrak{W}$	Direct sum $V \cup W$, where $V \cap W = \emptyset$.
$\mathcal{P}(\mathfrak{M})$	Powerset of set $\mathfrak{M}$.
$\mathfrak{K}^{m \times n}$	$m \times n$-matrix with elements from ring $\mathfrak{K}$.
$\text{rk}(M)$	Rank of matrix M.
$\langle \nu, r \rangle$	Access instance – see Section 2.3.
$OpndSel(i, \nu)$	Occurrence instance representing operand i of the operation ν. See definition 2.21 in Section 2.3.
m_∞	Big parameter ($m_\infty = \infty$) [Fea03] – Section 2.3.
m_c	Constant parameter known to be 1 ($m_c = 1$) – see Section 2.3.
$a\%b$	a modulo b ($a\%b = a - \left\lfloor \frac{a}{\lvert b \rvert} \right\rfloor \cdot b$).
$\Delta^i, \Delta^f, \Delta^a, \Delta^o$	Input-, flow-, anti- or output-dependence relation (defined in Section 2.3.2). In the reduced form, these dependences are usually defined by piecewise linear h-transformations of the form $$H_\Delta : \mathfrak{D} \subseteq \text{im}(\Delta) \to \mathfrak{DI} : \alpha \mapsto \begin{cases} M_{H_{\Delta,1}} \cdot \alpha & \text{if } \alpha \in \mathfrak{D}_1 \\ \vdots \\ M_{H_{\Delta,n}} \cdot \alpha & \text{if } \alpha \in \mathfrak{D}_n \end{cases}.$$ However, in this work, such a relation is usually represented by a family of n linear h-transformations of the form $$\begin{array}{c} H_{\Delta,1} : \mathfrak{D}_1 \subseteq \text{im}(\Delta) \to \mathfrak{DI} : \alpha \mapsto M_{H_{\Delta,1}} \cdot \alpha \\ \vdots \\ H_{\Delta,n} : \mathfrak{D}_n \subseteq \text{im}(\Delta) \to \mathfrak{DI} : \alpha \mapsto M_{H_{\Delta,n}} \cdot \alpha \end{array}$$ with $\mathfrak{D}_i \cap \mathfrak{D}_j = \emptyset$ for $i \neq j \in \{1, \ldots, n\}$.
Δ^e	Equivalence propagating dependences (see Section 3.2).
$\circ_-$	Subset composition of two relations (see Definition 2.36).
$\circ_+$	Superset composition of two relations (see Definition 2.32).

In addition, we obey some general naming conventions:

- General sets start with capital German letters: $\mathfrak{A}, \mathfrak{B}, \mathfrak{C}, \ldots$ Sets defining linear subspaces are usually named $\mathfrak{U}, \mathfrak{V}, \mathfrak{W}, \ldots$
- (Scalar) variables (whose type may be *anything*, including a vector type, if we do not use the fact that the variable is actually a vector) start with lower case Latin letters: $a, b, c, \ldots$ Usually, we further discern:
  - Align dummies are usually named $i, j, k, \ldots$.
  - Indices (in the program text) are usually named $i, i_1, i_2, \ldots$
  - Parameters (in the program text) are usually named $m, m_1, m_2, \ldots$
  - The number of indices in the source space of a program fragment considered is n_{src}, the number of index variables in the target space (after space-time mapping) is n_{tgt}.
  - The number of parameters in a program fragment considered is n_{blob} – we suppose that the target program fragment contains as many parameters as the source program fragment.[1]
  - Occurrences are usually named $o, o_1, o_2, \ldots$
  - The number of dimensions of a processor mesh or its corresponding template or of a processor array is $\#pdim$. The sum of all dimensions of all processor meshes is $\#pdims$.
  - The extent of dimension i of a processor mesh or a processor array is $\#\text{cpu}\,\texttt{[}i\texttt{]}$, and the number of physical processors in that mesh is $\#\text{cpu} = \prod_{i=1}^{\#pdim} \#\text{cpu}\,\texttt{[}i\texttt{]}$.
  - Lower bound, upper bound, and stride of a loop are usually named l, u, and s, respectively.
  - Dimensionalities of linear subspaces and polyhedra are usually written $q, r, s, \ldots$
  - Statements in the program text are denoted $S, T, \ldots$
- Terms (expressions) that constitute the program text (which is more general than a statement) are denoted by lower case German letters: $\mathfrak{a}, \mathfrak{b}, \mathfrak{c}, \ldots$
- Vectors start with lower case Greek letters: $\alpha, \beta, \gamma, \ldots$ In general, occurrence instances will be named $\alpha, \beta, \ldots$, while more general vectors that are not necessarily occurrence instances start somewhat higher up in the alphabet $(\mu, \nu, \psi, \ldots)$.
- Functions start with capital Latin letters: $F, G, H, \ldots$.
- Arrays in program fragments presented are also seen as functions; however, they are usually named $A, B, C, \ldots$ Arrays representing templates and processor arrays are usually named $T, P, \ldots$
- Functions in program fragments considered that are seen as (prefix, infix, or postfix) operators – i.e., functions that are used mainly inside of a larger expression in the program text rather than for output through some output arguments – may also be denoted by $\odot, \otimes, \ldots$
- Linear functions and linear relations start with capital Greek letters: $A, B, \Gamma, \ldots$ Special cases are:
  - The h-transformation, defined in Section 2.3, is usually denoted $H, H_1, H_2, \ldots$ Note that in this case, there is no difference between the capital Greek letter "eta" (H) and the capital Latin "h" (H). Of course, the name actually stands for "h-transformation".

[1] Actually, this implies that we do not introduce additional parameters in the course of parallelization and optimization. Although our methods *do* need parameters that do not occur in the source program. However, we may introduce these parameters in the source program fragment during the first step of program analysis, so that we can assume the number of parameters to remain constant.

- The schedule of a set of occurrence instances is usually denoted by Θ.
- The placement of a set of occurrence instances is usually denoted by Φ.
- The space-time mapping of a set of occurrence instances is usually denoted by T – note that, again, there is no difference between the capital Greek letter "tau" (T) and the capital Latin "t" (T). The name actually stands for "transformation".

A genuine exception is the projection $\pi(\cdot)$ defined above, since the capital Greek letter Π usually denotes multiplication, while $\pi(\cdot)$ denotes projection, and we will not break with these traditions.

- Matrices start with capital Latin letters – usually M. A matrix that defines a linear function $\Psi : \nu \mapsto M_\Psi \cdot \nu$ is usually denoted as M_Ψ. Similarly, a matrix that defines a polyhedron $\mathfrak{P}$ or a linear relation P by its roots ($\mathfrak{P} = \{\nu \,|\, M_\mathfrak{P} \cdot \nu \geq 0\}$, $P = \{(\mu, \nu) \,|\, M_P \cdot (\mu, \nu)^T = 0\}$) is usually denoted $M_\mathfrak{P}$, and M_P, respectively.
- For better presentation, matrices and vectors are accessed like arrays. We write $M\,[i,j]$ for the element in the i-th row of the j-th column of M. Correspondingly, the row vector consisting of the i-th row of M is $M\,[i,\cdot]$, and the j-th column is $M\,[\cdot,j]$.
- In the algorithms shown, we actually do not discern between a list, a vector, and a one-dimensional array, and sizes of objects of these structures can be obtained by a function size; elements of such an object *List* can be referenced by *List* [i]; the first element of an array or list is always 1. In addition, we can search for an element x in a list of array A via $x \in A$, and it is always possible to append an element x to a list or array A by using the function append(A, x) or by assigning to an undefined array position or by concatenating a list containing only that element via a concatenate function: concatenate$(A, (x))$. The empty list is represented by $\emptyset$.

Other variable names in algorithms presented in this work quite often consist of more than one letter; these variables usually start with a lower case Latin letter, and no type information can be deduced from their name. In addition, it is not always simple to choose the correct representation for a given object, since several characterizations of the same object may be interesting at a given time – for example, one may want to see some $a \in \mathbb{Q}$ as both a rational number, and as the one-dimensional vector $(a) = \alpha \in \mathbb{Q}^1$. Often, the excessive use of functions in order to represent such a conversion obscures the facts even further. We chose, not to use explicit conversions where they can be avoided; instead the name of an object should let the reader determine at least its main purpose in the given context.

Bibliography

[AB03] C. Alias and D. Barthou. On the recognition of algorithm templates. *Electr. Notes Theor. Comput. Sci*, 82(2), 2003. http://www1.elsevier.com/gej-ng/31/29/23/133/53/show/Products/notes/index.htt#007.

[AB05] C. Alias and D. Barthou. Deciding where to call performance libraries. In J. C. Cunha and P. D. Medeiros, editors, *Euro-Par 2005*, volume 3648 of *Lecture Notes in Computer Science*, pages 336–345. Springer, 2005. http://dx.doi.org/10.1007/11549468_39.

[ACIK95] C. Ancourt, F. Coelho, F. Irigoin, and R. Keryell. A linear algebra framework for static HPF code distribution. Technical Report A-278-CRI, Centre de Recherche en Informatique, Ecole Normale Superieure de Lyon, 46, Allee d'Italie, 69364 Lyon Cedex 07, France, November 1995.

[ACK81] F. E. Allen, J. Cocke, and K. Kennedy. *Reduction of Operator Strength*, pages 79–101. Prentice-Hall, Englewood Cliffs, NJ, USA, 1981.

[ACK87] R. Allen, D. Callahan, and K. Kennedy. Automatic decomposition of scientific programs for parallel execution. In *POPL*, pages 63–76, 1987.

[Agr99] G. Agrawal. A general interprocedural framework for placement of split-phase large latency operations. *IEEE Transactions on Parallel and Distributed Systems*, PDS-10(4):394–413, April 1999.

[AJMCY98] V. Adve, G. Jin, J. Mellor-Crummey, and Q. Yi. High performance fortran compilation techniques for parallelizing scientific codes. In *Proceedings of Supercomputing '98 (CD-ROM)*, November 1998.

[Ali05] C. Alias. *Program Optimization by Template Recognition and Replacement.* PhD thesis, University of Versailles, December 2005.

[AMC98] V. Adve and J. Mellor-Crummey. Using integer sets for data-parallel program analysis and optimization. In *Proceedings of the SIGPLAN'98 Conference on Programming Language Design and Implementation (PLDI)*, pages 186–198, June 1998.

[AMC01] V. Adve and J. Mellor-Crummey. Advanced code generation for High Performance Fortran. In S. Pande and D. P. Agrawal, editors, *Compiler Optimizations for Scalable Parallel Systems – Languages, Compilation Techniques, and Run Time Systems*, LNCS 1808, chapter 16, pages 553–596. Springer-Verlag, 2001.

[AMCS97] V. Adve, J. Mellor-Crummey, and A. Sethi. HPF analysis and code generation using integer sets. Technical report, Department of Computer Science, Rice University, April 1997. CS-TR97-275, http://citeseer.ist.psu.edu/adve97hpf.html.

[ASD95] G. Agrawal, J. H. Saltz, and R. Das. Interprocedural partial redundancy elimination and its application to distributed memory compilation. In *SIGPLAN Conference on Programming Language Design and Implementation*, pages 258–269, 1995. http://citeseer.ist.psu.edu/article/agrawal95interprocedural.html.

[ASU86] A. V. Aho, R. Sethi, and J. D. Ullman. *Compilers – Principles, Techniques, and Tools.* Addison-Wesley, 1986.

[Bal98] A. D. Balsa. F-CPU architecture: Register organization, August 1998. `http://f-cpu.tux.org/original/F-mem.php3`.

[Ban93] U. Banerjee. *Loop Transformations for Restructuring Compilers: The Foundations.* Series on Loop Transformations for Restructuring Compilers. Kluwer, 1993.

[Ban94] U. Banerjee. *Loop Parallelization.* Series on Loop Transformations for Restructuring Compilers. Kluwer, 1994.

[Bas03] C. Bastoul. Efficient code generation for automatic parallelization and optimization. In *ISPDC'2 IEEE International Symposium on Parallel and Distributed Computing*, pages 23–30, Ljubjana, October 2003.

[Bas04] C. Bastoul. Code generation in the polyhedron model is easier than you think. In *13th International Conference on Parallel Architectures and Compilation Techniques*, pages 7–16, Juan-les-Pins, September 2004. IEEE Computer Society Press.

[BCF97] D. Barthou, J.-F. Collard, and P. A. Feautrier. Fuzzy dataflow analysis. *J. Parallel and Distributed Computing*, 40(2):210–226, February 1997.

[BCG+03] C. Bastoul, A. Cohen, A. Girbal, S. Sharma, and O. Temam. Putting polyhedral loop transformations to work. In *LCPC'16 International Workshop on Languages and Compilers for Parallel Computers, LNCS 2958*, College Station, October 2003.

[BCL82] B. Buchberger, G. E. Collins, and R. Loos, editors. *Computer Algebra – Symbolic and Algebraic Computation.* Springer-Verlag, 2nd edition, 1982. In cooperation with Rudolf Albrecht.

[Bel03] V. Beletskyy. Finding synchronization-free parallelism for non-uniform loops. In *Computational Science – ICCS 2003, part II*, LNCS 2658, pages 925–934. Springer-Verlag, May 2003.

[BFR01] D. Barthou, P. Feautrier, and X. Redon. On the equivalence of two systems of affine reccurence equations. Technical Report RR-4285, INRIA, October 2001.

[BFR02] D. Barthou, P. Feautrier, and X. Redon. On the equivalence of two systems of affine recurrence equations (research note). *Lecture Notes in Computer Science*, 2400:309–313, 2002. `http://link.springer-ny.com/link/service/series/0558/bibs/2400/24000309.htm`.

[BHS+95] D. Bailey, T. Harris, W. Saphir, R. van der Wijngaart, A. Woo, and M. Yarrow. *The NAS Parallel Benchmarks 2.0.* NAS Systems Division, NASA Ames Research Center, Moffett Field, CA, 1995.

[BN98] F. Baader and T. Nipkow. *Term Rewriting and All That.* Cambridge University Press, New York, 1998.

[BS91] I. Bronstein and K. Semendjajew. *Taschenbuch der Mathematik*, volume 1. B.G. Teubner Verlagsgesellschaft, Stuttgart, Leipzig, und Verlag Nauka, Moskau, 25th edition, 1991.

[BZ94] T. Brandes and F. Zimmermann. ADAPTOR—a transformation tool for HPF programs. In K. M. Decker and R. M. Rehmann, editors, *Programming Environments for Massively Distributed Systems*, pages 91–96. Birkhäuser, 1994.

[CC96] A. Cohen and J.-F. Collard. Fuzzy array data-flow analysis, Part II: Recursive programs. Technical Report 96-036, PRiSM, 1996.

[CF93] J.-F. Collard and P. Feautrier. Automatic generation of data parallel code. In H. Sips, editor, *Proc. of the Fourth International Workshop on Compilers for Parallel Computers*, pages 321–332. TU Delft, December 1993.

[CG97] J.-F. Collard and M. Griebl. Array dataflow analysis for explicitly parallel programs. *Parallel Processing Letters*, 7(2):111–131, June 1997.

[CG99] J.-F. Collard and M. Griebl. A precise fixpoint reaching definition analysis for arrays. In L. Carter and J. Ferrante, editors, *Languages and Compilers for Parallel Computing, 12th International Workshop, LCPC'99*, LNCS 1863, pages 286–302. Springer-Verlag, 1999.

[Che64] N. V. Chernikova. Algorithm for finding a general formula for the non-negative solutions of system of linear equations. *U.S.S.R. Computational Mathematics and Mathematical Physics*, 4(4):151–158, 1964.

[Che65] N. V. Chernikova. Algorithm for finding a general formula for the non-negative solutions of system of linear inequalities. *U.S.S.R. Computational Mathematics and Mathematical Physics*, 5(2):228–233, 1965.

[Che68] N. V. Chernikova. Algorithm for discovering the set of all solutions of a linear programming problem. *U.S.S.R. Computational Mathematics and Mathematical Physics*, 8(6):282–293, 1968.

[Che86] P. L. Chenadec. *Canonical Forms in Finitely Presented Algebras.* Pitman Publishing, London, 1986.

[CK89] D. Callahan and K. Kennedy. Compiling programs for distributed-memory processors. *J. Supercomputing*, 2(2):151–169, October 1989.

[CL99] A. Cohen and V. Lefebvre. Storage mapping optimization for parallel programs. In P. Amestoy, P. Berger, M. J. Daydé, I. S. Duff, V. Frayssé, L. Giraud, and D. Ruiz, editors, *Euro-Par 2005*, volume 1685 of *Lecture Notes in Computer Science*, pages 375–382. Springer, 1999. `http://link.springer.de/link/service/series/0558/bibs/1685/16850375.htm`.

[Cla96] P. Clauss. Counting solutions to linear and nonlinear constraints through ehrhart polynomials: Applications to analyze and transform scientific programs. In *ACM Int. Conf. on Supercomputing.* ACM, May 1996. `http://icps.u-strasbg.fr/pub-96/pub-96-03.ps.gz`.

[Cla97] P. Clauss. Handling memory cache policy with integer points countings. In *Euro-Par'97*, LNCS 1300, pages 285–293, August 1997. `http://icps.u-strasbg.fr/pub-97/pub-97-07.ps.gz`.

[CM00] P. Clauss and B. Meister. Automatic memory layout transformations to optimize spatial locality in parameterized loop nests. In *4th Annual Workshop on Interaction between Compilers and Computer Architectures, INTERACT-4*, January 2000.

[CMZ92] B. Chapman, P. Mehrota, and H. Zima. Programming in Vienna Fortran. *Scientific Programming*, 1(1):31–50, 1992.

[Coh99] A. Cohen. *Program Analysis and Transformation: From the Polytope Model to Formal Languages.* PhD thesis, Laboratoire PRiSM, Université de Versailles, December 1999.

[Deh90] B. Dehbonei. *Etude de la génération de code et de l'analyse interprocédurale au sein d'un environment de programmation parallèle.* PhD thesis, L'Université Pierre et Marie Curie, Paris, 1990.

[DL97] J.-L. Dekeyser and C. Lefebvre. HPF-Builder: A visual environment to transform Fortran 90 codes to HPF. *The International Journal of Supercomputer Applications and High Performance Computing*, 11(2):95–102, Summer 1997. `http://citeseer.ist.psu.edu/315235.html`.

[DR95] M. Dion and Y. Robert. Mapping affine loop nests: New results. In B. Hertzberger and G. Serazzi, editors, *High-Performance Computing & Networking (HPCN'95)*, LNCS 919, pages 184–189. Springer-Verlag, 1995.

[DRV00] A. Darte, Y. Robert, and F. Vivien. *Scheduling and Automatic Parallelization.* Birkhäuser, Boston, 2000.

[DV94] A. Darte and F. Vivien. Automatic parallelization based on multi-dimensional scheduling. Technical Report 94-24, Laboratoire de l'Informatique du Parallélisme, Ecole Normale Supérieure de Lyon, September 1994.

[Fab97] P. Faber. Transformation von Shared-Memory-Programmen zu Distributed-Memory-Programmen. Master's thesis, Fakultät für Mathematik und Informatik, Universität Passau, November 1997. `http://www.fmi.uni-passau.de/loopo/doc/faber-d.ps.gz`.

[Fea88] P. Feautrier. Parametric integer programming. *Operations Research*, 22(3):243–268, 1988.

[Fea91] P. Feautrier. Dataflow analysis of array and scalar references. *Int. J. Parallel Programming*, 20(1):23–53, February 1991.

[Fea92a] P. Feautrier. Some efficient solutions to the affine scheduling problem. Part I. One-dimensional time. *Int. J. Parallel Programming*, 21(5):313–348, 1992.

[Fea92b] P. Feautrier. Some efficient solutions to the affine scheduling problem. Part II. Multidimensional time. *Int. J. Parallel Programming*, 21(6):389–420, 1992.

[Fea93] P. Feautrier. Toward automatic partitioning of arrays on distributed memory computers. In *International Conference on Supercomputing*, pages 175–184, 1993. `http://citeseer.ist.psu.edu/feautrier93toward.html`.

[Fea03] P. Feautrier. *Solving Systems of Affine (In)Equalities: PIP's User's Guide*, 2003. Additions by Jean-François Collard and Cédric Bastoul.

[FGL01a] P. Faber, M. Griebl, and C. Lengauer. A closer look at loop-carried code replacement. In *Proc. GI/ITG PARS'01*, PARS-Mitteilungen Nr.18, pages 109–118. Gesellschaft für Informatik e.V., November 2001.

[FGL01b] P. Faber, M. Griebl, and C. Lengauer. Issues of the automatic generation of HPF loop programs. In S. P. Midkiff, J. E. Moreira, M. Gupta, S. Chatterjee, J. Ferrante, J. Prins, W. Pugh, and C.-W. Tseng, editors, *13th Workshop on Languages and Compilers for Parallel Computing (LCPC 2000)*, LNCS 2017, pages 359–362. Springer-Verlag, 2001.

[FGL01c] P. Faber, M. Griebl, and C. Lengauer. Loop-carried code placement. In R. Sakellariou, J. Keane, J. Gurd, and L. Freeman, editors, *Euro-Par 2001: Parallel Processing*, LNCS 2150, pages 230–234. Springer-Verlag, 2001.

[FGL03] P. Faber, M. Griebl, and C. Lengauer. Replicated placements in the polyhedron model. In H. Kosch, L. Böszörményi, and H. Hellwagner, editors, *Euro-Par 2003: Parallel Processing*, Lecture Notes in Computer Science 2790, pages 303–308. Springer-Verlag, 2003.

[FGL04] P. Faber, M. Griebl, and C. Lengauer. Polyhedral loop parallelization: The fine grain. In M. Gerndt and E. Kereku, editors, *Proc. 11th Workshop on Compilers for Parallel Computers (CPC 2004)*, Research Report Series, pages 25–36. LRR-TUM, Technische Universität München, July 2004.

[Flo62] R. W. Floyd. Shortest path. *Comm. ACM*, 5(6):345, June 1962.

[Fos94] I. Foster. *Designing and Building Parallel Programs.* Addison-Wesley, 1994.

[GC95] M. Griebl and J.-F. Collard. Generation of synchronous code for automatic parallelization of `while` loops. In S. Haridi, K. Ali, and P. Magnusson, editors, *EURO-PAR'95: Parallel Processing*, LNCS 966, pages 315–326. Springer-Verlag, 1995.

[GGL06] A. Größlinger, M. Griebl, and C. Lengauer. Quantifier elimination in automatic loop parallelization. *Journal of Symbolic Computation*, 41(11):1206–1221, November 2006. doi:10.1016/j.jsc.2005.09.012.

[GL97] M. Griebl and C. Lengauer. The loop parallelizer LooPo—Announcement. In D. Sehr, U. Banerjee, D. Gelernter, A. Nicolau, and D. Padua, editors, *Languages and Compilers for Parallel Computing (LCPC'96)*, LNCS 1239, pages 603–604. Springer-Verlag, 1997. More details at `http://www.infosun.fmi.uni-passau.de/cl/loopo`.

[GLW98] M. Griebl, C. Lengauer, and S. Wetzel. Code generation in the polytope model. In *Proc. Int. Conf. on Parallel Architectures and Compilation Techniques (PACT'98)*, pages 106–111. IEEE Computer Society Press, 1998.

[GR05] J. Gummaraju and M. Rosenblum. Stream Programming on General-Purpose Processors. In *MICRO 38: Proceedings of the 38th annual ACM/IEEE international symposium on Microarchitecture*, Barcelona, Spain, November 2005.

[GR06] G. Gupta and S. Rajopadhye. Simplifying reductions. In J. G. Morrisett and S. L. P. Jones, editors, *Proceedings of the 33rd ACM SIGPLAN-SIGACT Symposium on Principles of Programming Languages, POPL 2006, Charleston, South Carolina, USA, January 11-13, 2006*, pages 30–41. ACM, 2006. `http://doi.acm.org/10.1145/1111037.1111041`.

[Gri96] M. Griebl. *The Mechanical Parallelization of Loop Nests Containing* while *Loops.* PhD thesis, University of Passau, 1996. also available as technical report MIP-9701, `http://www.uni-passau.de/~griebl/thesis.html`.

[GTA06] M. I. Gordon, W. Thies, and S. Amarasinghe. Exploiting coarse-grained task, data, and pipeline parallelism in stream programs. In *Proceedings of the Twelfth International Conference on Architectural Support for Programming Languages and Operating Systems (ASPLOS 2006)*, October 2006.

[GWI92] R. Gómez, J. Winicour, and R. Isaacson. Evolution of scalar fields from characteristic data. *Journal of Computational Physics*, 98(1):11–25, 1992.

[Hes06] P. Hester. Presentation at 2006 amd analyst day. website, December 2006. `http://www.amd.com/us-en/assets/content_type/DownloadableAssets/Dec-06A-DayPhilHester.pdf`, visited on 2007-05-19.

[Hig93] High Performance Fortran Forum. High Performance Fortran Language Specification. Final Version 1.0, Department of Computer Science, Rice University, May 1993.

[Hig97] High Performance Fortran Forum. High Performance Fortran language specification. Version 2.0, Department of Computer Science, Rice University, January 1997.

[HK91] D. Hofbauer and R.-D. Kutsche. *Grundlagen des maschinellen Beweisens.* Vieweg, Braunschweig, 2nd edition, 1991.

[HKT91] S. Hiranandani, K. Kennedy, and C.-W. Tseng. Compiler optimizations for Fortran D on MIMD distributed-memory machines. In *Proc. Supercomputing'91*, pages 86–100. IEEE Computer Society Press, 1991.

[HKT94] S. Hiranandani, K. Kennedy, and C.-W. Tseng. Evaluating compiler optimizations for Fortran D. *J. Parallel and Distributed Computing*, 21(1):27–45, April 1994.

[HM95] T. Haupt and M. Miller. Gravitational wave extraction – a benchmark? website, 1995. `http://www.npac.syr.edu/projects/bbh/PITT_CODES/intro/pitt.ps`, visited on 2007-07-02.

[Hoe07] J. Hoeflinger. Extending OpenMP to clusters, 2007. `http://cache-www.intel.com/cd/00/00/28/58/285865_285865.pdf`, visited on 2007-05-29.

[HT92] S. Hiranandani and K. K. C.-W. Tseng. Compiling Fortran D for MIMD Distributed Memory Machines. *Comm. ACM*, 35(8):66–80, August 1992.

[Int97] International Organization for Standardization (ISO). ISO/IEC 1539-1:1997, 1997. `http://www.iso.org/iso/iso_catalogue/catalogue_ics/catalogue_detail_ics.htm?csnumber=26933`, visited on 2008-02-01.

[KD95] K. Knobe and W. Dally. The subspace model: A theory of shapes for parallel systems, 1995. `http://citeseer.ist.psu.edu/knobe95subspace.html`.

[Kei97] H. Keimer. Datenabhängigkeitsanalyse zur Schleifenparallelisierung: Vergleich und Erweiterung zweier Ansätze. Master's thesis, Fakultät für Mathematik und Informatik, Universität Passau, January 1997. `http://www.fmi.uni-passau.de/loopo/doc/keimer-d.ps.gz`.

[KKS98] J. Knoop, D. Koschützki, and B. Steffen. Basic-block graphs: Living dinosaurs? In *Compiler Construction*, LNCS 1383, pages 65–79. Springer, 1998.

[KLS+94] C. H. Koelbel, D. B. Loveman, R. S. Schreiber, G. L. Steele, Jr., and M. E. Zosel. *The High Performance Fortran Handbook.* Scientific and Engineering Computation. MIT Press, 1994.

[KMP+96a] W. Kelly, V. Maslov, W. Pugh, E. Rosser, T. Shpeisman, and D. Wonnacott. *The Omega Calculator and Library*, 1996.

[KMP+96b] W. Kelly, V. Maslov, W. Pugh, E. Rosser, T. Shpeisman, and D. Wonnacott. The Omega Library Interface Guide. Technical report, Department of Computer Science, Univ. of Maryland, College Park, April 1996.

[Kna28] B. Knaster. Un théorème sur les fonctions d'ensembles. *Annales de la Societé Polonaise de Mathematique*, 6:133–134, 1928.

[Kno97] K. Knobe. *The Subspace Model: Shape-based Compilation for Parallel Systems.* PhD thesis, Massachusetts Institute of Technology, 1997. `http://citeseer.ist.psu.edu/knobe97subspace.html`.

[KPRS94] W. Kelly, W. Pugh, E. Rosser, and T. Shpeisman. Transitive closure of infinite graphs and its applications. Technical Report UMIACS-TR-95-48, University of Maryland, College Park, MD 20742, April 1994.

[KRS94] J. Knoop, O. Rüthing, and B. Steffen. Optimal code motion: Theory and practice. *ACM Transactions on Programming Languages and Systems (TOPLAS)*, 16(4):1117–1155, July 1994. `http://www.acm.org/pubs/toc/Abstracts/0164-0925/183443.html`.

[Lam74] L. Lamport. The parallel execution of DO loops. *Comm. ACM*, 17(2):83–93, February 1974.

[Len93] C. Lengauer. Loop parallelization in the polytope model. In E. Best, editor, *CONCUR'93*, LNCS 715, pages 398–416. Springer-Verlag, 1993.

[LF98] V. Lefebvre and P. Feautrier. Automatic storage management for parallel programs. *Parallel Computing*, 24(3–4):649–671, May 1998. `http://www.elsevier.com/cas/tree/store/parco/sub/1998/24/3-4/1306.pdf`.

[LL97] A. W. Lim and M. S. Lam. Maximizing parallelism and minimizing synchronization with affine transforms. In *Proceedings of the Twenty-fourth Annual ACM Symposium on the Principles of Programming Languages*, Paris, 1997. `citeseer.ist.psu.edu/article/lim97maximizing.html`.

[LMC02] V. Loechner, B. Meister, and P. Clauss. Precise data locality optimization of nested loops. *J. Supercomputing*, 21(1):37–76, January 2002.

[Loe99] V. Loechner. Polylib: A library for manipulating parameterized polyhedra, 1999. `citeseer.ist.psu.edu/loechner99polylib.html`.

[MCA97] J. Mellor-Crummey and V. Adve. Simplifying control flow in compiler-generated parallel code (extended abstract). In *Proceedings of the Tenth International Workshop on Languages and Compilers for Parallel Computing*, Minneapolis, MN, 1997. Springer-Verlag. `http://citeseer.ist.psu.edu/87338.html`.

[Mes95] Message Passing Interface Forum. *MPI: A Message-Passing Interface Standard.* University of Tennessee at Knoxville, 1995.

[Mes97] Message Passing Interface Forum. *MPI-2: Extensions to the Message Passing Interface.* University of Tennessee at Knoxville, 1997.

[Mol82] D. I. Moldovan. On the analysis and synthesis of VLSI algorithms. *IEEE Transactions on Computers*, C-31(11):1121–1126, November 1982.

[Mor98] R. Morgan. *Building an Optimizing Compiler.* Butterworth-Heinemann, 1998.

[MP87] S. P. Midkiff and D. A. Padua. Compiler algorithms for synchronization. *IEEE Transactions on Computers*, 36(12):1485–1495, December 1987.

[MR79] É. Morel and C. Renvoise. Global optimization by suppression of partial redundancies. *Communications of the ACM*, 22(2):96–103, February 1979. Data Flow Analysis.

[MR91] P. Mehrotra and J. V. Rosendale. Programming distributed memory architectures using kali. In A. Nicolau, editor, *Advances in languages and compilers for parallel processing*, chapter 19. Pitman Publishing, London, 1991.

[MR98] M. Metcalf and J. Reid. *Fortran 90/95 explained.* Oxford science Publications, 3rd edition, 1998.

[Muc97] S. S. Muchnick. *Advanced Compiler Design and Implementation.* Morgan Kaufmann Publishers, Inc., San Francisco, CA, USA, 1997.

[Ope00a] OpenMP Architecture Review Board. *OpenMP: A Proposed Industry Standard API for Shared Memory Programming*, 2000. `http://computer.org/cise/cs1998/c1046abs.htm`.

[Ope00b] OpenMP Architecture Review Board. *OpenMP Fortran Application Program Interface*, 2000. `http://www.openmp.org/specs/`.

[Pug92] W. Pugh. The Omega test: a fast and practical integer programming algorithm for dependence analysis. *Comm. ACM*, 35(8):102–114, August 1992.

[Pug93] W. Pugh. The omega test: a fast and practical integer programming algorithm for dependence analysis, July 1993.

[QRW00] F. Quilleré, S. Rajopadhye, and D. Wilde. Generation of efficient nested loops from polyhedra. *Int. J. Parallel Programming*, 28(5):469–498, 2000.

[Qui87] P. Quinton. The systematic design of systolic arrays. In F. F. Soulié, Y. Robert, and M. Tchuenté, editors, *Automata Networks in Computer Science*, chapter 9, pages 229–260. Manchester University Press, 1987. Also: Technical Reports 193 and 216, IRISA (INRIA-Rennes), 1983.

[Qui04] M. J. Quinn. *Parallel Programming in C with MPI and OpenMP.* McGraw-Hill, 2004.

[Ram06] R. Ramanathan. Extending the world's most popular processor architecture. website, October 2006. `http://www.intel.com/technology/magazine/computing/new-instructions-1006.pdf`, visited on 2007-05-19.

[RKS98] O. Rüthing, J. Knoop, and B. Steffen. Code motion and code placement: Just synonyms? In C. Hankin, editor, *ETAPS'98*, LNCS 1381, pages 154–169. Springer-Verlag, 1998.

[RL77] J. R. Reif and H. R. Lewis. Symbolic evaluation and the global value graph. In *Conference Record of the 4th ACM SIGACT/SIGPLAN Symposium on Principles on Programming Languages*, pages 104–118, January 1977.

[Sch86] A. Schrijver. *Theory of Linear and Integer Programming.* Series in Discrete Mathematics. John Wiley & Sons, 1986.

[SHM97] D. B. Skillicorn, J. M. D. Hill, and W. F. McColl. Questions and Answers about BSP. *Scientific Programming*, 6(3):249–274, 1997. `citeseer.ist.psu.edu/skillicorn96question.html`.

[SKR90] B. Steffen, J. Knoop, and O. Rüthing. The value flow graph: A program representation for optimal program transformations. In N. D. Jones, editor, *3rd European Symposium on Programming (ESOP'90)*, LNCS 432, pages 389–405. Springer-Verlag, 1990.

[Slo99] A. Slowik. *Volume Driven Selection of Loop and Data Transformations for Cache-Coherent Parallel Processors.* PhD thesis, Universität-GH Paderborn, 1999.

[SM98] P. Suresh and R. Moona. PERL - a registerless architecture. In *Proceedings of 5th international conference on High Performance Computing, HiPC 98*, December 1998.

[SPB93] E. Su, D. J. Palermo, and P. Banerjee. Automating parallelization of regular computations for distributed-memory multicomputers in the PARADIGM compiler. In A. N. Choudhary and P. B. Berra, editors, *Proc. Int. Conf. on Parallel Processing (ICPP'93), Vol. II: Software*, pages 30–38. CRC Press, 1993.

[Sun90] V. S. Sunderam. PVM: a framework for parallel distributed computing. *Concurrency, Practice and Experience*, 2(4):315–340, 1990. `citeseer.ist.psu.edu/sunderam90pvm.html`.

[Tar55] A. Tarski. A lattice-theoretical fixpoint theorem and its applications. *Pacific Journal of Mathematics*, 5(2):285–309, 1955. `http://projecteuclid.org/Dienst/UI/1.0/Summarize/euclid.pjm/1103044538`.

[TKA02] W. Thies, M. Karczmarek, and S. Amarasinghe. StreamIt: A language for streaming applications. In *Proceedings of the 2002 International Conference on Compiler Construction (CC 2002)*, April 2002.

[Usm87] R. A. Usmani. *Applied Linear Algebra.* Marcel Dekker Inc., 1987.

[Val90] L. G. Valiant. A bridging model for parallel computation. *Comm. ACM*, 33(8):103–111, August 1990.

[vE98] R. van Engelen. CTADEL*: A Generator of Efficient Numerical Codes.* PhD thesis, Rijksuniversiteit te Leiden, November 1998. `http://www.cs.fsu.edu/~engelen/thesis.ps.gz`.

[War62] S. Warshall. A theorem on boolean matrices. *J. Supercomputing*, 9(1):11–12, January 1962.

[Wet95] S. Wetzel. Automatic code generation in the polyhedron model. Master's thesis, Fakultät für Mathematik und Informatik, Universität Passau, November 1995. `http://www.fmi.uni-passau.de/loopo/doc/wetzel-d.ps.gz`.

[Wie95] C. Wieninger. Automatische Methoden zur Parallelisierung im Polyedermodell. Master's thesis, Fakultät für Mathematik und Informatik, Universität Passau, May 1995. `http://www.fmi.uni-passau.de/loopo/doc/wieninger-d.ps.gz`.

[Wil93] D. K. Wilde. A library for doing polyhedral operations. Technical Report RR-2157, INRIA, 1993.

[Wit18] L. Wittgenstein. *Tractatus Logico-Philosophicus.* Project Gutenberg, `http://www.gutenberg.org/etext/5740`, 1918. visited on 2007-05-29.

[Wol95] M. Wolfe. *High Performance Compilers for Parallel Computing.* Addison-Wesley, 1995.

[Won01] D. Wonnacott. Extending scalar optimizations for arrays. In S. Midkiff, J. Moreira, M. Gupta, S. Chatterjee, J. Ferrante, J. Prins, W. Pugh, and C.-W. Tseng, editors, *13th Workshop on Languages and Compilers for Parallel Computing (LCPC 2000)*, LNCS 2017, pages 97–111. Springer-Verlag, 2001.

[Won05] T. Wondrak. Verteilung von allgemeinen Berechnungsinstanzen in automatisch generierten Schleifen. Master's thesis, Fakultät für Mathematik und Informatik, Universität Passau, November 2005. `http://www.infosun.fim.uni-passau.de/cl/arbeiten/wondrak-d.pdf`.

[Xue93] J. Xue. A new formulation of mapping conditions for the synthesis of linear systolic arrays. In L. Dadda and B. W. Wah, editors, *Proc. Int. Conf. on Application Specific Array Processors*, pages 297–308. IEEE Computer Society Press, 1993.

[Xue94] J. Xue. Automating non-unimodular loop transformations for massive parallelism. *Parallel Computing*, 20(5):711–728, May 1994.

[Xue96] J. Xue. Transformations of nested loops with non-convex iteration spaces. *Parallel Computing*, 22(3):339–368, March 1996.

[ZCF+99] G. Zhang, B. Carpenter, G. Fox, X. Li, and Y. Wen. The HPspmd model and its java binding. In R. Buyya, editor, *High Performance Cluster Computing*, volume 2, Programming and Applications, pages 291–309. Prentice Hall PTR, Upper Saddle River, NJ, 1999. Chap. 14.

Index

NOTES

www.ingramcontent.com/pod-product-compliance
Lightning Source LLC
Chambersburg PA
CBHW080404060326
40689CB00019B/4130
9781409266129